H. James Price
Texas A&M University

Student Companion Guide
To Accompany

Principles of Genetics
Fourth Edition

D. Peter Snustad
University of Minnesota

Michael J. Simmons
University of Minnesota

WILEY
JOHN WILEY & SONS, INC.

Cover Photo: ©Dr. Jeremy Burgess/SPL/Photo Researchers, Inc.

To order books or for customer service please, call 1-800-CALL WILEY (225-5945).

ISBN-13   978- 0-471-73830-5
ISBN-10   0-471-73830-1

Printed in the United States of America

10 9 8 7 6 5 4 3 2 1

Printed and bound by Bind Rite Graphics

TABLE OF CONTENTS

Part I

Part II

To The Student

This companion to *Principles of Genetics*, 4th Edition, by Peter Snustad and Michael Simmons, is written to be a hands-on workbook to help you learn and reinforce terminology and concepts, develop problem-solving skills, and challenge your mind. Learning introductory genetics requires both communication of knowledge and solution of problems. Solving genetics problems requires the ability to analyze and interpret data. To communicate your newly acquired knowledge involves learning new terminology. Genetics becomes fascinating and interesting as your confidence in understanding and using your knowledge grows.

How to Use This Student Companion Guide

You should use this Student Companion Guide after you have read the textbook and attended the lectures on each topic. The guide is divided into two parts. **Part I** contains chapters designed to help review concepts and develop problem-solving skills. **Part II** contains the complete answers and solutions to the "Questions and Problems" at the end of the chapters in *Principles of Genetics*, 4th Edition.

In Part I, each chapter begins with an outline of the corresponding chapter in *Principles of Genetics*, 4th edition. A summary of **Important Concepts** follows to help you organize your studies.

Next are lists of **Important Terms** and **Important Names**. In the space allotted, you are asked to define concisely what each term means or to identify the contribution(s) made by an individual or group. A few short and well-worded sentences are better than rambling answers. More importantly, try to understand the real importance of the terms, and the contributions of the scientists.

The **Testing Your Knowledge** section is designed to allow you to evaluate your understanding of genetic concepts and terminology and to help develop problem-solving skills. As you progress through this section of each chapter, the questions become more complex and involve more problem working. The more problems you work, the better you will become at problem solving. Space is allotted in the workbook for your calculations.

A **Thought Challenging Exercise** is included in each chapter. It is intended to stimulate additional thought and discussion. I encourage you to discuss these with your fellow students.

A **Summary of Key Points** section lists the Key Points presented in the text, thereby providing another review of important concepts.

Answers to Questions and Problems are provided. The **Approaches to Problem Solving** section is designed to help you develop problem-solving skills. In many cases, a problem is much easier to work if it is broken down into simpler components. This approach is emphasized throughout.

Now that you have the resources to study, i.e., your professor, *Principles of Genetics*, 4th Edition, this Student Companion Guide and your inquisitive mind, it is time to embark on your fascinating study of genetics. Study hard, devote time to study each day, and watch your interest and knowledge grow. I encourage you to study in small groups after you have used this guide and have attempted to work all the problems. Explaining concepts and discussing solutions to problems helps the learning process.

Acknowledgments

I thank Merillat Staat of John Wiley & Sons, Inc., for her editorial assistance. I appreciate most my wife Patricia for her understanding and patience during the many evenings and weekends that were used to prepare this *Student Companion Guide*.

PART I

1

The Science of Genetics

IMPORTANT CONCEPTS

A. The science called genetics emerged to explain similarities and differences among organisms and focuses on the material that influences how organisms develop, function and behave.
    1. The hereditary material that passes from generation to generation must have three properties.
        a. It must be able to replicate so that copies can be transmitted from parents to offspring.
        b. It must contain information to guide the development, functioning and behavior of an organism.
        c. It must be able to change.

B. Genetics is rooted in the research of Gregor Mendel who discovered how traits are inherited.
    1. Mendel carried out hybridization studies involving peas. He postulated the existence of hereditary factors (genes) associated with traits and how they were inherited.
    2. Mendel's discoveries published in 1866 were not much noticed until 1900 after which the science of genetics was born.

C. After the birth of genetics, scientists were interested in the chemical nature of a gene.
    1. By the middle of the twentieth century, genes were shown to consist of nucleic acids.
    2. The big breakthrough came in 1953 when James Watson and Francis Crick discovered how nucleotides (the building blocks of nucleic acids) are organized within DNA.
    3. By the end of the twentieth century, sequence of bases in DNA of several organisms including human beings had been determined.
    4. The paragon of all sequencing programs is the Human Genome project.

5. Genes can now be studied at the molecular level with relative ease, and vast numbers of genes can be studied simultaneously. Researchers are now able to construct and scan enormous databases containing DNA sequences to address questions concerning genetics.

6. The study of the structure and function of DNA and chromosomes is called genomics.

D. The central dogma in biology is that information flows from DNA to RNA to proteins.

1. DNA molecules contain information encoded in the sequence of nucleotides to direct the activities of cells and to guide the development, functioning, and behavior of the organisms that contain these cells.

    a. Genes contain the instructions for the synthesis of specific proteins.

    b. Proteins are made up of specific sequences of amino acids. The sequence of amino acids in a protein determines its function.

    c. The flow of information from DNA to protein involves an intermediate molecule of RNA. This RNA called messenger RNA is a copy of a gene and directs protein synthesis.

    d. The collection of all the different proteins in an organism is referred to as the proteome.

E. DNA replication is an extraordinarily accurate process, but it is not perfect. Mistakes in replication and damage to DNA may result in deletion, duplication, or rearrangement of individual or stretches of nucleotides. These are called mutations.

1. Mutant genes may have their encoded information altered or disrupted.

2. The process of mutation introduces genetic variability into a population.

F. Charles Darwin and Alfred Wallace proposed that variation makes it possible for species to change, i.e., to evolve over time.

1. Darwin and Wallace gave credence to the concept that all living organisms are related by virtue of decent from a common ancestor.

2. Researchers can establish the historical relationships among organism (phylogenies) by comparing DNA sequences.

    a. Construction of phylogenetic trees is an important part of the study of evolution.

G. Genetic analysis can be conducted at different levels.

1. The oldest type of genetic analysis involves the inheritance of traits when different strains are hybridized. This is referred to as transmission genetics.

2. A second type of genetic analysis focuses on the molecular makeup of the genetic material.

3. A third approach focuses on entire populations of organisms.

H. Genetics is applied in agriculture, medicine, and many other human endeavors.

1. Plant and animal breeders utilize classical genetic approaches as well as molecular techniques to introduce genes from other species into crop plants and livestock.

    a. Plants and animals that have been altered by the introduction of foreign genes are called GMOs (genetically modified organisms).

2. Classical genetics has provided physicians with a long list of diseases that are caused by mutant genes.

    a. Many hospitals and clinics have genetic counselors who are trained to advise people about risks of inheriting or transmitting genetic diseases.

3. Advances in molecular genetics are providing new ways of detecting mutant genes in individuals.

4. Molecular genetics provides new ways to treat diseases.

I. Modern societies depend heavily on the technology that emerges from research in the basic sciences.

1. Genetics has contributed to technologies to provide food and health care.

2. Discoveries in genetic research have nourished countless business ventures in the biotechnology industry.

3. Evidence based on analysis of DNA sequences is routinely used to test for paternity, to convict the guilty and exonerate the innocent of crimes they are accused, to authenticate claims of inheritances, and to identify the dead.

J. The virus ΦX174, which infects E. coli, was pivotal in the development of techniques to sequence DNA.

1. In 1977, Frederick Sanger and his colleagues reported a new technique for sequencing DNA.

    a. This chain-terminating technique was a big improvement over all previous methods that had been used to sequence DNA.

    b. In 1978 they published the complete sequence of the ΦX174 genome to be 5,386 nucleotides.

c. Sanger's chain-termination technique became the standard method to sequence DNA, and the various genome sequencing projects including the human genome project would not have occurred without it.

IMPORTANT TERMS

In the space allotted, concisely define each term.

genetics:

genes:

alleles:

nucleic acids:

nucleotide:

ribonucleic acid:

deoxyribonucleic acid:

genome:

human genome project:

genomics:

messenger RNA (mRNA):

translocation:

proteomics:

central dogma of molecular biology:

reverse transcriptase:

mutations:

phylogeny:

classical genetics:

transmission genetics:

GMOs:

genetic counselor:

IMPORTANT NAMES

In the space allotted, concisely state the major contribution made by the following individual.

Gregor Mendel:

James Watson and Francis Crick:

Charles Darwin and Alfred Wallace:

Sir Archibald Garrod:

Frederick Sanger:

TESTING YOUR KNOWLEDGE

In this section, fill in the blank or answer the question in the space allotted.

1. The fundamental genetics laws were first discovered by _____.

2. The study of the relationship between the units of inheritance and the physical appearance of an organism falls into the discipline of _____.

3. The study of the biochemical nature of genes and how genes express their encoded information is called _____.

4. What are the three properties required of genes?

5. What is the central dogma of molecular biology? What is an exception to this concept?

6. What were the three major milestones that have occurred in genetics?

7. The area of genetics that includes the study of allelic frequencies in a population and the forces that cause change in theses frequencies over time is called _____.

8. What are at least three major areas where applications of genetics have benefited humans?

THOUGHT CHALLENGING EXERCISE

Before you continue further into your study of genetics, prepare a list of the ways in which you think genetics has directly and indirectly influenced your life. Save this list. At the end of your genetics course, prepare another list of how genetics has affected your life. Prepare the last list without referring to the first one. Then compare the two lists.

SUMMARY OF KEY POINTS

Genetics is the study of the hereditary material.

The hereditary material explains both similarities and differences among organisms.

Gregor Mendel postulated the existence of genes to explain how traits are inherited.

Alleles, the alternative forms of genes, account for heritable differences among individuals.

James Watson and Francis Crick elucidated the structure of DNA, a macromolecule composed of two complementary chains of nucleotides.

DNA is the hereditary material in all life forms except some types of viruses, in which RNA is the hereditary material.

The Human Genome Project determined the sequences of nucleotides in the DNA of the human genome.

Sequencing of the DNA of a genome provides the data to identify and catalogue all the genes of an organism.

When DNA replicates, each strand of a duplex molecule serves as the template for the synthesis of a complementary strand.

When genetic information is expressed, one strand of a gene's DNA duplex is used as a template for the synthesis of a complementary strand of RNA.

For most genes, RNA synthesis (transcription) generates a molecule (the RNA transcript) that becomes a messenger RNA (mRNA).

Coded information in an mRNA is translated into a sequence of amino acids in a polypeptide.

Mutations can alter the DNA sequence of a gene.

The genetic variability created by mutation is the basis for biological evolution.

Evolution depends on the occurrence, transmission, and spread of mutant genes in groups of organisms.

DNA sequence data provide a way of studying the historical process of evolution.

In classical genetic analysis, genes are studied by following the inheritance of traits in crosses between different strains of an organism.

In molecular genetic analysis, genes are studied by isolating, sequencing, and manipulating DNA and by examining the products of gene expression.

In population genetic analysis, genes are studied by assessing the variability among individuals in a group of organisms.

Discoveries in genetics are changing procedures and practices in agriculture and medicine.

Advances in genetics are raising ethical, legal, political, social, and philosophical questions.

ANSWERS TO QUESTIONS

1) Gregor Mendel **2)** classical genetics **3)** molecular genetics **4)** The three properties required of genes are the ability to replicate; to store information needed to direct cellular activities and guide development; and to change (mutate) **5)** The central dogma of molecular biology is that information flows from DNA to RNA to protein. An exception to the central dogma is the activity of an enzyme called reverse transcriptase that synthesizes DNA from RNA. **6)** The three milestones of genetics were the discovery of rules governing the inheritance of traits in organisms; the identification of the material (DNA) responsible for this inheritance, and the elucidation of its structure; and the comprehensive analysis of the hereditary material. **7)** population genetics **8)** The application of genetics has benefited humans in many areas including plant and animal breeding, medicine, commercial application of biotechnology, and its use in positively identifying individuals for legal proceedings.

2

Cellular Reproduction and Model Genetic Organisms

CELLS AND CHROMOSOMES
    The Cellular Environment
    Prokaryotic and Eukaryotic Cells
    Chromosomes: Where Genes Are Located
    Cell Division

MITOSIS

MEIOSIS
    Meiosis I
    Meiosis II and the Outcomes of Meiosis

GENETICS IN THE LABORATORY: AN INTRODUCTION TO SOME MODEL RESEARCH ORGANISMS
    Escherichia coli, a bacterium
    Saccharomyces cerevisiae, Baker's Yeast
    Drosophila melanogaster, a Fruit Fly
    Arabidopsis thaliana, a Fast-growing Plant
    Homo sapiens, Our Own Species

A MILESTONE IN GENETICS: CULTURING HUMAN CELLS

IMPORTANT CONCEPTS

A. In both prokaryotic and eukaryotic cells, the genetic material is organized into chromosomes.
    1. Prokaryotic cells are the simpler than eukaryotic cells.
        a. They are usually less than a thousandth of a millimeter long and typically lack a complicated system of internal membranes and membranous organelles.
        b. There is no membrane-bound nucleus. Their DNA is concentrated in a special subcellular compartment.
        c. Examples of prokaryotic organisms are bacteria and the archaea.
    2. Eukaryotic cells are more complex.
        a. They contain a membrane-bound true nucleus which houses the chromosomes, that is comprised of DNA, RNA, and various proteins.
        b. They possess complicated systems of internal membranes, some of which are associated with conspicuous organelles.
        c. Examples of organelles are mitochondria, chloroplasts, endoplasmic reticulum, Golgi complex, lysosomes, and perioxisomes.
        d. All eukaryotic cells have a network of protein filaments, called a cytoskeleton, which gives the cell its shape, its ability to move, and its ability to move materials to specific locations within cells.

B. Chromosomes contain the genetic material.
    1. Most eukaryotic nuclei contain several different chromosomes that are larger and more complex than those found in prokaryotic cells.
    2. For most eukaryotes, somatic or body cells are diploid, i.e., they contain two copies of each chromosome. However, the sex cells or gametes contain just one copy of each chromosome, a condition called haploid.
    3. Eukaryotic chromosomes can be examined using light microscopy, whereas prokaryotic chromosomes can be only be visualized by electron microscopy.

4. Eukaryotic chromosomes seen during cell division using light microscopy are made up of two sister chromatids held together by a centromere.
2. A part of the centromere called the kinetochore functions in chromosome movement during the cell cycle.
C. Each time a eukaryotic cell divides it goes through a series of phases that collectively form the cell cycle. The progression of phases is G1 → S → G2→ M. Duplication of the DNA and other chromosomal material occurs in S. The mother cell divides during the M phase. G1 and G2 are "gaps" between S and M phases.
D. The two types of eukaryotic cell division are mitosis and meiosis.
   1. The result of mitosis is two daughter cells that are identical to each other and to the parent cell.
      a. The DNA of the chromosomes replicates and the synthesis of a variety of proteins necessary for mitosis occurs in interphase of mitosis. The individual chromosomes at this time are too extended and thin to be identified and collectively are called chromatin.
      b. During mitosis chromosomes condense. In prophase each chromosome consists of two rod-shaped identical sister chromatids that are connected at a centromere. As prophase progresses, the nuclear membrane and nucleolus break up, the chromosomes move toward the equator of the cell, and microtubules assemble.
      c. Attachment of spindle microtubules to the kinetochores marks the beginning of metaphase. The chromosomes are fully contracted at metaphase and line up on an equatorial plane called the metaphase plate.
      d. Sister chromatids of individual chromosomes are separated from each other during anaphase and move toward opposite poles of the cell.
      e. Telophase begins when the chromosomes reach the poles. Decondensation of the chromosomes and restoration of the internal organelles and nuclear membrane are characteristic of telophase.
      f. The daughter nuclei produced by division of a mother cell are genetically and chromosomally identical to each other and to the mother cell that underwent mitosis.
      g. When mitosis is complete, the daughter cell nuclei are partitioned into different cells by a process called cytokinesis.
   2. Meiosis is a process involving one round of DNA replication during premeiotic interphase followed by two cell divisions that reduces the diploid state ($2n$) to the haploid state ($1n$). The resulting haploid cells either directly become gametes, or divide to produce cells that later become gametes.
      a. Prophase I of meiosis is divided into five stages, leptonema, zygonema, pachynema, diplonema, and diakinesis.
      b. Leptonema follows the premeiotic interphase. Duplicate chromosomes condense out of the diffuse chromatin network.
      c. During zygonema homologous chromosomes pair by a process called synapsis.
      d. At pachynema, each member of the paired chromosomes consists of two sister chromatids. Paired homologs are called a bivalent. A breakage and reunion occurs during crossing over that leads to recombination of genetic material between paired chromosomes.
      e. During diplonema paired chromosomes separate slightly, but are still held together by chiasmata. A chiasma results from crossing over and involves only two of the four chromosomes of a tetrad.
      f. At diakinesis spindle microtubules penetrate into the nuclear space and attach to the kinetochore of the chromosomes. The bivalents move toward a central plane of the cell that is perpendicular to the axis of the spindle apparatus.
      g. At metaphase I the homologous pairs (bivalents) line up on the equatorial plane, the paired chromosomes oriented toward opposite poles of the spindle.
      h. During anaphase I the paired chromosomes separate from each other and migrate toward opposite poles.
      i. Telophase I begins when the separating chromosomes have reached the opposite poles. The spindle apparatus is disassembled, chromosomes decondense, and a nucleus is formed around the chromosomes at each pole.
      j. During prophase II, the chromosomes condense and become attached to a new spindle apparatus.

k. Chromosomes line up on the equatorial plane at metaphase II.
l. At anaphase II, the centromeres split and the sister chromatids migrate to opposite poles.
m. During telophase II, the separated chromatids (chromosomes) have reached the poles and daughter nuclei form around them. Each daughter nucleus contains a haploid set of chromosomes.
n. Unlike the products of mitosis, the cells that result from meiosis are not genetically identical. Meiosis results in recombination of the maternal and paternal genetic material into haploid meiotic products. Recombination results from both independent assortment of chromosome pairs and from crossing over between non-sister homologous chromatids.
o. A potentially almost infinite variety of new genetic combinations can result from meiosis and fertilization.
4. Meiosis occurs in all sexually reproducing eukaryotes.
a. The cells of primitive eukaryotes such as unicellular algae and fungi are typically haploid. Under appropriate conditions sexual cells form by mitosis and fuse to form a diploid zygote. The zygote usually proceeds into meiosis without intervening mitotic growth. The resulting haploid spores divide by mitosis to perpetuate the predominant haploid stage of the life cycle.
b. In plants, meiosis produces haploid spores that in turn undergo mitotic divisions and differentiate into gametophytes. The haploid gametophytes produce the haploid gametes. Fertilization produces a diploid zygote that divides and differentiates into the sporophyte.
c. In animals, the haploid products of meiosis develop directly into gametes. This process called gametogenesis occurs in gonads, ovaries (oogenesis) of females and testes (spermatogenesis) of males.
E. Several research organisms listed below are well suited as models for genetic research on microorganisms, plants, and animals.
1. *Escherichia coli,* a bacterium
a. *E. coli* is a bacterium that occurs in the intestines of animals including humans. It can be grown on a simple medium, and it is amenable to biochemical and genetic analyses.
b. The DNA of the *E. coli* genome consists of 4.6×10^6 nucleotide pairs that are arranged in a single chromosome with a circumference of 1.4 mm. The genome has been sequenced and is thought to contain 4,288 protein-encoding genes.
c. *E. coli* can be used to study the biology and genetics of viruses (bacteriophages) that infect them.
2. *Saccharomyces cerevisiae,* baker's yeast
a. Yeast is typically a unicellular fungi used as a leavening agent is making bread. It has developed as a model organism because it can be grown on a simple medium, large numbers of cells can be rapidly obtained from a single mother cell, and mutant strains with different growth characteristics can be readily isolated.
b. The yeast genome consists of 12×10^6 nucleotide pairs distributed among its 16 chromosomes, and it is thought to contain 6,268 genes. It was the first eukaryotic genome to be analyzed completely.
c. Yeast reproduces both sexually and asexually.
3. *Drosophila melanogaster,* a fruit fly
a. A mature *D. melanogaster* is about 2mm long. It reproduces prolifically, is easy to rear in the laboratory and has a relatively short life cycle.
b. Thousands of mutant strains have been obtained affecting characters including eye color and anatomy.
c. The genome of *D. melanogaster* has been sequenced and and contains 170×10^6 nucleotide pairs. It is estimated to contain 13,792 genes.
4. *Caenorhaditis elegans*
a. *C. elegans* is a nemotode that has played a major role in the genetic analysis of development.
b. Its genome has been sequenced and consists of 100×10^6 nucleotide pairs housing an estimated 20,516 genes.

5. *Arabidopsis thaliana*, a fast-growing plant
   a. *A thaliana* is a weedy member of the mustard family called the mouse ear cress. It is small with a relatively short generation time.
   b. The genome of *A thaliana* consists of 157×10^6 nucleotide pairs and is thought to contain 27,706 genes. Since many of these genes are present in agriculturally important crop species, genetic analysis of *Arabidopsis* is providing a framework to understand how genes function in important food crops.
6. *Homo sapiens*, our own species
   a. Since human beings cannot be subjected to genetic experimentation like *Drosophila* or yeast, *Homo sapiens* it is not a true model system. The desire to learn about the genetics of our own species is very strong, and great efforts have been expended to satisfy this desire in spite of many obstacles.
   b. Techniques such as cell culture and DNA cloning have added new dimensions to the study the genetic material of humans. Culturing of human cells allows genetic phenomena to be studied outside the human organism.
      (1) Human cell lines have been extensively used to study cell structure, metabolism, and growth, to localize genes to chromosomes, and to analyze the expression of genes and the functions of their products.
   c. The human genome contains 3.2×10^6 nucleotide pairs containing 20,000 to 25,000 genes.
   d. Recent focus has been on ways to isolate and culture human stem cells. Cells, obtained from certain types of adult tissues and from embryos, can be induced to differentiate in culture into specialized cells types and, therefore, have much potential medicinal value.

IMPORTANT TERMS

In the space allotted, concisely define each term.

cytoplasm:

organelles:

prokaryotic cell:

eukaryotic cell:

nucleus:

mitochondria:

chloroplast:

chromosome:

ribosomes:

plasmids:

diploid:

somatic:

haploid:

gametes:

germ line:

centromere:

cell cycle:

G_1 phase:

S phase:

G$_2$ phase:

M phase:

mitosis:

cytokinesis:

chromatin:

interphase:

sister chromatids:

microtubules:

prophase:

meiosis:

nucleolus:

kinetochore:

metaphase:

metaphase plate:

anaphase:

telophase:

meiosis:

homologues:

prophase I:

leptonema:

zygonema:

synapsis:

synaptonemal complex:

pachynema:

bivalent:

tetrad:

crossing over:

diplonema:

chiasmata:

diakinesis:

metaphase I:

anaphase I:

chromosome disjunction:

telophase I:

prophase II:

metaphase II:

anaphase II:

chromatid disjunction:

telophase II:

gametophyte:

sporophyte:

megaspore:

microspore:

alternation of generations:

gametogenesis:

model organism:

bacteriophages:

larva:

instar:

imago:

pupa:

imaginal discs:

anthers:

stamens:

ovary:

pistil:

megaspore mother cell:

secondary endosperm nucleus:

DNA cloning:

IMPORTANT NAMES

Gregor Mendel:

George Gey:

TESTING YOUR KNOWLEDGE

*In this section, answer the questions, fill in the blanks, or solve the problems in the space allotted. Problems noted with an * are solved in the Approaches to Problem Solving section at the end of the chapter.*

1. A cross-shaped structure formed between nonsister homologous chromatids by crossing over that is visible during the diplonema stage of meiosis is called a _____.

2. The stage of meiosis at which synapsis occurs is called _____.

*3. Four gametes result from a cell that has undergone meiosis. If the nucleus of the cell at pachynema had 5 picograms of DNA, the total amount of DNA in each gamete must be _____ picograms.

4. Segregation of homologous chromosomes occurs during _____ of meiosis.

*5. A sunflower has a diploid chromosome number of $2n = 34$. How many chromatids are present in a secondary meiocyte at prophase II?

6. A pair of synapsed chromosomes in the first meiotic division is called a _____.

7. The failure of a homologous chromosome pair to segregate during meiosis, therefore resulting in meiotic products with either an extra or a missing chromosome is called _____.

8. The process of pairing of homologous chromosomes during meiosis is called _____.

*9. A human has 46 chromosomes in somatic (body) cells. How many chromatids are present in a secondary polar body nucleus?

10. Metaphase I of meiosis and mitosis differ in regard to the presence of _____.

11. For a human, the life cycle has the sequence
    (a) $1n = 23$ -> meiosis -> $2n = 46$ -> fertilization -> $1n = 23$
    (b) $2n = 46$ -> meiosis -> $1n = 23$ -> fertilization -> $2n = 46$
    (c) $1n = 23$ -> mitosis -> $2n = 46$ -> fertilization -> $1n = 23$
    (d) $2n = 46$ -> mitosis -> $1n = 23$ -> fertilization -> $2n = 46$

*12. A geneticist cytologically examined a corn plant and found that it had one member of chromosome # 10 possessing a very large terminal knob and the other member lacking this knob. A knob is a darkly staining enlarged region on a chromosome. When the geneticist looked at the much larger chromosome # 1, it was observed that only one chromosome of this pair possessed a terminal knob.

(a) In regard to these two heteromorphic chromosome pairs, draw all possible types of meiotic products that can be produced by this plant.

(b) Can all of these types of gametes be produced by one meiosis? Why or why not?

*13. A particular animal species has four chromosomes ($2n = 4$). For each of the following, indicate whether the nucleus or cell is in mitosis or meiosis, the particular stage, and the substage if in meiosis.

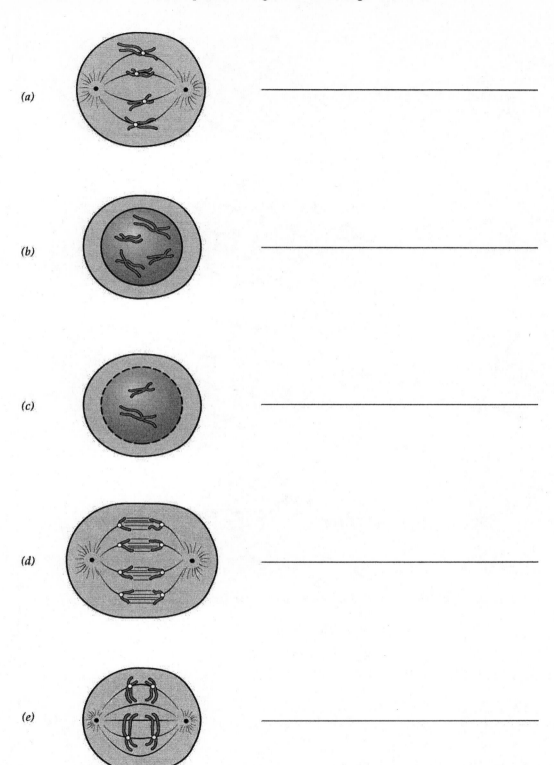

(a) _____

(b) _____

(c) _____

(d) _____

(e) _____

*14. Cells from a diploid organism ($2n = 4$) are shown undergoing division in the diagrams, (I), (II), (III), and (IV), shown below:

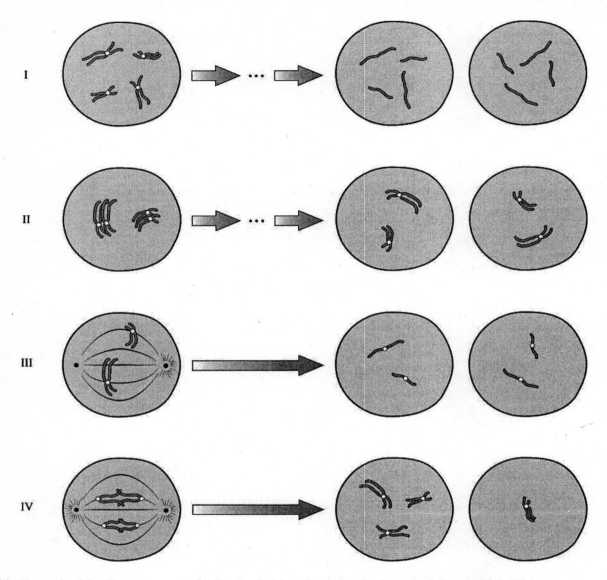

(a) For each of the diagrams, state whether the division shown is mitosis, meiosis I, or meiosis II?

Division I: _____   Division III: _____

Division II: _____   Division IV: _____

(b) Which diagrams show synapsed chromosomes?

(c) Which of the above diagrams starts with a haploid cell?

(d) In which diagram did segregation of both chromosome pairs occur?

(e) What is the term that best describes the phenomenon that occurred in the division of diagram IV?

15. For each of the following stated phenomena, indicate whether it occurs in mitosis and/or meiosis and the phase or subphase at which it occurs.

(a) Chromosomes line up with their centromeres on the equatorial plate.

(b) DNA replication occurs.

(c) Homologous chromosomes undergo synapsis.

(d) Centromeres divide and sister chromosomes move to opposite poles.

(e) The cell is haploid and the chromosomes consist of one chromatid.

(f) Homologous chromosomes segregate to opposite poles.

(g) Nonsister homologous chromosomes undergo crossing over.

(h) The first appearance of haploid cells, with each chromosome consisting of two chromatids.

(i) Chromosomes are maximally condensed.

(j) Bivalents line up with their centromeres on either side of the equatorial plate.

(k) Chromatids are first visually recognized through a light microscope.

(l) Two nuclei are formed, each identical to that which underwent division.

(m) Independent assortment of chromosomes occurs.

THOUGHT CHALLENGING EXERCISE

It is generally agreed that meiosis and fertilization evolved because they produce an enormous variety of new genetic combinations upon which evolutionary forces can act. Although most higher organisms have meiosis and fertilization as part of their reproductive cycle, some plants and animals skip meiosis and fertilization and reproduce parthenogenically by the division and differentiation of a cell produced asexually by mitosis. Discuss possible reasons for the evolution of parthenogenesis from sexually reproducing populations.

SUMMARY OF KEY POINTS

Cells, the basic units of all living things, are enclosed by membranes.

Chromosomes, the cellular structures that carry the genes, are composed of DNA, RNA, and protein.

In eukaryotes, chromosomes are contained within a membrane-bounded nucleus; in prokaryotes they are not.

Eukaryotic cells possess complex systems of internal membranes as well as membranous organelles such as mitochondria, chloroplasts, and endoplasmic reticulum.

Haploid eukaryotic cells possess one copy of each chromosome; diploid cells possess two copies.

Prokaryotic cells divide by fission; eukaryotic cells divide by mitosis and cytokinesis.

Eukaryotic chromosomes duplicate when a cell's DNA is synthesized; this event, which precedes mitosis, is characteristic of the S phase of the cell cycle.

As a cell enters mitosis, it duplicated chromosomes condense into rod-shaped bodies (prophase).

As mitosis progresses, the chromosomes migrate to the equatorial plane of the cell (metaphase).

Later in mitosis, the centromere which holds the sister chromatids of a duplicated chromosome together splits, and the sister chromatids separate (or disjoin) from each other (anaphase).

As mitosis comes to an end, the chromosomes decondense and a nuclear membrane reforms around them (telophase).

Each daughter cell produced by mitosis and cytokinesis has the same set of chromosomes; thus, daughter cells are genetically identical.

Diploid eukaryotic cells form haploid cells by meiosis, a process involving one round of chromosome duplication followed by two cell divisions (meiosis I and meiosis II).

During meiosis I, homologous chromosomes pair (synapse), exchange material (cross over), and separate (disjoin) from each other.

During meiosis II, chromatids disjoin from each other.

In many organisms, the haploid products of meiosis develop directly into gametes.

In plants, the products of meiosis divide mitotically to form haploid gametophytes.

The gametophyte phase of a plant's life cycle alternates with the sporophytic phase, which is diploid; meiosis occurs in the sporophyte.

The bacterium *E. coli* is the premier prokaryote for genetic analysis.

Model eukaryotes include yeast (*S. cerevisiae*), a fruit fly (*D. melanogaster*), and a fast-growing plant (*A. thaliana*).

Techniques such as cell culture and DNA cloning have made it possible to study the genetic material of human beings and many other organisms.

ANSWERS TO QUESTIONS AND PROBLEMS

1) chiasma **2)** zygonema **3)** 1.25 **4)** anaphase I **5)** 34 **6)** bivalent **7)** nondisjunction **8)** synapsis **9)** 23 **10)** paired chromosomes (bivalents) at MI of meiosis **11)** b **12)** see answer in next section **13) a.** mitosis, metaphase **b.** mitosis, prophase **c.** meiosis, prophase II **d.** mitosis, anaphase **e.** meiosis, anaphase I **14) a.** (I) mitosis; (II) meiosis I; (III) meiosis II; (IV) meiosis I **b.** II, IV; **c.** III; **d.** II; **e.** nondisjunction. **15) a.** mitosis, metaphase; meiosis, metaphase II **b.** interphase before mitosis and meiosis **c.** meiosis, prophase I, zygonema **d.** mitosis, anaphase; meiosis, anaphase II **e.** meiosis, telophase II **f.** meiosis, anaphase I **g.** meiosis, prophase I, pachynema **h.** meiosis, telophase I **i.** metaphase of mitosis; metaphase I and metaphase II of meiosis **j.** meiosis, metaphase I **k.** meiosis, leptonema; mitosis, prophase **l.** mitosis, telophase **m.** meiosis, anaphase I

Comments on the thought challenging exercise:

In plants, parthenogenesis generally occurs following hybridization of divergent populations. The parthenogenetic individuals often have a broader range of adaptation than either parent and occupy different ecological niches. If the heterozygosity of the hybrids results in highly adaptive individuals, then ways to restrict recombination and pass the adaptive genotypes to progeny would have a selective advantage. Parthenogenesis is a way of eliminating meiosis and thereby maintaining the genotype of the mother. Many other mechanisms have evolved that limit the recombination of chromosomes while still utilizing meiosis and fertilization. Also, organisms that reproduce by parthenogenesis generally retain a low frequency of sexual reproduction.

APPROACHES TO PROBLEM SOLVING

3. The cell at pachynema is diploid and has 5 pg of DNA. Each chromosome consists of two chromatids. At anaphase I the bivalents segregate to opposite poles into secondary meiocytes with a complete set of chromosomes, each consisting of two chromatids This reduces the DNA amount to 2.5 pg or 1/2 that of the cell at pachynema. The anaphase of meiosis II involves a division of the centromeres and the separation of sister chromatids to opposite poles. This reduces the DNA amount by 1/2 again and, thus, produces gametes with 1.25 pg DNA.

5. The first meiotic division reduces the chromosome number from $2n = 34$ to $1n = 17$. Since prophase II chromosomes still have two chromatids each, there is a total of 34 chromatids.

9. The first meiotic division reduces the chromosome number from $2n = 46$ to $1n = 23$. A secondary polar body is a product of a second meiotic division during oogenesis. Therefore, a secondary polar body is haploid, $1n = 23$, with each chromosome having just one chromatid. There are 23 chromatids present.

12. The cell that will undergo meiosis has the following chromosome makeup in regard to chromosomes 1 and 10.

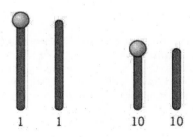

(a) The independent assortment of the two heteromorphic chromosome pairs results in the following four kinds of gametes produced in equal frequencies.

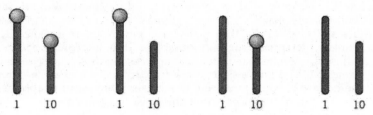

(b) The four types of gametes could be produced from one meiosis only if crossing over simultaneously occurred between the knob and the centromere in each of the heteromorphic chromosome pairs. If so, then both chromosomes in each of the two secondary meiocytes would consist of one knobbed and one unknobbed chromatid as drawn below.

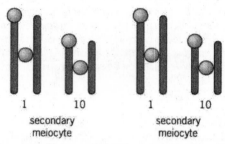

The random disjunction of the chromatids at anaphase II could result in all four gametic types being produced.

13. (a) This figure shows four chromosomes lined up with their centromeres on the equatorial plate. Since the diploid number of this cell is $2n = 4$, it must be a mitotic metaphase.

(b) This figure shows four unpaired chromosomes, each chromosome consisting of two chromatids. The nuclear membrane in still intact and spindle fibers have not formed. Therefore, this cell is prophase of mitosis.

(c) Two different chromosomes are apparent in the nucleus. Since this is the haploid number, the cell must have at least completed the first meiotic division. Each chromosome still consists of two chromatids. Therefore, the nucleus must be at prophase II of meiosis.

(d) Four chromosomes have migrated toward, but have not reached, each pole. Therefore, the nucleus is at anaphase of mitosis.

(e) This drawing depicts two chromosomes migrating to each pole. Since the diploid chromosome number is $2n = 4$, this must be a meiotic anaphase. Each chromosome consists of two chromatids, indicating that this must be anaphase I.

14. (a) I. The first cell has the diploid number ($2n = 4$) of chromosomes. Each chromosome consists of two chromatids, and the chromosomes are not paired. The initial cell must be at mitotic prophase. The two chromatids of each chromosome have migrated to separate cells during the division depicted, as would be expected of a mitotic division.
II. This figure starts with paired chromosomes at pachynema (prophase I of meiosis) and ends with the haploid cells at prophase of meiosis II.
III. The initial cell shown has the haploid number of chromosomes ($n = 2$). Each chromosome consists of two chromatids and is lined up on the equatorial plate. This cell is in metaphase II, this diagram depicts a cell in meiosis II.
IV. The initial cell has paired chromosomes (at metaphase I) that display terminalized

chiasmata. Therefore, this nucleus is in the first meiotic division.

(b) The chromosomes in II and IV are paired or synapsed.

(c) The division shown in III has two chromosomes, which is the haploid number. Therefore, the division shown starts with a haploid cell.

(d) Since the members of each chromosome pair have gone to different poles in diagram II, it depicts the segregation of both chromosome pairs.

(e) In diagram IV, both members of the larger chromosome pair have ended up in the same cell after the first meiotic division. This illustrates the phenomenon of nondisjunction.

3

Mendelism: The Basic Principles of Inheritance

IMPORTANT CONCEPTS

A. Through carefully designed hybridization experiments with the garden pea and numerical analyses of progeny, Gregor Mendel deduced fundamental laws of transmission genetics.
   1. The principle of segregation states that alleles of a gene pair segregate from each other during gamete formation so that gametes have only one member of a gene pair.
   2. The principle of independent assortment states that alleles of different genes segregate, or assort, independently of each other, and recombine at random at fertilization. In other words, the segregation of one gene pair does not interfere with or influence the segregation of other gene pairs.
   3. Mendel also discovered another principle involving gene function, referred to as dominance.
B. Mendel's discoveries went essentially unnoticed until 35 years later.
   1. In 1900, three scientists (Correns, deVries, and von Tschermak-Seysenegg) independently rediscovered Mendel's laws.
   2. The discipline of genetics was born in 1900 when William Bateson championed the cause of Mendelian genetics and introduced it to the English-speaking world.
C. The results of genetic crosses can be predicted by the Punnett square, forked line, or probability methods.
D. Since deviations from expected genetic ratios result from elements of chance, statistical methods such as the Chi-square analysis are used to determine the "goodness of fit" of data.
E. The probabilities of various combinations of events in groups of a specific size can be calculated by the binomial expansion $(p + q)^n$.
   1. If the probabilities of p and q are known, then $(p + q)^n$ represents the probabilities of all combinations of events p and q in a sample size of n.
   2. In F_2 and testcross progeny, the genotypes resulting from segregation and independent assortment of heterozygous gene pairs are binomially distributed.
F. Genetics in humans has traditionally been aided by pedigree analysis.

1. Pedigrees are diagrams that show relationships among members of a family and the occurrence of the trait(s) of interest among these family members.

IMPORTANT TERMS

In the space allotted, concisely define each term.

genetics:

true-breeding:

cross-fertilized:

dominant:

recessive:

monohybrid cross:

gene:

allele:

homozygous:

heterozygous:

genotype:

phenotype:

segregate:

parental:

F_1:

F_2:

principle of dominance:

principle of segregation:

dihybrid cross:

principle of independent assortment:

Punnett square method:

backcross:

testcross:

probability:

Chi-square (X^2):

degrees of freedom:

pedigree:

multiplicative rule:

additive rule:

binomial distribution:

IMPORTANT NAMES

In the space allotted, concisely state the major contribution made by the individual.

Gregor Johann Mendel:

Hugo de Vries:

Carl Correns:

Eric von Tschermak-Seysenegg:

William Bateson:

Wilhelm Johannsen:

TESTING YOUR KNOWLEDGE

*In this section, answer the questions, fill in the blanks, and solve the problems in the space allotted. Problems noted with an * are solved in the Approaches to Problem Solving section at the end of the chapter.*

1. Alternative forms of genes are called _____.

2. The expression of one allele to the exclusion of the other allele of a heterozygous gene pair is called _____.

3. Name the three rediscoverers of Mendel's laws.

4. The person who gave genetics its name was _____.

5. Which one of the following is <u>true</u> when genotype *Cc Dd* undergoes independent assortment?

    (a) the parental combinations will be most frequent among the gametes.
    (b) four different gametic genotypes will occur in equal frequency.
    (c) the gametes will be either *CD* or *cd*.
    (d) the segregation of one gene pair will determine how the other gene pair segregates.

*6. In maize, a color gene and a height gene control the following phenotypes: *CC* and *Cc*, purple; *cc*, white; *TT*, tall; *Tt*, intermediate; *tt*, short. If a dihybrid is self-fertilized, what is the resulting phenotypic ratio?

    (a) 1:2:1:2:4:2:1:2:1   (b) 2:3:6:3:2   (c) 3:3:3:3:3:1   (d) 3:6:3:1:2:1 (e) 9:3:3:1

*7. From a cross, *Aa Bb cc Dd Ee ff* x *AA Bb Cc Dd ee Ff*, what is the probability of obtaining the genotype *Aa bb Cc DD Ee ff* in the progeny?

*8. What is the probability of rolling a combination totaling eight on two dice thrown simultaneously?

*9. If both parents were known to be carriers (*Aa*) for the recessive albino allele, what is the chance of three albino and two normal children if they are to have five children?

10. In a family of six children, what is the probability that 4 will be girls and 2 will be boys?

11. A man and wife are both heterozygous for the recessive albino allele. What is the probability, if they parent six children, that three will be albino and three will be normal?

*12. A phenotypically normal couple has one albino child. What is the probability that their other phenotypically normal child is a heterozygote?

     (a) 1/4   (b) 1/3   (c) 1/2   (d) 3/4   (e) 2/3

13. A mouse of the *CcDd* genotype is testcrossed. What is the probability that the first three offspring are dihybrid females? What is the probability that the first four offspring will be homozygous?

*14. Two strains of sorghum (one homozygous for all dominant alleles and the other homozygous for all recessive alleles at the same six genes) are crossed. The six gene pairs at which they differ segregate independently. The F_1 hybrid is selfed to produce F_2s.

(a) How many kinds of gametes can be produced by the F_1 plants?

(b) What is the number of possible genotypes among the F_2s?

(c) What proportion of the F_2 progeny is expected to be homozygous at all six gene pairs?

(d) What proportion of the F_2 progeny is expected to show all the dominant phenotypes?

(e) What proportion of the F_2 progeny is expected to be homozygous for all dominant alleles?

*15. In squash, fruit color is determined by one gene, and fruit shape by another independently segregating gene. The white fruit allele (W) is dominant to yellow fruit allele (w) and disk-shaped fruit allele (D) is dominant to round-shaped fruit allele (d). From progenies listed in the following table, provide probable genotypes for the parents of each cross.

Parents	Phenotypes of Progeny			
	white disk	white round	yellow disk	yellow round
(a) white, round x white, round	0	31	0	9
(b) white, disk x yellow, round	52	48	0	0
(c) white, disk x white, round	31	29	9	10
(d) white, disk x white, disk	88	30	27	10
(e) white, disk x yellow, round	16	14	13	15

*16. The following is a legend for maize.

$v^+\_$ = nonvariegated leaves $b^+\_$ = yellow seed
$v\ v$ = variegated leaves $b\ b$ = bronze seed

$g^+\_$ = nonglossy leaves $D\_$ = dwarf
$g\ g$ = glossy leaves d^+d^+ = tall

$l^+\_$ = liguled $w+\_$ = starchy endosperm
$l\ l$ = liguleless $w\ w$ = waxy endosperm

From a cross:

$$vv\ g^+g\ ll\ b^+b\ Dd^+\ w^+w\ \text{(female)}\ X\ v^+v\ gg\ ll\ b^+b\ Dd^+\ w^+w\ \text{(male)}$$

(a) How many different gametes can be formed by the male parent of the above cross?

(b) How many different genotypes are possible in the progeny of the above cross?

(c) How many different phenotypes are possible in the progeny of the above cross?

(d) What is the probability of obtaining the following genotype from the above cross?

$$vv\ gg\ ll\ b^+b\ Dd^+\ ww$$

(e) What is the probability of obtaining the following phenotype in the progeny of the above cross?

nonvariegated, glossy, liguleless, bronze, dwarf, waxy

*17. The following is a legend for chickens.

$B\ B$ = black	$C\ C$ = embryonic lethal	$P\ \_$ = pea comb
$B\ b$ = blue	$C\ c$ = creeper	$p\ p$ = single comb
$b\ b$ = white	$c\ c$ = normal	
$M\ \_$ = bearded	$S\ \_$ = silky feathers	$W\ \_$ = white skin
$m\ m$ = beardless	$s\ s$ = normal feathers	$w\ w$ = yellow skin

From a cross:

$Bb\ Cc\ pp\ Mm\ ss\ Ww$ (female) x $Bb\ Cc\ Pp\ Mm\ Ss\ Ww$ (male)

(a) How many different gametes can be formed by the female parent of the above cross?

(b) How many genotypes are possible in the **living** progeny of the above cross?

(c) How many phenotypes are possible in the **living** progeny of the above cross?

(d) What is the probability of obtaining the following genotype in the **living** progeny from the above cross?

$Bb\ Cc\ pp\ MM\ Ss\ ww$

(e) What is the probability of obtaining the following phenotype in the **living** progeny of the above cross?

black, non-creeper, pea comb, bearded, silky, yellow skin

*18. To this point we have been considering the basic F_2 phenotypic ratios of 3:1, 1:2:1, and 9:3:3:1. In later chapters you will learn that modifications of these ratios may occur due to various types of gene interaction. In a plant species, red, yellow and white seed colors are found. A plant from a line breeding true for red seeds crossed to a plant from a line breeding true for white seeds produces F_1 progeny that are all red. The 820 plants of the F_2 generation consisted of 450 red seeded, 160 yellow seeded, and 210 white seeded individuals. Using the Chi-square procedure and the following Chi-square table, test whether these data better fit a 2 red: 1 yellow: 1 white or a 9 red: 3 yellow: 4 white ratio.

Table of Chi-square (X^2)[a]

Degrees of Freedom	Probability (P)						
	0.99	0.95	0.80	0.50	0.20	0.05	0.01
1	0.0002	0.004	0.064	0.455	1.642	3.841	6.635
2	0.020	0.103	0.446	1.386	3.219	5.991	9.210
3	0.115	0.352	1.005	2.366	4.642	7.815	11.345
4	0.297	0.711	1.649	3.357	5.989	9.488	13.277
5	0.554	1.145	2.343	4.351	7.289	11.070	15.086
6	0.872	1.635	3.070	5.348	8.558	12.592	16.812

[a]Selected data from R. A. Fisher and F. Yates, *Statistical Tables for Biological, Agriculture and Medical Research.* Oliver & Boyd, London, 1943.

*19. The following is a pedigree for a rare trait:

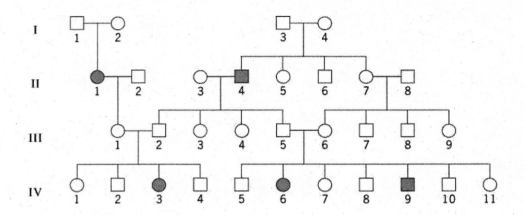

(a) What is the most probable mode of inheritance for the trait indicated above?

(b) If individuals III-3 and III-8 marry, what is the probability of them having a child with the trait?

(c) If individuals IV-1 and IV-10 marry, what is the probability of them having a child who expresses the trait?

*20. The following is a pedigree for a rare trait:

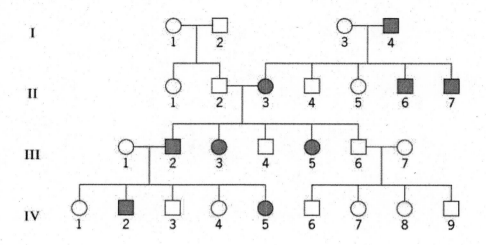

(a) What is the most probable mode of inheritance for the trait indicated above?

(b) If individuals IV-4 and IV-6 marry, what is the probability of them having a child homozygous for the allele determining the trait?

(c) If individuals IV-5 and IV-6 marry, what is the probability of them having a child with the trait?

THOUGHT CHALLENGING EXERCISE

A beginning genetics student gathered data from many phenotypically normal couples to whom one or more albinos had been born. The pooled data involving their children showed 61 normal and 39 albino children. These data were subjected to Chi-square analysis to test a "goodness of fit" to a 3:1 ratio. Much to the student's surprise, the ratio was not consistent with the data ($X^2 = 10.41$; df = 1; P < 0.01). What factor(s) could have contributed to this highly significant variation from a 3:1 ratio of normal to albino in children of heterozygous parents?

SUMMARY OF KEY POINTS

Mendel studied the inheritance of seven different traits in garden peas, each trait being controlled by a different gene.

Mendel's research led him to formulate three principles of inheritance: (1) the alleles of a gene are either dominant or recessive; (2) different alleles of a gene segregate from each other during the formation of gametes; and (3) the alleles of different genes assort independently.

The outcome of a cross can be predicted by the systematic enumeration of the genotypes using a Punnett square.

When more than two genes are involved, the forked-line or probability method is used to predict the outcome of a cross.

The chi-square statistic is calculated as $X^2 = \sum$(observed number – expected number)^2/expected number, with the sum computed over all categories comprising the data.

Each chi-square statistic is associated with an index, the degrees of freedom, which is equal to the number of data categories minus one.

Pedigrees are used to identify dominant and recessive traits in human families.

The analysis of pedigrees allows genetic counselors to assess the risk that an individual will inherit a particular trait.

ANSWERS TO QUESTIONS AND PROBLEMS

1) alleles 2) dominance 3) Correns, deVries, von Tschermak-Seyenegg 4) Bateson 5) b 6) d 7) 1/256 8) 5/36 9) 45/512 10) 15/64 11) 135/1024 12) e 13) 1/512, 1/256 14) a. 64; b. 729; c. 1/64; d. 729/4096; e. 1/4096 15) a. *Ww dd* x *Ww dd*; b. *WW Dd* x *ww dd*; c. *Ww Dd* x *Ww dd*; d. *Ww Dd* x *Ww Dd*; e. *Ww Dd* x *ww dd* 16) a. 16; b. 108; c. 32; d. 1/64; e. 3/256 17) a. 16; b. 216; c. 96;

d. 1/192; **e.** 1/256 **18)** data fit a 9:3:4 ratio; **19) a.** autosomal recessive; **b.** 1/8; **c.** 1/9; **20) a.** autosomal dominant; **b.** 0; **c.** 1/2.

Some ideas concerning thought challenging exercise:

The first thought may be that one genotype shows less viability than the others, e.g., *aa*. But if albinos had a higher death rate than normal, the ratio would be skewed in excess to 3:1. A more careful investigation indicates a bias toward albinos due to the method of identifying heterozygous couples; the couples were identified by having produced one or more albino children. Therefore, the heterozygous couples producing only normal children went unidentified. If the albino children that led the investigator to the family were excluded from the data, a 3:1 normal:albino ratio should be obtained.

APPROACHES TO PROBLEM SOLVING

6. This problem can be worked by the probability method by analyzing the inheritance of one gene pair at a time and then multiplying their probabilities. From the $Cc\,Tt$ selfed plant a progeny will be either purple, $C\_$ (p = 3/4), or white, cc (p = 1/4). A progeny will be either Tall, TT (p = 1/4), intermediate, Tt (p = 2/4), or short, tt (p = 1/4). Since the genes segregate independently of each other, the expected phenotypic ratio of the progeny can be calculated by assembling all combinations of phenotypes and multiplying their individual probabilities.

   > purple (3/4) and tall (1/4) = 3/16
   > purple (3/4) and intermediate (2/4) = 6/16
   > purple (3/4) and short (1/4) = 3/16
   > white (1/4) and tall (1/4) = 1/16
   > white (1/4) and intermediate (2/4) = 2/16
   > white (1/4) and short (1/4) = 1/16

   The phenotypic ratio is 3:6:3:1:2:1

7. Assume that the six gene pairs segregate independently. To work the problem, it is best to break it down into a series of monohybrid crosses and then multiply the probabilities of obtaining the individual genotypes, e.g., the probability of obtaining Aa from $Aa \times AA$ is 1/2; bb from $Bb \times Bb$ is 1/4, etc.

   $$\begin{array}{cccccc} Aa & bb & Cc & DD & Ee & ff \\ 1/2 \ \times & 1/4 \ \times & 1/2 \ \times & 1/4 \ \times & 1/2 \ \times & 1/2 = 1/256 \end{array}$$

8. There are five random combinations on two dice that equal eight; 2 and 6, 6 and 2, 5 and 3, 3 and 5, and 4 and 4. Each combination has a 1/6 x 1/6 or 1/36 probability of occurring. Since each combination is a mutually exclusive event, the individual probabilities are summed. The answer is 5/36.

9. This problem may be solved by use of the binomial expansion $(B + b)^5$. The terms of the expanded binomial represent all combinations of events B (normal phenotype, P = 3/4) and b (albino, P =1/4) in a sample size of 5. The term utilized is $10\,B^2b^3$. The probability is $10(3/4)^2(1/4)^3 = 45/512$.

A way to expand a binomial and assign the coefficient to each term without the aid of Pascal's triangle is shown using the example of $(B + b)^n$, where n = 5. The first term is the first letter of the binomial to the power of n, or B^5 in this example. For each subsequent term, the exponent of the first letter is decreased by a power of 1, and the exponent of the second letter increases by one. The six terms of the expansion of $(B +b)^5$ are:

$$B^5 \ + \ B^4b^1 + \ B^3b^2 \ + \ B^2b^3 \ + \ B^1b^4 \ + \ b^5$$

The coefficients are symmetrically distributed. The coefficient of the first and last terms is 1.

The coefficient of the second term is "n" or $5B^4b^1$. The proceeding coefficient is determined by multiplying the exponent of the first letter in the current term by the coefficient of the term, and dividing by the term number. In our example, the coefficient of the term following $5B^4b^1$ is $\frac{5 \times 4}{2} = 10$, and the term is $10 B^3b^2$. The coefficients of the six terms are 1, 5, 10, 10, 5, 1.

A way to represent one term of an expanded binomial involves the use of the factorial method and the formula

$$P = \frac{N!}{x! (N-x)!} p^x q^y$$

The best way to start is to define the terms of the factorial. From the example in this problem

P = probability
N = sample size (5)
x = number of event one (2 normal)

y = number of event two (3 albino)
p = probability of event one (3/4)
q = probability of event two (1/4)

Entering these numbers into the terms of the equation results in

$$P = \frac{5!}{2! (5-2)!} (3/4)^2(1/4)^3 = \frac{5 \times 4 \times 3 \times 2 \times 1}{2 \times 1 \times 3 \times 2 \times 1} (9/16)(1/64) = 45/512.$$

12. The genotypes and their probabilities produced by $Aa \times Aa$ mating is 1/4 AA: 2/4 Aa: 1/4 aa. Since the child in question is normal, it is not aa. Therefore, it has twice the probability of being heterozygous Aa than homozygous AA. The probability of being heterozygous, if normal, is 2/3.

14. (a) The F_1 is heterozygous at six gene pairs, $Aa\ Bb\ Cc\ Dd\ Ee\ Ff$. Assume that each gene pair assorts independently. The segregation of each heterozygous gene pair produces 2 kinds of gametes. The total number of gametes produced by the F1 plant is therefore $2 \times 2 \times 2 \times 2 \times 2 \times 2 = 64$. A formula that denotes the number of gametes formed is $(2)^n$, where "n" is the number of heterozygous gene pairs.

(b) The F_1 is heterozygous at six gene pairs. Selfing of the F1 results in three genotypes in the F_2 for each gene pair, e.g., AA, Aa, aa. The formula $(3)^n$ can be used to determine the number of genotypes possible in the F_2 generation, i.e., $3^6 = 729$.

(c) Half of the F_2 progeny are expected to be homozygous for each gene pair, e.g, **1/4 AA**, 2/4 Aa **1/4 aa.** Therefore, the proportion of F_2 progeny expected to be homozygous at all six gene pairs is $(1/2)^6$ or 1/64.

(d) Three-quarters of the F_2 progeny are expected to show the dominant phenotype at each gene pair, e.g., **1/4 AA, 2/4 Aa,** 1/4 aa. The proportion of the F_2 progeny expected to show all the dominant phenotypes is $(3/4)^6$ or 729/4096.

(e) The probability that the F_2 progeny will be homozygous for any dominant allele, e.g., AA, is 1/4. The proportion of progeny expected to be homozygous for all six dominant alleles is $(1/4)^6$ or 1/4096.

15. (a) The progeny segregate white to yellow in a 3:1 ratio. Round and disk do not segregate. The parents must be $Ww\ dd \times Ww\ dd$.

(b) Disk and round segregate in the progeny, while white and yellow do not segregate. The parents must be $WW\ Dd \times ww\ dd$.

(c) White and yellow segregate in a 3:1 ratio and disk and round segregate in a 1:1 ratio. The parents must be $Ww\ Dd \times Ww\ dd$.

(d) Both gene pairs are segregating to give a 9:3:3:1 ratio. The parents are dihybrids of the genotype *Ww Dd* x *Ww Dd*.

(e) Both gene pairs are segregating in a 1:1 ratio. The parents must be *Ww Dd* x *ww dd*.

16. (a) The male parent is heterozygous at 4 gene pairs. Therefore $(2)^4$ or 16 kinds of gametes can be produced.

(b) To determine the number of genotypes and phenotypes possible in the progeny, we analyze this multiple hybrid as a series of individual monohybrid crosses and multiply the probabilities.

vv x v^+v -> v^+v and vv = 2 genotypes and 2 phenotypes
g^+g x gg -> g^+g and gg = 2 genotypes and 2 phenotypes
ll x ll -> ll = 1 genotype and 1 phenotype
b^+b x b^+b -> b^+b^+, b^+b, and bb = 3 genotypes and 2 phenotypes
Dd^+ x Dd^+ -> DD, Dd^+, and d^+d^+ = 3 genotypes and 2 phenotypes
w^+w x w^+w -> w^+w^+, w^+w, and ww = 3 genotypes and 2 phenotypes

The number of genotypes possible in the progeny is 2 x 2 x 1 x 3 x 3 x 3 = 108.

(c) The number of different phenotypes possible in the progeny is 2 x 2 x 1 x 2 x 2 x 2 = 32.

(d) To calculate the probability of obtaining the $vv\ gg\ ll\ b^+b\ Dd^+\ ww$ genotype, determine the probability of obtaining from the parents the genotype at each gene pair, and multiply these. This is 1/2 x 1/2 x 1 x 1/2 x 1/2 x 1/4 = 1/64.

(e) Multiply the probabilities of obtaining the six respective individual phenotypes, i.e., 1/2 x 1/2 x 1 x 1/4 x 3/4 x 1/4 = 3/256.

17. This problem is worked by the same procedure as problem 16, except that there is a lethal segregating in the cross. The fact that the *CC* genotype and phenotype don't appear in the living progeny must be taken into account when calculating genotypic and phenotypic ratios.

(a) 2^4 = 16

(b) 3 x 2 x 2 x 3 x 2 x 3 = 216

(c) 3 x 2 x 2 x 2 x 2 x 2 = 96

(d) 1/2 x **2/3** x 1/2 x 1/4 x 1/2 x 1/4 = 1/192

(e) 1/4 x **1/3** x 1/2 x 3/4 x 1/2 x 1/4 = 1/256

18. Chi-square analysis for 2:1:1 ratio:

	observed (O)	expected (E)	(O - E)	$(O - E)^2$	$(O - E)^2/E$
red	450	410	40	1,600	3.9
yellow	160	205	- 45	2,025	9.9
white	210	205	5	25	0.1
	820	820			X^2 = 13.9

Obtain the probability from the chi-square table with d.f. = 2. P = < 0.01
This means that the probability of obtaining deviations from the expected ratio, due to chance alone, as large as were observed, is less than 1/1000. Therefore, a 2:1:1 ratio is not consistent with the data.

Chi-square analysis for 9:3:4 ratio:

	observed (O)	expected (E)	(O - E)	(O - E)^2	(O - E)^2/E
red	450	461	- 11	121	0.262
yellow	160	154	6	36	0.234
white	210	205	5	25	0.122
	820	820			$X^2 = 0.618$

Obtain the probability from the chi-square table with d.f. = 2. $0.70 > P > .50$
This means that the probability of obtaining deviations from the expected ratio as large as observed is between 50% and 70%. Therefore, a 9:3:4 ratio is compatible with the data.

19. (a) The inheritance of the trait appearing in the pedigree is most probably due to a recessive allele. This assessment is based upon several features of the pedigree, i.e., the trait appears to occur equally in both males and females; individuals with the trait are not present every generation; and the trait appears among the progeny of some parents who are both phenotypically normal.

(b) Individual III-3 is heterozygous (*Aa*) for the recessive allele. She inherited the recessive allele from her father who displayed the recessive phenotype and thus was homozygous (*aa*). III- 6 is heterozygous (*Aa*) because she had two homozygous recessive offspring. It follows that she inherited the recessive allele from II-7 (*Aa*). III-8 is a son of II-7 and there is a 1/2 chance that he is heterozygous (*Aa*). The probability of having a child with the trait is the probability of III-3 being heterozygous (P = 1) times the probability of III-8 being heterozygous (P = 1/2) times 1/4 = 1/8.

(c) For IV-1 and IV-10 to have a child that shows the trait (homozygous recessive, aa) requires that both be heterozygous (*Aa*). Since IV-1 has a homozygous recessive sibling (*aa*) and phenotypically normal parents (*A_*), both parents must be heterozygous (*Aa*). It follows that IV-1 has a 2/3 chance of being heterozygous (see problem 12 solution above). For the same reasons, IV-10 has a 2/3 chance of being heterozygous for the recessive allele. The probability of IV-1 and IV-10 having a child with the trait (*aa*) is 2/3 x 2/3 x 1/4 = 1/9.

20. (a) The mode of inheritance of the trait shown in the pedigree is most probably due to an autosomal dominant allele. This was concluded because the trait (progressing upward through the pedigree from affected individuals) is not skipping generations and it occurs equally in both males and females.

(b) Since individuals IV-4 and IV-6 do not express the trait and, therefore, do not have the dominant allele, they cannot have a child homozygous for the allele determining the dominant trait.

(c) Individual IV-5 shows the dominant trait. Her mother was unaffected (*aa*) and her father showed the dominant phenotype and was heterozygous (*Aa*). IV-5 must be heterozygous (*Aa*) for the dominant allele. IV-6 is phenotypically normal and does not carry a dominant allele. If IV-5 (*Aa*) and IV-6 (*aa*) marry, the probability of producing a child (*Aa*) with the trait is 1/2.

4

Extensions of Mendelism

ALLELIC VARIATION AND GENE FUNCTION
    Incomplete Dominance and Codominance
    Multiple Alleles
    Allelic Series
    Testing Gene Mutations for Allelism
    Variation Among the Effects of Mutations
    Genes Function to Produce Polypeptides
    Why Are Some Mutations Dominant and Others Recessive?

GENE ACTION: FROM GENOTYPE TO PHENOTYPE
    Influence of the Environment
    Environmental Effects on the Expression of Human Genes
    Penetrance and Expressivity
    Gene Interactions
    Epistasis
    Pleiotropy

A MILESTONE IN GENETICS: GARROD'S INBORN ERRORS OF METABOLISM

IMPORTANT CONCEPTS

A. Mendel's principle of dominance has been modified by the discovery of incompletely dominant and codominant alleles.
    1. Incomplete or partial dominance is when a heterozygous allelic pair results in a phenotype intermediate to that of the respective homozygous genotypes.
    2. Codominance is when both members of a heterozygous allelic pair are expressed.
B. Alternative forms of genes are called alleles.
    1. Some genes exist in multiple allelic states.
    2. Most mutant alleles are recessive to the wild-type alleles because they result in a loss of gene function; however, some mutant alleles are dominant to the wild-type allele because they supersede or interfere with its function.
    3. Recessive mutations can be tested for allelism by combining them in the same individual. If a mutant phenotype occurs in the hybrid, the mutants are alleles of the same gene. If a dominant phenotype occurs in the hybrid, then the mutations are generally nonallelic.
    4. Visual mutations affect some aspect of the phenotype.
    5. Sterile mutations limit reproductive ability.
    6. Lethal mutations kill their carriers.
C. The function of most genes is to produce a polypeptide which may be an enzyme.
    1. Enzymes catalyze specific steps in biochemical pathways.
D. Gene action can be influenced by environmental factors.
E. Different genes may interact to determine a phenotype.
    1. Epistatic interactions, in which a gene has an overriding effect on the phenotype, may indicate that the genes involved control different steps in a pathway.
        a. Epistasis can result in many different F_2 dihybrid ratios, depending upon the type of dominance displayed. Some common F_2 dihybrid ratios involving full dominance are 9:3:4 (recessive epistasis), 12:3:1 (dominant epistasis), 9:7 (duplicate recessive epistasis), 15:1 (duplicate dominant epistasis), and 13:3 (both dominant and recessive epistasis).
F. The action of a gene may affect more than one aspect of the phenotype. This phenomenon is called pleiotropy.

G. When individuals do not show a trait even though they have the appropriate genotype, the trait exhibits reduced or incomplete penetrance.
H. The term expressivity is used if a trait is not manifested uniformly among individuals that show it.
I. In the early twentieth century, a British physician Archibald E. Garrod observed that certain abnormalities caused by recessive alleles ("inborn errors of metabolism") were associated with metabolic abnormalities.
   1. Garrod's work led to the concept that metabolism is under genetic control.
      a. Genes encode enzymes that catalyze specific biochemical reactions.

IMPORTANT TERMS

In the space allotted, concisely define each term.

complete or full dominance:

partial, incomplete, or semidominance:

codominance:

multiple alleles:

wild-type:

polymorphic:

mutation:

visible mutations :

lethal mutations:

incomplete penetrance:

loss-of-function alleles:

gain-of-function mutations:

expressivity: _____

epistasis:

pleiotropic:

IMPORTANT NAMES

In the space allotted, concisely state the major contribution made by the individual.

William Bateson:

Archibald Garrod:

George Beadle and Edward Tatum:

TESTING YOUR KNOWLEDGE

*In this section, answer the questions, fill in the blanks, and solve the problems in the space allotted. Problems noted by an * are solved in the Approaches to Problem Solving section at the end of the chapter:*

1. The expression of both alleles of a heterozygous gene pair is called _____.

2. The percentage of individuals of a genotype that show the expected phenotype is called _____.

3. The degree to which a genotype is expressed in the phenotype is called _____.

4. The multiple phenotypic effect of a gene is called _____.

5. The suppression of the expression of a gene by another nonallelic gene is called _____.

6. How many different genotypes can occur in a population for a gene that exists in 6 different allelic forms?

7. Epistasis influences:

   (a) genotypic ratios
   (b) segregation of alleles
   (c) the kinds of gametes formed
   (d) the phenotypic ratio
   (e) none of the above

8. The phenomenon occurring when a single-gene heterozygous genotype results in a phenotype intermediate to that of the two respective homozygous genotypes is called _____.

9. When black mice of a true-breeding (homozygous) line are crossed with mice from a true-breeding white line, all the F_1s are black. The F_2 generation consists of about 9 black: 3 brown: 4 white. This is an example of _____ (be specific).

10. You are studying a disease in humans caused by a recessive allele and find that not all individuals who inherit two recessive alleles have the expected disease symptoms. Which genetic term is used to describe this phenomenon?

*11. A recessive trait (*aa*) in mice has a penetrance of 0.75. From the cross *Aa* X *Aa*, what is the expected proportion of normal to affected progeny?

*12. A mouse with the wild-type phenotype:

   (a) must be homozygous for recessive alleles at every locus
   (b) must be homozygous for dominant alleles at every locus
   (c) must have a dominant allele at each locus, but does not have to be homozygous
   (d) may be heterozygous at every locus and still be wild-type
   (e) none of the above

*13. You have obtained a series of mutants, all affecting the same phenotype, e.g., coat color. Each mutant, when crossed to a wild-type, genetically behaves as a single gene recessive mutation. Explain how you might determine whether these mutations are of the same gene or of more than one gene.

*14. In British cattle there are phenotypes with normal long legs called Kerry and others with extremely shortened legs called Dexter. The Kerry cattle, when crossed to each other, produce only progeny with the Kerry phenotype. Dexter cattle, when crossed with Kerry cattle, produce the Kerry and Dexter phenotypes in a 1:1 ratio. It is not possible to obtain true breeding strains of the Dexter phenotype. When Dexter are crossed with Dexter, about 1/4 of the calves are the Kerry phenotype, 1/2 are the Dexter phenotype, and 1/4 are spontaneously aborted after about seven months of gestation. These calves are born dead and have extremely shortened legs and nose. Their superficial resemblance to a bulldog led to the name of bulldog calves. Interpret the genetics of the Kerry and Dexter phenotypes.

*15. An insect from a strain, breeding true for white eyes, was crossed to an insect from a strain breeding true for red eyes. The F_1s all had red eyes. The F_1s were crossed to produce an F_2 generation that consisted of 175 red-eyed: 62 cream-eyed: 81 white-eyed.

(a) The results illustrate the phenomenon of _____.

(b) Provide genotypes for the parents, F_1s and F_2s of this cross.

(c) Illustrate a biochemical pathway that best explains the steps in pigment production, and indicate the step affected by each gene.

*16. Now assume when you crossed the red-eyed and white-eyed insects that the F_1s all had white eyes, and the F_2s consisted of 243 white-eyed: 58 red-eyed: 21 cream-eyed individuals.

(a) The results illustrate the phenomenon of _____.

(b) Provide genotypes for the parents, F_1s and F_2s of this cross.

(c) Illustrate a biochemical pathway that best explains the steps in pigment production, and indicate the step affected by each gene.

*17. In a particular species of fish, a region proximal on the tail may be unspotted or possess one spot, two spots, three spots, or a crescent marking. The following table lists a series of fish crosses and their offspring in regard to tail markings.

Cross	Parents	Offspring
A	one-spot X one-spot	95 one-spot
B	crescent X crescent	101 crescent
C	unspotted X unspotted	98 unspotted
D	one-spot X one-spot	74 one-spot, 26 unspotted
E	crescent X crescent	69 crescent, 22 unspotted
F	crescent X one-spot	49 crescent, 23 one-spot, 27 unspotted
G	crescent X two-spot	50 crescent, 24 two-spot, 25 unspotted
H	two-spot X one-spot	98 three-spot
I	three-spot X three-spot	55 three-spot, 27 one-spot, 28 two-spot

Analyze these data and indicate all possible genotypes for the following phenotypes (use logical allelic symbols of your choice).

unspotted   _____

one-spotted   _____

two-spotted   _____

three-spotted   _____

crescent   _____

*18. A geneticist crossed two true-breeding lines of a pet bird. One line "chirped" and had maroon feathers. The other line "screeched" and had brown feathers. All the F_1s had maroon feathers and chirped. The F_1s were crossed and the F_2 data were distributed in the following ratio.

27/64   maroon feathers, chirper
12/64   brown feathers, chirper
9/64   maroon feathers, screecher
9/64   red feathers, chirper
4/64   brown feathers, screecher
3/64   red feathers, screecher

Provide a legend for the above inheritance and indicate the specific type of gene interaction (if any) that occurs in this cross.

*19. In wild sunflowers, populations occur that have yellow flowers with black centers (wild-type), and yellow flowers with yellow centers. The following summarizes a series of crosses involving true breeding sunflower plants:

Cross #1   P   wild-type X yellow-centered (plant 1)

                F_1 black-centered

                F_2 75 black-centered, 26 yellow-centered

Cross #2   P   wild-type X yellow-centered (plant 2)

                F_1 black-centered

                F_2 62 black-centered, 21 yellow-centered

Cross #3   P   yellow-centered (plant 1) X yellow-centered (plant 2)

                F_1 black-centered

                F_2 56 black-centered, 44 yellow-centered

(a) Using genetic symbols of your choice, provide a legend for the inheritance of color of flower centers in the sunflower.

(b) What specific kind of interaction is apparent from the results?

THOUGHT CHALLENGING EXERCISE

The himalayan phenotype in rabbits consists of a white body with dark extremities, i.e., the tips of the nose, ears and legs. This phenotype results from a single gene recessive mutation of a dominant wild-type allele. The wild-type allele determines a fully colored rabbit. The himalayan mutation also occurs in Siamese cats and leads to the seal point phenotype of a lightly pigmented body and darkly pigmented extremities. It has been postulated that both the wild-type allele and the himalayan mutation produces an enzyme that catalyzes pigment formation, but that the himalayan allele produces a defective enzyme that is functionally temperature sensitive. How might you test the hypothesis that the expression of the himalayan allele is temperature sensitive?

SUMMARY OF KEY POINTS

Genes often have multiple alleles.

Mutant alleles may be dominant, recessive, incompletely dominant or codominant.

If a hybrid that inherited a recessive mutation from each of its parents has a mutant phenotype, then the recessive mutants are alleles of the same gene; if the hybrid has a wild phenotype, then the recessive mutations are alleles of different genes.

Most genes encode polypeptides.

In homozygous condition, recessive mutations often abolish or diminish polypeptide activity.

Some dominant mutations produce a polypeptide that interferes with the activity of the polypeptide encoded by the wild-type allele of the gene.

Gene action is affected by biological and physical factors in the environment.

Two or more genes may determine a trait.

A mutant allele is epistatic to a mutant allele of another gene if it has an overriding effect on the phenotype.

A gene is pleiotropic if it influences many different phenotypes.

ANSWERS TO QUESTIONS AND PROBLEMS

1) codominance **2)** penetrance **3)** expressivity **4)** pleiotropy **5)** epistasis **6)** 21 **7)** d
8) partial or incomplete dominance **9)** recessive epistasis **10)** reduced penetrance **11 to 19)** see *Approaches to Problem Solving* section.

Some ideas concerning thought challenging exercise:

One way to test the hypothesis that the himalayan allele is a temperature sensitive mutant of the wild-type allele would be to shave some of the fur from the back of the cat and to put a cool pack on it for several days as the hair grows back. If the allele is temperature sensitive, the fur should grow back fully pigmented. Another test of the hypothesis would be to shave a patch of dark colored fur from the tail and cover the area with a thick patch to elevate the temperature. If the allele is temperature sensitive, it should not be expressed at the higher temperature and the fur should grow back lightly pigmented. Of course, more sophisticated experiments could be designed involving cell extracts to measure the rate of conversion of a precursor compound to melanin pigment *in vitro*.

APPROACHES TO PROBLEM SOLVING

11. From the mating, *Aa* X *Aa*, there is a .25 probability of obtaining "aa". The "aa" genotype displays reduced penetrance (.75). Therefore, the probability of obtaining an individual expressing the trait is (.25)(.75) = .1875 or 3/16. The ratio of normal to affected progeny is expected to be 13/16: 3/16 = 13:3.

12. A phenotypically wild-type mouse will be expressing the most common, non-selected phenotype. Some of the wild-type traits will be determined by dominant alleles, others by recessive alleles. Therefore, the answer is (e) none of the above.

13. If all of the recessive mutant alleles are of the same gene, they should yield a mutant phenotype when combined in a hybrid. A dominant hybrid phenotype results from complementation of alleles of different genes. Additional evidence that the mutants are all of the same gene is obtained if only monohybrid ratios are observed in F_2 progeny of crosses between the mutant individuals. If a dihybrid ratio is observed in the F_2 generation of a cross, then the parents must have differed at two gene pairs.

14. The cross of Dexter X Kerry gives a 1:1 ratio, suggesting that a heterozygous gene pair is segregating in one parent. Kerry X Kerry crosses always produce Kerry offspring. Dexter X Dexter crosses produce 1/4 Kerry: 1/2 Dexter: 1/4 bulldog calves. This indicates that the Dexter phenotype is heterozygous for the Dexter and Kerry alleles. Dexter is lethal in the homozygous condition resulting in bulldog calves, but acts like a dominant allele in the heterozygous condition.

15. The 175 red, 62 cream, and 81 white progeny in the F_2 reduces to a 9/16: 3/16: 4/16 ratio. The 16 in the denominator indicates that two gene pairs are segregating in the F_1, and therefore, the F_1 is a dihybrid. The F_1 can be arbitrarily given the genotype of *AaBb*. The phenotypes of the F_2 can be interpreted in terms of a modified dihybrid ratio:

$$9 \ A\_ B\_$$
$$3 \ A\_ b\,b$$
$$3 \ a\,a\,B\_$$
$$1 \ a\,a\,b\,b$$

The dominant alleles "*A*" and "*B*" interact to produce the red phenotype which is expected to occur in a frequency of 9/16. The white phenotype results from recessive epistasis of "*aa*", and comprises 4/16 of the progeny. A colored phenotype can occur only if the genotype is *A_*. The cream allele "*b*" is a recessive mutant of the red allele "*B*". The cream phenotype results from the genotype *A_ bb* which occurs 3/16 of the time.

(a) The results illustrate the phenomenon of recessive epistasis (9:3:4 ratio).

(b) The genotypes for the parents, F_1s and F_2s of this cross are:

P   red (*AABB*) X white (*aabb*)

F_1 red (*AaBb*)

F_2 A_ B_   = red
             A_ b b = cream
             a a B_   = white
             a a b b = white

(c) Illustrate a biochemical pathway that best explains the steps in pigment production, and indicate the step affected by each gene.

Dominant alleles code for enzymes which carry out specific steps in biochemical pathways. Recessive alleles are mutants of the dominant allele that code for a nonfunctional or partially functional enzyme or no enzyme at all. Homozygosity for a recessive allele may result in a block in a biochemical pathway because of production of a nonfunctional enzyme. With these concepts in mind, a biochemical pathway can be illustrated as follows:

```
          Gene A              Gene B
            ⇓                   ⇓
         enzyme A            enzyme B
colorless  ────────→  cream  ────────→  red
precursor            pigment           pigment
```

16. The 243 white: 58 red: 21 cream reduces to a 12/16: 3/16: 1/16 ratio or 12: 3: 1. The 16 in the denominator indicates that the F_1 is segregating for two gene pairs. It follows that the F_1 can be written as *Aa Bb*. The F_2 can be analyzed in respect to a modified 9:3:3:1 dihybrid ratio.

        9   A_ B_
        3   A_ b b
        3   a a B_
        1   a a b b

If dominant epistasis is assigned to the white determining "*A*" allele, then anytime "*A*" is present the eye color will be white, regardless of the genotype at the second gene pair. The insect must be "*aa*" for color to be expressed. The allele "*B*" results in red eye color only if the insect is "*aa*". The recessive cream eye color phenotype results if the insect is *aa bb*.

(a) The results illustrate the phenomenon of dominant epistasis (12:3:1 ratio).

(b) Provide genotypes for the parents, F_1s and F_2s of this cross.

P   red (*aa BB*) X white (*AAbb*)

F_1 white (*Aa Bb*)

F_2 A_ _ _   = white
             a a B_   = red
             a a b b = cream

(c) Illustrate a biochemical pathway that best explains the steps in pigment production, and indicate the step affected by each gene.

A hypothetical biochemical pathway can be illustrated by having an enzyme, coded by allele. "B", convert a cream colored precursor to the red pigment. Allele "b" does not code for a functional enzyme. In our model, an enzyme coded by allele "A" is much more active than the enzyme coded by allele "B", and when present all of the cream colored pigment is converted to a colorless compound and white eyes result. The alleles at the second gene pair are expressed only if the insect is "aa" at the first gene pair, assuming that the "a" allele does not code for a functional enzyme.

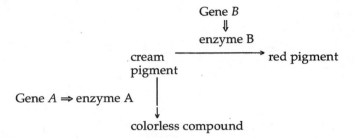

17. To approach this problem it is best to first look at the ratios of the crosses. In doing this it is seen that all data can be reduced to 3:1, 1:2:1, or 1:1 ratios, which all result from monohybrid crosses. Therefore, multiple alleles of a single gene are segregating in these crosses.

Crosses A, B, and C indicate that one-spot, crescent, and unspotted can be in the form of homozygous genotypes. This is concluded from the fact that no segregation of phenotypes is seen in these crosses.

Cross D indicates that the one-spot allele (arbitrarily designated as t^1) is dominant to the unspotted allele t^0 since a 3:1 ratio of one-spot to unspotted is observed among the progeny.

Cross E indicates that crescent (t^C) is dominant to unspotted.

Cross F suggests that the genotype of crescent is $t^C t^0$ and the genotype of one-spot is $t^1 t^0$. Since the crescent phenotype makes up 1/2 of the progeny, crescent must be dominant to both the unspotted and one-spot allele and one-spot is dominant to unspotted.

Cross G likewise indicates that the crescent fish is heterozygous for crescent and unspotted, and the two-spot fish is heterozygous for the two-spot and the unspotted alleles.
This cross, $t^C t^0$ X $t^2 t^0$ produces 1/2 crescent ($t^C t^2$ and $t^C t^0$), 1/4 two-spot ($t^2 t^0$) and 1/4 unspotted ($t^0 t^0$). The crescent allele is dominant to the two-spot allele. The two-spot allele is dominant to the unspotted allele.

Cross H suggests that three-spot results from the heterozygous condition for two-spot and one-spot.

Cross I confirms that the genotype of three-spot is $t^1 t^2$. Three spot ($t^1 t^2$) X three-spot ($t^1 t^2$) results in 1/2 three-spot ($t^1 t^2$), 1/4 one-spot ($t^1 t^1$) and 1/4 two-spot ($t^2 t^2$)

All possible genotypes are indicated for the following phenotypes:

unspotted	$t^0 t^0$
one-spotted	$t^1 t^1$, $t^1 t^0$
two-spotted	$t^2 t^2$, $t^2 t^0$
three-spotted	$t^1 t^2$
crescent	$t^C t^C$, $t^C t^0$, $t^C t^1$, $t^C t^2$

18. A superficial glance at the ratios with 64 in the denominator indicates that three gene pairs are segregating in this cross. It is easiest to analyze this cross by breaking it into simpler components. Therefore, first consider the inheritance of chirping and screeching.

$$P \qquad \text{chirper} \ X \ \text{screecher}$$

$$F_1 \qquad \text{chirper}$$

$$F_2 \qquad 27/64 + 12/64 + 9/64 = 48/64 \text{ chirpers}$$

$$9/64 + 4/16 + 3/64 = 16/64 \text{ screechers}$$

The 3:1 ratio indicates that chirper is dominant to screecher.

Next, consider the inheritance of maroon, red, and brown.

$$P \qquad \text{maroon feathers} \ X \ \text{brown feathers}$$

$$F_1 \qquad \text{maroon feathers}$$

$$F_2 \qquad 27/64 + 9/64 \quad = 36/64 \text{ maroon}$$

$$9/64 + 3/64 \quad = 12/64 \text{ red}$$

$$12/64 + 4/64 \quad = 16/64 \text{ brown}$$

The 36:12:16 or 9:3:4 ratio indicates that two genes are involved in inheritance of feather color and also that recessive epistasis occurs for the brown allele.

The legend is:

chirper	$= C \_$	maroon	$= R\_ \ B\_$
screecher	$= c \ c$	red	$= r \ r \ B\_$
		brown	$= \_ \_ \ b \ b$

19. Crosses #1 and #2 indicate that the two yellow-center strains differ from the wild-type (black center) at one gene pair. Black is dominant to yellow as deduced from the 3:1 ratio in each cross. Further analysis indicates that the yellow-center alleles of cross #1 and cross # 2 are of two different gene pairs because they complement each other to give the wild-type in the F_1 hybrid. The dihybrid nature of this cross is seen when the F_2 data of 56 black-centered to 44 yellow-centered are reduced to a 9/16 to 7/16 ratio. This is a duplicate recessive epistatic ratio. Therefore, the yellow-centered alleles of each gene pair display recessive epistasis. The legend for this inheritance is:

$$B \_ \ C\_ \quad = \text{black center}$$
$$\_ \_ \ c \ c \quad = \text{yellow center}$$
$$b \ b \ \_ \_ \quad = \text{yellow center}$$

5

The Chromosomal Basis of Mendelism

IMPORTANT CONCEPTS

   A. Each species has a characteristic chromosome number.
   1. A human has 23 pairs of chromosomes ($2n = 46$).
   2. A *Drosophila* has four pairs of chromosomes ($2n = 8$).
   B. The discovery of sex chromosomes paralleled the re-discovery of Mendel's work at the turn of the twentieth century.
   1. E. B. Wilson and others observed that the differences in the sexes were confined to the single pair of sex chromosomes, and that the behavior of these chromosomes during meiosis could account for the inheritance of sex.
   a. Such observations supported the emerging *chromosome theory of inheritance* which postulated that genes were found on chromosomes and the behavior of chromosomes during meiosis may explain Mendel's principles of segregation and independent assortment.
   2. In humans and *Drosophila*, females have two X chromosomes and males have an X and a Y chromosome.
   3. Chromosomes in the genome other than sex chromosomes are called autosomes.
   C. Thomas Hunt Morgan discovered that the white-eye gene in *Drosophila* behaved genetically as if it were located on the X chromosome.
   1. Morgan's student, Calvin Bridges, proved that the X chromosome carried the white-eye gene by demonstrating that exceptions to the pattern of X-linked inheritance were caused by nondisjunction of the X chromosomes during meiosis.
   2. Morgan's group discovered that many genes could be assigned to the X chromosome of *Drosophila*.

3. Morgan's group also identified many genes located on autosomes.

D. The early studies of *Drosophila* greatly strengthened the view that all genes were located on chromosomes and that Mendel's principles were explainable by the transmissional properties of chromosomes during meiosis.

E. The inheritance of X-linked traits conforms to the following pattern.
 1. A male progeny receives a Y chromosome from his father and an X chromosome from his mother. Therefore, an X-linked trait cannot be transmitted from father to son.
 2. A female receives one X chromosome from her mother and the other from her father.
 3. In pedigrees, an X-linked recessive trait is expressed more often in males than females. It typically skips generations; being passed from male to heterozygous daughters to about half of grandsons.
 4. For a female to express a sex-linked recessive trait, she must be homozygous; receiving an X chromosome carrying the recessive allele from both her mother and father.
 5. A sex-linked dominant trait will be passed from an affected male to all his daughters.
 6. A female, heterozygous for the dominant trait, will pass it on to about half of her progeny, regardless of sex.

F. In humans, hemophilia and color blindness are due to recessive mutations of X-linked genes.
 1. These traits are more prevalent in males than females because males are hemizygous for the X chromosome; affected males always inherit the mutant allele from their heterozygous mothers.
 2. For a female to be homozygous for a sex-linked recessive trait requires that her father and mother both carry the mutant allele.
 3. The fragile X syndrome is caused by an X-linked dominant allele with reduced penetrance.

G. Only a few Y-linked genes have been discovered in humans.
 1. A gene for the testis-determining factor (*TDF*), required for male sexual differentiation, is located on the Y chromosome; without this factor, a human embryo develops into a female.

H. Mechanisms of sex determination vary among organisms.
 1. In *Drosophila*, sex determination is based on the ratio of X chromosomes to sets of autosomes. A ratio of 1.0 or greater determines a female, a ratio of 0.5 or less results in a male, and a ratio between 0.5 and 1.0 leads to an intersex.
 2. In birds and moths females are the heterogametic sex (ZW) and males are the homogametic sex (ZZ).
 3. Diploid honeybees are females and haploids are males.
 4. In some insects, such as grasshoppers, the Y chromosome has been eliminated completely; females are XX and males are XO.

I. Dosage compensation is a phenomenon in which the activity of a gene is increased or decreased according to the number of copies of that gene in the cell.
 1. Dosage compensation for X-linked genes in *Drosophila* is achieved by hyperactivation of a single X chromosome in males.
 2. In mammals, dosage compensation is achieved by randomly inactivating one of the two X chromosomes in females.
    a. The inactivated X chromosome takes the form of a heterochromatic Barr body and is late in replicating its DNA in the S phase of the cell cycle.
    b. In chromosomally abnormal mammals, including humans, with more than two X chromosomes, all the X chromosomes, except one, form a Barr body.

IMPORTANT TERMS

In the space allotted, concisely define each term.

chromatin:

euchromatin:

heterochromatin:

haploid:

haploid genome (n):

diploid ($2n$):

X chromosome:

Y chromosome:

sex chromosomes:

autosomes:

hemizygote:

chromosome theory of heredity:

disjoin:

nondisjunction:

hemophilia:

testis determining factor (*TDF*):

sex determining region Y (*SRY*):

testicular feminization:

heterogametic:

homogametic:

hyperactivation:

genetic mosaics:

Barr body:

IMPORTANT NAMES

In the space allotted, concisely state the major contribution made by each individual.

Thomas Hunt Morgan:

Edmund Beecher Wilson:

C. E. McClung:

N. M. Stevens:

W. S. Sutton:

Calvin B. Bridges:

Mary Lyon:

Murray Barr:

TESTING YOUR KNOWLEDGE

*In this section, answer all questions, fill in the blanks, and solve the problems in the space allotted. Problems noted with an * are solved in the Approaches to Problem Solving Section at the end of the chapter.*

1. The position a gene occupies on a chromosome is called a _____.

2. The X chromosome in mammals that is seen as a heterochromatin spot in an interphase nucleus is called a _____.

3. The person who first conclusively proved that a gene was located on a chromosome was _____.

4. In humans, about 1/500 phenotypically normal males have an extra Y chromosome, i.e., XYY. Account for the origin of the XYY karyotype.

5. A mammal with the 48 chromosomes, XXXY, would have _____ Barr bodies in somatic cell nuclei.

6. The person who discovered sex linkage in *Drosophila* was _____.

*7. A male is hemizygous for a sex-linked dominant allele, causing a serious abnormality that shows 60% penetrance. What is the probability that if he has a daughter that she will be afflicted with the abnormality? What is the probability that he will have an afflicted son?

8. Why do nuclei of somatic cells of female, but not male, mammals possess a Barr body?

*9. A normal female has a father with a skin disorder called *ichthyosis*. This is caused by a sex-linked recessive allele that results in severely scaled skin.

(a) What is the probability that she will have an affected son if she marries a normal male?

(b) What is the probability that she will have an affected son if she marries a male with *ichthyosis*?

(c) What is the probability that she will have an affected daughter if she marries a normal male?

(d) What is the probability that she will have an affected daughter if she marries a male with *ichthyosis*?

*10. In *Drosophila*, waxy wings are inherited as a sex-linked recessive trait, and hairy body is inherited as an autosomal dominant trait. Indicate the F_1 and F_2 phenotypic ratios expected from a cross between a waxy wing male and a hairy body female.

*11. Provide a legend for the inheritances of the following cross of chickens:

P   white skin, non-barred feathers, female X yellow skin, barred feathers, male

F_1 both males and females have barred feathers and white skin

F_2 3/16 barred feathers, white skin females
1/16 barred feathers, yellow skin females
3/16 non-barred feathers, white skin females
1/16 non-barred feathers, yellow skin female
6/16 barred feathers, white skin males
2/16 barred feathers, yellow skin males

*12. What is the most probable mode of inheritance for the rare trait indicated in the following pedigree?

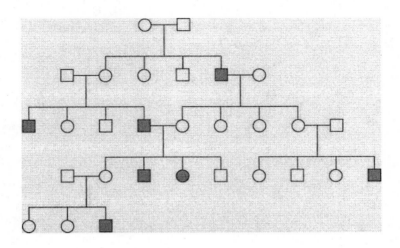

*13. What is the most probable mode of inheritance for the rare trait indicated in the following pedigree?

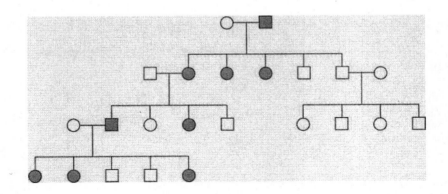

*14. What is the most probable mode of inheritance for the rare trait indicated in the following pedigree?

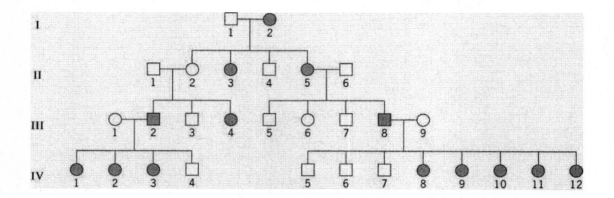

*15. A wildlife photographer who was also a geneticist observed one population of mountain sheep with blue eyes and horns, and another population with brown eyes and no horns. Several brown-eyed hornless females were mated with horned, blue-eyed males. The F_1 progeny consisted of only horned, brown-eyed males and hornless, brown-eyed females. Matings among these F_1s produced the following F_2 progeny:

3/16	horned, brown-eyed males
1/8	horned, brown-eyed females
1/16	hornless, brown-eyed males
3/8	hornless, brown-eyed females
3/16	horned, blue-eyed males
1/16	hornless, blue-eyed males

Using allelic symbols of your choice, provide a legend for the above inheritance. State the specific types of inheritance involved in the cross.

*16. In a *Drosophila* cross involving a straight-winged, round-eyed female and a curved-wing oval-eyed male, all the F_1 flies had straight wings and round eyes. The F_2 generation was as follows:

305 straight-winged, round-eyed males
607 straight-winged, round-eyed females
299 straight-winged, oval-eyed males
103 curved-winged, round-eyed males
201 curved-winged, round-eyed females
 99 curved-winged, oval-eyed males

What were the genotypes of the parents and the F_1s of the above cross?

*17. You have selected a laboratory population of mice consisting of black and orange males and tortoise shell females. You make the observation that the individual tortoise shell females have mated with both black and orange mice in the cage. What type of litter would unambiguously demonstrate that a tortoise shell mouse actually conceived by more than one male?

THOUGHT CHALLENGING EXERCISE

A lawyer friend asks your opinion on whether or not he should pursue a legal case requested of a client. He explains to you that a woman has given birth to a son that has been medically diagnosed with Lesch-Nyhan syndrome. She claims that her husband worked in a chemical plant for ten years before they met and was chronically exposed to vinyl chloride (a suspected mutagen). The couple has it set in their minds that the exposure to vinyl chloride induced a mutation in the man that in turn was passed on to their son. Based upon genetic knowledge, what would be your advice to the lawyer?

SUMMARY OF KEY POINTS

Individual chromosomes become visible during cell division; between divisions they form in a diffuse network of fibers called chromatin.

Diploid somatic cells have twice as many chromosomes as haploid gametes.

The cells of males and females may have different numbers of X and Y sex chromosomes; however, the number of autosomes in these cells is the same.

Genes are located on chromosomes.

The disjunction of chromosomes during meiosis is responsible for the segregation and independent assortment of genes.

Nondisjunction during meiosis leads to abnormal numbers of chromosomes in gametes, and ultimately, in zygotes.

Disorders such as hemophilia and color blindness, which are caused by recessive X-linked mutations, are more common in males than in females.

In humans the Y chromosome carries fewer genes than the X chromosome.

In humans pseudoautosomal genes are located on both the X and Y chromosomes.

In humans sex is determined by a dominant effect of the SRY gene on the Y chromosome; the product of this gene, the testis-determining factor (TDF), causes a human embryo to develop into a male.

In *Drosophila*, sex is determined by the ratio of X chromosomes to sets of autosomes (X:A). For X:A $\leq$ 0.5, the fly develops as a male, for X:A $\geq$ 1.0, it develops as a female, and for 0.5 < X:A < 1.0, it develops as an intersex.

In honeybees, sex is determined by the number of chromosome sets; haploid embryos develop into males and diploid embryos develop into females.

In *Drosophila*, dosage compensation for X-linked genes is achieved by hyperactivating the single X chromosome in males.

In mammals, dosage compensation for X-linked genes is achieved by inactivating one of the two X chromosomes in females.

ANSWERS TO QUESTIONS AND PROBLEMS

1) locus **2)** Barr body **3)** Calvin Bridges **4)** nondisjunction of the Y chromosomes in the second meiotic division of males to yield a YY sperm **5)** 2 **6)** Morgan **7)** 0.60, 0 **8)** it is a means of dosage compensation in females that have 2 X chromosomes **9) a.** 1/4; **b.** 1/4; **c.** 0; **d.** 1/4 **10)** F_1 - both males and females are hairy, F_2 - 3/16 hairy males, 1/16 normal males, 3/16 waxy hairy males. 1/16 waxy males, 3/8 hairy females, 1/8 normal females. **11)** $W \_$ = white, *ww* = yellow for both males and females (white is dominant to yellow, autosomally inherited); $Z^B Z^B$ and $Z^B Z^b$ = barred males, $Z^b Z^b$ = non-barred male, $Z^B W$ = barred female, $Z^b W$ = nonbarred female (barred is a Z-linked dominant allele) **12)** sex-linked recessive **13)** sex-linked dominant **14)** sex-linked dominant with reduced penetrance **15)** *HH* = horned males and females, *Hh* = horned males and hornless females, *hh* = hornless males and females (horned is dominant in males and recessive in females, or sex-influenced); $X^B X^B$, $X^B X^b$ = brown-eyed female, $X^b X^b$ = blue-eyed female, $X^B Y$ = brown-eyed male, $X^b Y$ = blue-eyed male (brown is dominant to blue and sex-linked) **16)** parents - $c^+ c^+$ $o^+ o^+$ female X *cc* *o* Y male; F_1s, - $c^+ c$ $o^+ o$ females and $c^+ c$ $o^+ Y$ males (curved is an autosomal recessive allele; oval is a sex-linked recessive allele), **17)** a litter possessing a black female and also an orange female.

Ideas concerning the thought challenging exercise:

Lesch-Nyhan syndrome is caused by a sex-linked recessive allele. You would tell the lawyer that the son inherited the mutant allele from his heterozygous mother. It would be best to diagram the inheritance of a sex-linked trait. Indicate that males receive their X chromosome only from their mothers. Show why the son could not have received a mutant allele on his X chromosome from his father. The lawyer would then realize that there were no grounds for the claim that the vinyl chloride induced the mutation that the son inherited, because he inherited it from his mother.

APPROACHES TO PROBLEM SOLVING

7. A male will pass his X chromosome to all his daughters. Therefore, if the gene shows 60% penetrance, 0.60 of the females should be afflicted with the abnormality. None of the sons will be afflicted because they inherit their X chromosome from their mother.

9.   Approach this problem by characterizing the normal female. Since her father had *ichthyosis*, he was of the genotype i Y. The normal daughter must be heterozygous, I i, because she inherited one of her X chromosomes from her father.

(a) Males receive their X chromosome from their mother. Therefore, the normal heterozygous female, *Ii*, would have a 1/2 or 50% chance that her male children would be hemizygous, i Y, for the recessive allele. Since there is a 1/2 chance that a child will be a male, the probability of having an affected son is $(1/2(1/2) = 1/4$.

(b) Since it is the female parent that contributes the X chromosome to her sons, the genotype of the father is inconsequential in this case. The probability is still 1/4.

(c) For a daughter to be homozygous for a sex-linked recessive allele, her father must have been. hemizygous and her mother at least heterozygous for the allele. Since the normal male cannot carry a sex-linked recessive allele, assuming full penetrance, no affected daughters will result from this mating.

(d) If the heterozygous woman, I i, marries a male with *ichthyosis*, i Y, then there is a 50% chance that each daughter will be affected. Since there is a 1/2 probability of having a female child, the probability of having an affected daughter is $(1/2)(1/2) = 1/4$.

10. Approach this problem by first denoting the genotypes at both genes of the parents.

P   waxy male X hairy female
$wY\ h^+h^+$ $w^+w^+\ HH$

Then, proceed to the F_1 generation. Remember males get their X chromosome from their mother and their Y chromosome from their father, and females get an X chromosome from both their mother and father.

F_1 $w^+Y\ Hh^+$ males $w^+w\ Hh^+$ females

Next, indicate the F_2 generalized genotypes that will result in each phenotype. Calculate the probability of obtaining the genotype at each gene pair and multiply the individual probabilities. his gives the expected ratio of the corresponding phenotype as indicated below.

F_2 $w^+\_\ H\_$ probability = 1/2 x 3/4 = 3/8 hairy females
        $w^+\_\ h^+h^+$ probability = 1/2 x 1/4 = 1/8 wild-type females
        $w^+Y\ H\_$ probability = 1/4 x 3/4 = 3/16 hairy males
        $w^+Y\ h^+h^+$ probability = 1/4 x 1/4 = 1/16 wild-type males
        $wY\ H\_$ probability = 1/4 x 3/4 = 3/16 waxy, hairy males
        $wY\ h^+h^+$ probability = 1/4 x 1/4 = 1/16 waxy males

11. This problem is approached by the typical method of analyzing one set of traits at a time. First analyze the inheritance of white and yellow skin.

P   white skin X yellow skin

F_1 all white skin

F_2 white skin females, 3/16 + 3/16 = 6/16
         yellow skin females, 1/16 + 1/16 = 2/16   3:1 white:yellow

         white skin males, 6/16
         yellow skin males, 2/16   3:1 white:yellow male

Since white skin and yellow skin segregate in a 3:1 ratio in both sexes, one autosomal gene is involved with the white allele (*W*) being dominant to yellow (*w*).

Next, analyze the inheritance of barred and non-barred.

P   non-barred female X barred male

F_1 all barred feathered

F_2 barred females, $3/16 + 1/16 = 4/16$
           non-barred females, $3/16 + 1/16 = 4/16$ 1:1 barred: non-barred

           barred males, $6/16 + 2/16 = 8/16$
           non-barred males, none

The 1:1 segregation ratio of barred and non-barred in the F_2 females, in addition to the F_2 males being all barred, indicate a sex-linked trait. Barred is dominant to non-barred. The females show the 1:1 F_2 ratio and therefore are the heterogametic sex. This is a Z Z and Z W sex determining mechanism.

The legend is as follows:

          $W\_$ = white skin $Z^B\,W$ = barred feathered female
          $w\,w$ = yellow skin $Z^b\,W$ = nonbarred feathered female
                                    $Z^B\,Z^B,\ Z^B\,Z^b$ = barred feathered male
                                      $Z^b\,Z^b$ = non-barred feathered male

12. Since the trait occurs mostly in males and generally skips a generation, thereby appearing in grandfathers and grandsons, it fits the pattern of a sex-linked recessive trait. The only female in the pedigree with the trait had an affected father and a mother who was a carrier (heterozygous) for the trait.

13. Notice that this trait does not skip a generation when proceeding from an affected individual on the bottom and ascending to the top of the pedigree. Therefore, it must be determined by a dominant allele. Next, it must be determined whether it is sex-linked or autosomal. The definitive parts of the pedigree are that all affected males pass the trait to all of their daughters, and to none of their sons. This is the expected pattern of inheritance for a dominant sex-linked trait.

14. This pedigree is similar to the one in problem 13, in that males pass the trait on to their daughters. There is an exception in individual II-2. The simplest interpretation is that this trait is due to a sex-linked dominant trait with incomplete penetrance. That assumption is supported by the fact that individual II-2 has affected children.

15. The cross is analyzed one set of traits at a time as follows:

        P   brown-eyed females X blue-eyed males

        F_1 all brown-eyed

        F_2 brown-eyed females, $1/8 + 3/8 = 4/8$
                 brown-eyed males, $3/16 + 1/16 = 4/16$
                 blue-eyed males, $3/16 + 1/16 = 4/16$ 1:1 brown:blue

All F_2 females are brown-eyed, but the F_2 males segregate brown and blue eyes in a 1:1 ratio. These F_2 ratios are expected of a sex-linked trait with brown being dominant to blue, and the male being the heterogametic sex.

        P   hornless female X horned males

        F_1 all males horned, all females hornless

        F_2 horned males, $3/16 + 3/16 = 6/16$
                 hornless males, $1/16 + 1/16 = 2/16$ 3:1 horned:hornless

horned females, 1/8
hornless females, 3/8   1:3 horned:hornless

The 3:1 F$_2$ ratio in each sex suggests that a single gene pair is segregating with simple dominance of one allele. However, the reversed 3:1 ratios indicate that horned is dominant in males and hornless is dominant in females. Alleles showing reversal of dominance related to the sex of the individual are called *sex influenced*.

A legend is:

	females		males
HH	horned		HH, Hh
Hh, hh	hornless		hh
$X^B X^B, X^B X^b$	brown		$X^B Y$
$X^b X^b$	blue		$X^b Y$

16. For this problem, we again use the approach of breaking the more complex set of data into simpler components. Let's arbitrarily start with the inheritance of straight and curved wings.

P   straight-winged female X curved-winged male

F$_1$ all straight-winged

F$_2$ straight-winged males, 305 + 299 = 604
curved-winged males, 103 + 99 = 202   3:1 straight:curved

straight-winged females, 607
curved-winged females, 201   3:1 straight:curved

The 3:1 ratio of straight to curved in the F$_2$ generation indicates that one gene is involved in the inheritance of straight and curved. The straight allele (c^+) is dominant to the curved allele (c).

P   round-eyed female X oval-eyed male

F$_1$ all round-eyed

F$_2$ round-eyed males, 305 + 103 = 408
oval-eyed males, 299 + 99 = 398   1:1 round:oval

round-eyed females, 607 + 201 = 808
oval-eyed females, none

The F$_2$ 1:1 ratio of round to oval eyes in males and only round eyes in F$_2$ females suggest that the oval eye locus is sex-linked with round (o^+) dominant to oval (o).

The genotypes of the parents were: female, $c^+c^+ o^+o^+$; male, $cc\ o$ Y.

The genotypes of the F$_1$s were: females, $c^+c\ o^+o$; males, $c^+c\ o^+$Y.

17. A tortoise shell mouse is always a female and heterozygous for the sex-linked alleles, orange (c^o) and black (c^b). The random lyonization of one of the X chromosomes in each cell at an early stage of embryonic development and the retaining of the inactivation of the same X chromosome in all cells of a lineage result in a mosaic of patches of orange and black fur. If the tortoise shell mouse ($c^o c^b$) conceived only by a black male (c^b Y), then the female progeny could be tortoise shell ($c^o c^b$) or black ($c^b c^b$). If she conceived only by an orange male (c^o Y), then the female progeny could be either tortoise shell ($c^o c^b$) or orange ($c^o c^o$). Therefore, a litter having both an orange female and a black female would indicate that the female tortoise shell mouse had conceived by two different males.

6

Variation in Chromosome Number and Structure

IMPORTANT CONCEPTS

A. Cytogenetic analysis of stained mitotic chromosomes is generally performed at late prophase or metaphase.
    1. Specific treatments with stains such as quinicrine and Giemsa create a characteristic banding pattern along the chromosomes.
    2. The human karyotype consists of 23 pairs of chromosomes, numbered (approximately) from the largest to smallest.
    3. Each chromosome is characterized by a short (p) and a long (q) arm.
    4. Cytological analysis can reveal variation in chromosome number and structure.

B. Polyploidy is the presence of chromosome sets above the diploid (2n) level, e.g., triploidy (3n), tetraploidy (4n), hexaploidy (6n).
    1. Many polyploids are sterile or only show partial fertility because their chromosomes segregate irregularly during meiosis.
        a. The presence of incomplete sets of chromosomes upsets the genetic balance and causes inviability of gametophytes in plants, and inviability of zygotes or abnormal development of embryos in animals.
    2. Alloploids, produced by doubling the chromosome number of chromosomally sterile hybrids, are often fertile because each chromosome has a homolog with which to pair.
    3. Cytogeneticists have been able to induce polyploidy by treating organisms with the spindle-poisoning drug colchicine.

C. A special form of polyploid tissue called polyteny occurs in salivary glands of *Drosophila* larvae.
    1. In polytene nuclei, sister chromatids remain together through many rounds of DNA synthesis, forming large, banded, polytene chromosomes.
        a. These interphase chromosomes are excellent materials for cytogenetic analysis.

D. Aneuploidy involves the over- or under-replication of particular chromosomes or chromosome

segments.
1. Aneuploidy in animals was first detected in *Drosophila* and involved the sex chromosomes.
2. Aneuploidy in plants was originally discovered in mutant strains of *Datura*.
3. Aneuploidy also occurs in humans.
    a. The most common aneuploidy is trisomy for chromosome 21 (Down syndrome).
    b. Triplo-X and XXY (Klinefelter syndrome) are other viable human trisomics.
    c. The only viable human monosome is the 45, X karyotype (XO or Turner syndrome).
4. Deletions or duplications of chromosome segments may also cause aneuploidy.
    a. A deletion of the short arm of human chromosome 5 (5p⁻) causes the *cri-du-chat* syndrome.
    b. In *Drosophila*, a duplication of a segment of the X chromosome causes the Bar eyes phenotype.
E. The gross structure of chromosomes may be altered by inversions and translocations.
1. Inversions reverse the order of genes within a segment of a chromosome.
    a. An inversion that includes the centromere is called pericentric.
    b. Inversions in a chromosome arm that don't involve the centromere are called paracentric.
2. Reciprocal translocations interchange segments between two nonhomologous chromosomes.
    a. In translocation heterozygotes, chromosome disjunction during meiosis may produce aneuploid gametes that cause reduced fertility.
3. Compound chromosomes, such as attached-X, are formed by the fusion of homologous chromosome segments.
4. Robertsonian translocations are formed by fusions between the centromeres of nonhomologous chromosomes.

IMPORTANT TERMS

In the space allotted, concisely define each term.

cytogenetics:

Q banding:

G banding:

R banding:

C banding:

karyotype:

polyploidy:

univalent:

trivalent:

apomixis:

allopolyploids:

autopolyploids:

endomitosis:

polytene:

chromocenter:

aneuploidy:

trisomies:

hypoploid:

hyperploid:

Down syndrome:

Klinefelter syndrome:

monosomy:

Turner syndrome:

somatic mosaics:

gyandromorphs:

deletion (deficiency):

cri-du-chat syndrome:

duplication:

pericentric inversion:

paracentric inversion:

translocation:

reciprocal translocation:

adjacent disjunction:

alternate disjunction:

compound chromosome:

isochromosome:

attached-X chromosome:

Robertsonian translocation:

amniocentesis:

chorionic biopsy:

IMPORTANT NAMES

In the space allotted, concisely state the major contribution made by the individual.

C. B. Bridges:

Albert Blakeslee and John Belling:

F. W. Robertson:

Joe Hin Tjio and Albert Levan:

TESTING YOUR KNOWLEDGE

*In this section, answer the questions, fill in the blanks, or solve the problem in the space allotted. Problems noted with an * are solved in the Approaches to Problem Solving section at the end of the chapter.*

1. Trisomy for chromosome 21 in humans results in _____ syndrome.

2. An inversion that does not include the centromere is called _____.

3. An aberrant chromosome number in which a normally diploid cell has three copies of one chromosome and two copies of all the others is called _____.

4. An organism with four sets of homologous chromosomes is called an _____.

5. A reciprocal exchange of non-homologous chromosomes is called a _____.

6. A phenotypically normal man or woman with only 45 chromosomes and carrying a 14/21 translocation has about a 10% to 15% chance that each child will be afflicted with _____ syndrome.

7. An aberrant chromosomal condition in which a normally diploid cell has one copy of a chromosome and two copies of all others is called _____.

8. An inversion that includes the centromere is called _____.

9. An organism with four sets of chromosomes, two of one genome and two of a different genome, is called an _____.

10. A child trisomic for chromosome 18 has _____ syndrome.

11. An individual possessing three sets of chromosomes is called _____.

12. How many chromosomes would be found in a root tip nucleus of a hexaploid plant, derived from two plants, a diploid with $2n = 14$, and a tetraploid with $2n = 4x = 30$?

13. Which chromosome aberration in the heterozygous condition may lead to an acentric fragment and a bridge at anaphase I of meiosis?

14. A child trisomic for chromosome 13 has _____ syndrome.

15. A deletion of the short arm of one member of chromosome 5 results in _____ syndrome.

16. If a female human has only one X chromosome, an abnormal phenotype results called _____ syndrome.

17. If a human male has two X chromosomes and a Y chromosome, an abnormal phenotype results that is called _____ syndrome.

*18. A couple gives birth to an extremely mentally and physically retarded child. Upon karyotyping of the parents, it is discovered that the man is heterozygous for a normal chromosome and a chromosome carrying a large inversion.

(a) What is the likely cause of the phenotypically abnormal child?

(b) What advice would you give the parents on the probability of them having a second abnormal child?

*19. Two wild bluebell species, when crossed, produced a vigorous but sterile hybrid with large, horticulturally desirable flowers. You want to commercially develop a plant having the showy flowers of the hybrid. Unfortunately, the wild species grow so poorly under cultivated conditions that mass production of hybrid seed is unfeasible. Upon chromosomal analysis of the wild species you find that one has a somatic chromosome number of $2n = 10$, and the other has a somatic chromosome of $2n = 6$. Diagram the easiest procedure by which you may potentially derive a fertile, large-flowered variety for commercial seed production. Indicate the chromosome composition of the derived variety.

*20. Strains 1 and 2 are homozygous for the following respective chromosomal arrangement and alleles.

1. L   M   N   O   P   Q   2.   l   m   p   o   n   q

   L   M   N   O   P   Q   l   m   p   o   n   q

(a) Diagram pairing of the chromosomes during meiosis of the hybrid obtained from crossing strains 1 and 2. Indicate a chiasma in the region between the O and P loci.

(b) Indicate the genotype of the meiotic products formed by the hybrid, if a single crossover occurred between the O and P loci.

(c) Which meiotic products would result in inviable gametophytes if the organism were a plant, or abnormal or inviable zygotes upon fertilization if the organism were an animal? Why?

*21. The New World allotetraploid cotton *Gossypium hirsutum* ($4x = 2n = 52$) has 13 relatively large pairs and 13 smaller pairs of chromosomes. Thirteen of the chromosome pairs apparently were derived from the Old World *G. herbaceum*, which possesses $2n = 26$ large chromosomes. The other thirteen pairs were apparently derived from the New World *G. raimondii*, or a very similar species, which has $2n = 26$ small chromosomes.

(a) If *G. hirsutum* were crossed with *G. herbaceum*, how many bivalents and univalents and their sizes would you expect to see in the hybrid at metaphase I of meiosis?

(b) If *G. hirsutum* were crossed with *G. raimondii*, how many bivalents and univalents and their sizes would you expect to see in the hybrid at metaphase I of meiosis?

(c) How many bivalents would you expect to observe at metaphase I of meiosis in a hybrid between *G. raimondii* and *G. herbaceum*? How many univalents?

*22. In the Jimson weed, trisomic plants can be found for each of the twelve chromosomes. From various crosses of plants (female is always written first), disomic or trisomic for the chromosome on which the locus for purple (p^+) and white (p) flower color resides, indicate the expected proportions of purple and white flower plants among the progeny. Female gametes may carry an extra chromosome, but pollen possessing an extra chromosome is inviable.

(a) p^+p^+p X $p\,p$

(b) p^+p^+p X p^+p^+p

(c) p^+p^+p X p^+p

(d) $p^+p\,p$ X p^+p

(e) $p^+p\,p$ X $p^+p\,p$

*23. The following diagram depicts a pair of mouse chromosomes.

A B C D E F G

A B C D C D E F G

(a) What type of chromosomal aberration is depicted in this heterozygous chromosome pair?

(b) Diagram pairing of these chromosomes at pachynema.

*24. Show how two nonhomologous chromosomes can be rearranged so that genes, formerly on different chromosomes, share the same chromosome. Assume that the resulting individual is fully fertile and has only bivalents formed in meiosis.

*25. What is the expected phenotypic ratio in the progeny of a cross involving two autotetraploids of the genotype AaAa (assume random chromosome segregation)?

*26. What is the expected phenotypic ratio of a cross between two allotetraploids, $A_1a_1 A_2a_2$ X $a_1a_1 a_2a_2$ (the subscripts designate chromosomes of the two different genomes)?

*27. What are the expected phenotypic and genotypic ratios of a cross between two allotetraploids, $A_1A_1 a_2a_2$ X $A_1A_1 a_2a_2$ (the subscripts designate chromosomes of the two different genomes)?

*28. Illustrate how you can get an *aa* gamete from an autotetraploid of the genotype *AAAa*.

THOUGHT CHALLENGING EXERCISE

In some crop plants such as tomato ($2n = 24$), plants can be maintained that are individually trisomic for one of the twelve chromosomes. The twelve different trisomics can be easily identified by their specific abnormal effect on the phenotype. Describe a procedure, using trisomic plants, by which you could assign the newly discovered mutant gene in tomato to a specific chromosome.

SUMMARY OF KEY POINTS

Cytogenetic analysis usually focuses on chromosomes in dividing cells.

Dyes such as quinacrine and Giemsa create banding patterns that are useful in identifying individual chromosomes within a cell.

A karyotype shows the photographed chromosomes of a cell arranged for cytogenetic analysis.

Polyploidy contain extra sets of chromosomes.

Many polyploids are sterile because their multiple sets of chromosomes segregate irregularly in meiosis.

Polyploids produced by chromosome doubling in interspecific hybrids may be fertile if their constituent genomes segregate independently.

In some somatic tissues — for example, the salivary gland of *Drosophila* larvae — successive rounds of chromosome replication occur without intervening cell divisions and produce large polytene chromosomes that are ideal for cytogenetic analysis.

In trisomy, such as Down Syndrome in humans, three copies of a chromosome are present; in a monosomy, such as Turner Syndrome in humans, only one copy of a chromosome is present

Aneuploidy may involve the deletion or duplication of a chromosome segment.

An inversion reverses the order of genes in a segment of a chromosome.

A translocation interchanges segments between two nonhomologous chromosomes.

Compound chromosomes result from the fusion of homologous chromosomes, or from the fusion of the arms of homologous chromosomes.

Robertsonian translocations result from the fusion of nonhomologous chromosomes

ANSWERS TO QUESTIONS AND PROBLEMS

1) Down 2) paracentric 3) trisomic 4) autotetraploid 5) reciprocal translocation 6) Familial Down 7) monosomic 8) pericentric 9) allotetraploid 10) Edward 11) triploid 12) 44 13) paracentric

inversion **14)** Patau **15)** Cri-du-chat **16)** Turner **17)** Klinefelter **18-28)** see *Approaches to Problem Solving* section.

Ideas concerning thought challenging exercise:

A procedure to determine the chromosome on which the new mutant resides follows: 1) cross the mutant plant as a male parent to each of the twelve trisomic plants; 2) select a trisomic hybrid plant from the progeny of each cross; 3) testcross each trisomic plant; 4) determine the phenotypic ratio of mutant to normal in the testcross progeny. For eleven of the twelve trisomics selected, a 1:1 testcross ratio indicates disomic inheritance and the gene is not located on the chromosome that was trisomic. For one of the trisomics a distorted testcross ratio (2:1 to 5:1) will be observed that results from segregation of the heterozygous trisomic condition. This indicates that the gene is located on the chromosome that is trisomic in this cross.

APPROACHES TO PROBLEM SOLVING

18. (a) The key to the abnormal child resides in the fact that crossing over within the inversion loop of an inversion heterozygote will generate recombinant chromatids with terminal duplications and deficiencies. The duplicated/deficient chromosomes will cause genetic imbalance and hence abnormal development.

    (b) The parents should be advised that there is an increased risk of having a second abnormal child. However, a numerical probability cannot be assigned without knowing the frequency of recombination within the inversion loop.

19. The key to solving this problem is the sterility of the hybrid and the fact that the two species differ in chromosome number. This suggests that the sterility of the hybrid may result from lack of chromosome pairing due to divergence of the chromosomes. The doubling of the chromosome number by treatment with colchicine may lead to restoration of fertility because all chromosomes would now have a homolog. If so, then the resulting allotetraploid ($2n = 4x = 16$) could be mass produced from seeds.

20. (a) By analyzing the arrangement of chromosome regions in the two strains, it is seen that the sequence N to P is inverted in strain 2 relative to strain 1. Maximizing of pairing of homologous chromosome regions will lead to the formation of an inversion loop as depicted below at pachynema.

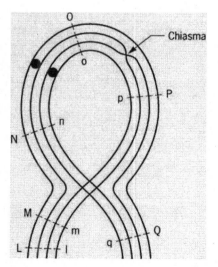

    (b) The crossover between O and P will lead to one chromatid of each chromosome having a duplicated and deficient end. At the end of meiosis, there will be four meiotic products; one with a normally arranged chromosome, one with the chromosome carrying the inversion, and two carrying a duplicated/deficient chromosome as follows:

L M N$_o$O P Q **L M N$_o$O** p **m l** q n$_o$o P **Q** l m p o$_o$n q

| Normal | Duplicated for m and l
Deficient for Q | Duplicated for Q
Deficient for l and m | Inversion |

(c) Due to the genetic imbalance, the meiotic products with duplicated/deficient chromosomes would result in inviable gametophytes in plants and inviable zygotes or abnormal embryonic development in animals.

21. (a) Since *G. hirsutum* contains 13 chromosomes similar to those of *G. herbaceum*, these chromosomes will pair in the triploid hybrid. Therefore, in the hybrid 13 larger bivalents and 13 smaller univalents would be seen at metaphase I of meiosis.

(b) *G. hirsutum* also has 13 chromosomes similar to those of *G. raimondii*. These would pair with the 13 chromosomes of *G. raimondii* in the hybrid forming 13 smaller bivalents; 13 larger univalents would also be observed.

(c) If no multivalents are observed in normal *G. hirsutum* (as is the case), and no trivalents are formed in the hybrids between *G. hirsutum* and the diploids, *G. raimondii* and *G. herbaceum*, then no pairing would be expected in the *G. raimondii* X *G. herbaceum* hybrid; 26 univalents and no bivalents should be observed.

22. To approach this problem, first determine the various genotypes of meiotic products formed by the random (two chromosomes to one pole and one chromosome to the other pole) segregation of the trisomic chromosomes. Also note that male gametophytes containing an extra chromosome grow slower and generally are not involved in fertilization of the egg. Therefore, the extra chromosome is transmitted only through the female parent.

(a) Let's consider the random segregation of chromosomes from the trisome p^+p^+p. To make it easier to follow, the three chromosomes will be denoted as 1, 2, and 3, i.e., $_1p^+{}_2p^+{}_3p$. Three random types of segregation occur during meiosis: 1 and 2 to one pole and 3 to the other pole; 2 and 3 to one pole and 1 to the other; 1 and 3 to one pole and 2 to the other. These segregations result in the four genotypes: p^+p^+, p^+p, p^+ and p in a 1:2:2:1 ratio. Since 1/6 of the gametes from this trisomic plant carry only p there should be a distorted testcross ratio of 5:1 purple to white. Since all female gametophytes carrying an extra chromosome may not survive and since trivalent association of the trisomic chromosomes does not occur all the time, the ratio is generally about 4:1 instead of 5:1.

(b) The female trisomic plant will produce gametophytes containing just p about 1/6 of the time. Since the male meiotic products containing an extra chromosome do not function in fertilization, the frequency of chromosomally normal male gametophytes containing just p will be 1/3. Therefore, in this cross the frequency of white flower plants will be $(1/6)(1/3) = 1/18$. The ratio of purple to white flower plants among the progeny will be 17:1.

(c) The frequency of p meiotic products from the trisomic plant is 1/6. The heterozygous disomic male, p^+p, produces 1/2 of the meiotic products carrying p. Therefore, the expected frequency of pp (white) offspring is expected to be 1/12. The ratio of purple to white is 11:1.

(d) The random segregation from the trisomic female p^+pp will produce the meiotic products pp, p^+p, p, and p^+, in a 1:2:2:1 ratio. Half of the meiotic products are of the genotype pp or p. The male disomic parent produces p^+ and p meiotic products in a 1:1 ratio. This cross segregates purple to white in a 3:1 ratio.

(e) Half of the meiotic products of the female parent are p or pp. The meiotic products with an extra chromosome do not function as viable gametophytes in the male parent. Therefore the male parent, p^+pp, will produce gametophytes carrying p and p^+ in a 2:1 ratio, and the expected ratio of purple to white from this cross is 2:1.

23. (a) An analysis of this pair of chromosomes reveals that the region C - D is represented twice in one of the chromosomes. Therefore, a duplication occurred in one chromosome.

   (b) The chromosomes will pair as homologously as possible. However the duplicated region will form a buckle as indicated below:

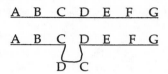

24. A translocation between nonhomologous chromosomes initially results in a translocation heterozygote as illustrated below:

A	A		W	W			A	W		A	W
B	B		X	X			B	X		B	X
C	C		Y	Y	reciprocal translocation occurs between two nonhomologous chromosomes		C	C		Y	Y
D	D		Z	Z			D	D		Z	Z

Alternate segregation from the cross-shaped quadrivalent that results from homologous pairing of the two normal and two translocated chromosomes will produce meiotic products containing either the two normal chromosomes or the two translocation chromosomes. These meiotic products are genetically balanced and have all the regions of the chromosomes present twice as in the normal cells, except that the linkage relationships have changed in the gametes that get the two translocated chromosomes. These are indicated below:

A	W		W	A
B	X		X	B
C	Y		C	Y
D	Z		D	Z

normal chromosomes in half of the gametes resulting from adjacent segregation

translocated chromosomes in half of the gametes resulting from adjacent segregation

Fusion of two gametes containing the two translocated chromosomes will result in a zygote homozygous for both of the translocated chromosomes that form two separate bivalents during meiosis. The chromosomal constitution of such a zygote is shown below:

W	W		A	A
X	X		B	B
C	C		Y	Y
D	D		Z	Z

The translocation homozygote has two new linkage groups that have brought together genes formerly on different chromosomes. Since it is homozygous for these rearranged chromosomes, two bivalents will form in meiosis and the plant will be fully fertile.

25. To start solving this problem, number the four homologous chromosomes of the autotetraploid that carry the A or a alleles as follows:

$$
\begin{array}{cccc}
1 & 2 & 3 & 4 \\
A & a & A & a
\end{array}
$$

Next, determine the genotypes of the meiotic products resulting from random, two-and-two segregation of the four homologous chromosomes. These are indicated below:

$$
\begin{array}{ccccccccc}
1 & 2 & \Leftrightarrow & 3 & 4 & \quad 1 & 3 & \Leftrightarrow & 2 & 4 & \quad 1 & 4 & \Leftrightarrow & 2 & 3 \\
A & a & & A & a & \quad A & A & & a & a & \quad A & a & & a & A
\end{array}
$$

As evident, the aa gametes will occur in a frequency of 1/6. The homozygous recessive genotype will occur in a frequency of $(1/6)(1/6) = 1/36$. Therefore, the phenotypic ratio of the dominant phenotype $A\_\_\_$ to the recessive phenotype $aaaa$ is 35:1.

26. Since this plant is an allotetraploid, the chromosomes designated 1 and 2 will not pair with each other. The two chromosomes designated 1 will pair and the two designated 2 will pair. Therefore, these chromosome pairs will undergo independent assortment. The $a_1 a_2$ gamete is expected to occur in a frequency of 0.25. From the testcross indicated, the frequency of the $a_1a_1\ a_2a_2$ progeny is 0.25. The expected phenotypic ratio is 3:1 dominant ($A\_\_\_$) to recessive ($aaaa$).

27. Because the chromosome pair designated 1 is homozygous for the dominant allele A_1 in both parents, and the second chromosome pair is homozygous for a_2, all gametes from each parent are $A_1 a_2$. All progeny are $A_1A_1\ a_2a_2$.

28. To obtain an aa gamete from an autotetraploid of the genotype $AAAa$ requires crossing over between the gene and the centromere with both recombinant chromosomes going to the same pole at anaphase I of meiosis. This secondary meiocyte will contain two recombinant chromosomes of the genotype Aa and Aa. This secondary meiocyte will proceed through the second meiotic division to produce two meiotic products. These meiotic products would be expected to be of the genotypes Aa and Aa 50% of the time and AA and aa 50% of the time. This is illustrated below.

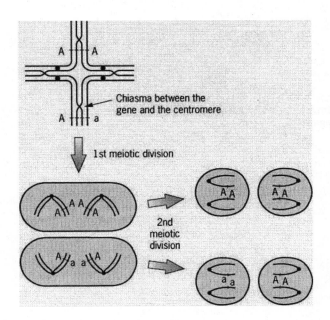

7

Linkage, Crossing Over, and Chromosome Mapping in Eukaryotes

IMPORTANT CONCEPTS

  A. Genes that are located on the same chromosome are called linked.
    1. Linked genes are an exception to the Principle of Independent Assortment.
    2. The intensity of linkage is measured by the frequency of recombination, which has a maximum value of 50%.
      a. The less often recombination occurs, the tighter is the linkage.
        (1) No recombination between linked genes results in complete linkage.
    3. Recombination is caused by crossing over between homologous chromosomes after the chromosomes have replicated.
      a. Crossing over apparently occurs during pachynema of meiosis.
      b. Each crossover event is a physical exchange between two of the four chromatids in a tetrad.
      c. Multiple crossovers involving different combinations of chromatids may occur along the length of a tetrad.
      d. Crossovers become visible as chiasmata from diplonema (pachynema in cytologically favorable material) through metaphase I of meiosis.
  B. The average number of crossovers on chromosome pairs during meiosis is a measure of genetic distance, calibrated in centiMorgans (cM), and is the basis for constructing chromosome maps.
    1. Map distance can be estimated by counting chiasmata in cytological preparations or by

calculating the frequency of recombination observed between genes in experimental crosses, e.g., in two- or three-point testcrosses.

2. An average of one chiasma is equivalent to a distance of 50 cM and one percent recombination is equivalent to one cM.

3. Due to formation of multiple crossovers, recombination frequencies greater than 20 - 25% usually underestimate genetic distance.
   a. Multiple crossovers are rare over short genetic distances. This is due to a phenomenon called interference.
      (1) Interference is crossing over at one point reducing the chance of another crossover occurring nearby.
      (2) The amount of interference is measured by the coefficient of coincidence.

4. A recombination map showing the positions of genes on a chromosome is colinear with the physical map of the chromosome; however, recombination distances between genes are not proportional to physical distances.
   a. The frequency and distribution of crossing over is influenced both by genetics and the physical nature of the chromosomes.

C. The assignment of genes to physical locations on chromosomes can be accomplished through a process called cytogenetic mapping.
   1. Cytogenetic mapping has been most thoroughly developed in *Drosophila* using polytene chromosomes.
      a. The physical mapping of genes may be done using deletions (deletion mapping) or duplications (duplication mapping).

D. Linkage between human genes can be detected by analysis of pedigrees or by using cultured somatic cell hybrids.

E. Recombination generates new combinations of alleles, some of which are favorable, upon which selection may act.

F. Recombination and the recovery of recombinant chromosomes are suppressed by heterozygosity for chromosomal rearrangements, especially inversions.

IMPORTANT TERMS

In the space allotted, concisely define each term.

locus:

linkage:

recombination:

chiasma (plural, chiasmata):

recombinants:

linkage phase:

tetrad:

centiMorgan (cM):

interference:

coefficient of coincidence:

cytogenetic mapping:

deletion mapping:

duplication mapping:

selectable genetic markers:

IMPORTANT NAMES

In the space allotted, concisely state the major contribution made by the individual or group.

Thomas Hunt Morgan:

Alfred H. Sturtevant:

W. Bateson and R. C. Punnett:

Harriet Creighton and Barbara McClintock:

Calvin B. Bridges and T. M. Olbrycht:

J. H. Renwick and S. D. Lawler:

G. Barski and B Ephrussi:

Oscar Miller:

Bruce T. Lahn and David C. Page:

Nancy Wexler:

TESTING YOUR KNOWLEDGE

In this section, answer the questions, fill in the blanks, or solve the problems in the space allotted. Problems noted with an * are solved in the Approaches to Problem Solving section at the end of the chapter.

1. The linkage phase in which the dominant alleles of two genes are on one homologous chromosome and the two respective recessive alleles are on the other homolog is called
    _____.

2. The phenomenon of a crossover at one point inhibiting the formation of another crossover in an adjacent region is called _____.

3. The term describing genes located on the same chromosome is _____.

4. The linkage phase in which a dominate allele of one gene and a recessive allele of a second gene are found on one homolog and the respective recessive and dominant alleles are on the other homolog is called _____.

5. If in the hybrid, *E f/e F*, <u>crossing over</u> occurs 40% of the time, then the frequency of the *E F* gamete will be _____.

*6. A dihybrid plant, *Cc Dd*, is self-fertilized and the double recessive, *cc dd*, phenotype occurs in 4% of the progeny.

   (a) What is the expected frequency of the *c D* gamete?

   (b) What is the linkage phase of this dihybrid plant?

*7. It is observed that two heterochromatic knobs on chromosome 6 of maize have a chiasma between them in 20% of the meiocytes studied. What is the genetic map distance between these knobs?

*8. A mouse has a chromosome number $2n = 40$. How many linkage groups does the mouse have?

9. A dihybrid plant, *Cc Dd*, is self-fertilized and the double recessive, *cc dd*, phenotype occurs in 16% of the progeny.

   (a) What is the expected frequency of the *c D* gamete?

   (b) What is the linkage phase of this dihybrid plant?

*10. In Upland cotton, the laciniate leaf and brown lint loci are 34 map units apart. What percentage of the meioses is expected to have a chiasma involving the region of the bivalent between these loci?

*11. The following are progeny of a heterozygous tomato plant, *R r Wf wf* (*R* = red fruit, *r* = yellow fruit; and *Wf* = yellow flowers, *wf* = white flowers) testcrossed to an *rr wfwf* plant: 77 red fruit, yellow flowers; 73 yellow fruit, white flowers; 28 red fruit, white flowers; 32 yellow fruit, yellow flowers.

   (a) What is the recombination frequency between the fruit color and flower color loci?

   (b) What is the linkage phase of the alleles in the dihybrid parent?

*12. In cotton, genes located on chromosome 15 give the following recombination frequencies as determined by two-point testcross mapping: L^0- Lg = 50%; L^0- sxl = 3%, s -cr = 0.2%; vf -s = .8%; sxl - Lg = 44%; Lg - vf = 5%; and Lg - cr = 6%. What is the relative order of these genes?

*13. A large inversion has been detected in pachytene nuclei of corn that involves about 25% of chromosome 2. A detailed analysis of pachytene chromosomes of plants heterozygous for this inversion showed that an inversion loop forms essentially in all of the meioses. Furthermore, a chiasma occurs in 54% of the inversion loops. How large is this inversion measured in cM map units?

*14. In cotton, the dwarf red mutation (Rd = dwarf red, rd = green) and the virescent mutation (v = virescent, V = nonvirescent) mapped about 10 cM apart. List the phenotypes and their frequencies if an $Rd\ V\ /\ rd\ v$ plant is allowed to self-pollinate.

*15. In *Drosophila*, rose eyes (rs), dichaete wings (Dt) and curled wings (cu) reside at the following positions on chromosome 3.

rs	Dt	cu
35	40	50

List all gametes and their frequencies that will be produced by a female fly of

the genotype $\dfrac{rs^+}{rs} \quad \dfrac{Dt^+}{Dt} \quad \dfrac{cu^+}{cu}$ assuming a coefficient of coincidence of 0.90.

*16. Rabbits fully heterozygous at three loci, $B\ b\ \ c^{ch}\ c^h\ \ S\ s$ (B = black, b = brown; c^{ch} = chinchilla, c^h = himalayan; S = long fur, s = short fur) are testcrossed and the 200 progeny are tabulated.

Phenotype			Number
black	himalayan	long	35
brown	chinchilla	long	36
black	himalayan	short	35
brown	chinchilla	short	37
black	chinchilla	long	13
brown	himalayan	long	14
black	chinchilla	short	13
brown	himalayan	short	17

(a) Which genes are linked? Which gene, if any, is not linked to the others?

(b) Indicate the map distance between all linked genes.

*17. The progeny of a *Drosophila* female (heterozygous at three loci, $y^+y\ ct^+ct\ w^+\ w$) crossed to a wild-type male are listed as follows:

	Phenotype	Number
females:	$y^+\ ct^+\ w^+$	2000
males:	$y^+\ ct^+\ w$	773
	$y\ \ ct\ \ w^+$	782
	$y\ \ ct^+\ w^+$	201
	$y^+\ ct\ \ w$	209
	$y^+\ ct\ \ w^+$	15
	$y\ \ ct^+\ w$	16
	$y\ \ ct\ \ w$	3
	$y^+\ ct^+\ w^+$	1

(a) On which chromosome do these genes reside?

(b) What is the gene order and gene arrangement in the trihybrid female parent?

(c) What are the map distances between these genes?

(d) Calculate the coefficient of coincidence and the value of interference for this cross.

*18. In tomato the genes for jointless (*j*), hairless (*hl*)and leafy (*lf*) are known to be linked on chromosome 5. Plants heterozygous for these three gene pairs, $j^+j\ hl^+hl\ lf^+lf$, were testcrossed and the progeny listed:

Phenotypes of progeny	Number
jointed, hairless, normal	142
jointless, hairy, leafy	139
jointed, hairy, leafy	9
jointless, hairless, normal	10
jointed, hairy, normal	96
jointless, hairless, leafy	103
jointless, hairy, normal	1
	500

(a) What is the correct sequence of genes in the linkage group?

(b) What is the correct linkage phase of the trihybrid parent?

(c) What are the map distances between the loci?

(d) What is the coefficient of coincidence and the value of interference for this cross?

*19. The following pedigree is for a family in which Duchenne's muscular dystrophy (DMD) and colorblindness are segregating. For this problem assume that the DMD and colorblindness loci are 10 cM apart on the X chromosome.

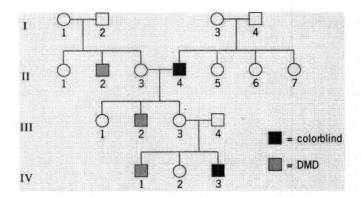

(a) What is the probability if III-3 and III-4 have another son that he will have DMD?

(b) What is the probability if III-3 and III-4 have another son that he will have both colorblindness and DMD?

THOUGHT CHALLENGING EXERCISE

Meiotic cytogenetic analysis of chromosome pairing of plant hybrids indicates that different populations and species often differ in regard to translocations in the homozygous state. Explain how the translocation became homozygous. How does a translocation affect linkage relationships? How might these be of selective advantage?

SUMMARY OF KEY POINTS

Linkage between two genes is detected as a deviation from expectations based on Mendel's Principle of Independent Assortment.

The frequency of recombination measures the intensity of linkage. In the absence of linkage, this frequency is 50 percent; for very tight linkage, it is close to zero.

Recombination is caused by the physical exchange between paired homologous chromosomes early in the prophase of the first meiotic division.

At any one point along a chromosome, the process of exchange (crossing over) involves only two of the four chromatids in a meiotic tetrad.

Late in prophase I, crossovers become visible as chiasmata.

The genetic maps of chromosomes are based on the average number of crossovers that occur during meiosis.

Genetic map distances are estimated by calculating the frequency of recombination between genes in experimental crosses.

Recombination frequencies less than 20 percent estimate map distance directly; however, recombination frequencies greater than 20 percent underestimate map distance because multiple crossover events do not always produce recombinant chromosomes.

An average of one chiasma during meiosis is equivalent to 50 centiMorgans of genetic map distance.

In *Drosophila*, genes can be localized on maps of the polytene chromosomes by combining recessive mutations with cytologically defined deletions and duplications.

A deletion will reveal the phenotype of a recessive mutation located between its endpoints, whereas a duplication will conceal the mutant phenotype.

Genetic and cytological maps are colinear, however genetic distances are not proportional to cytological distances.

Linkage between human genes can be detected by analyzing pedigrees.

A gene can be located on a specific human chromosome by fusing human cells with rodent cells and then analyzing the resulting somatic-cell hybrids for the presence of the gene or its product.

Recombination can bring favorable mutations together.

Chromosomal rearrangements, especially inversions, can suppress recombination.

Recombination is under genetic control.

ANSWERS TO QUESTIONS AND PROBLEMS

1) coupling 2) interference 3) linkage 4) repulsion 5) 10 6) a. 3; b. repulsion or *trans* 7) 10 cM 8) 20
9) a. 0.1; b. coupling or *cis* 10) $\cong 68\%$ 11) a. 0.2857 or 28.57%; b. coupling 12) L^0-sxl - cr - s - vf - Lg or
L^0- sxl - Lg - wf - s - cr 13. 27 cM 14 - 19) see *Approaches to Problem Solving* section.

Some ideas concerning thought challenging exercise:

A translocation becomes homozygous when two gametes carrying the two translocated chromosomes fuse. These chromosomes show normal bivalent pairing and are new linkage groups. The original translocation in the heterozygous condition may have kept highly adaptive allele combinations on the chromosomes from being recombined by independent assortment. The heterozygous plant displays semisterility (which would be disadvantageous) due to adjacent segregation of the translocation quadrivalents. The plants that are homozygous for the translocation would retain the advantageous alleles and would be fully fertile. Geographic isolation from the parental population would be favorable to maintain the translocation homozygosity, because it minimizes hybridization with the normal plants that would generate translocation heterozygotes.

APPROACHES TO PROBLEM SOLVING

6. (a) The homozygous recessive phenotype (*cc dd*) occurred in a frequency of 4% or 0.04. Since the *cc dd* genotype results from the fusion of two *c d* gametes, the frequency of the *c d* gamete must be 0.2, ($0.2 \times 0.2 = 0.04$) This frequency is less than 0.25 and thus *c d* is a recombinant. The other recombinant gamete *C D* also should occur in a frequency of 0.2. This leaves 0.6 of the gametes as nonrecombinants. *c D* is one of the two nonrecombinant types. Its frequency is 0.3.

    (b) From the calculations and reasoning involved in part (a), we can assign the linkage phase of the dihybrid parent as *c D/C d*. This is the repulsed or *trans* linkage phase.

7. Chiasmata represent crossovers. Because only two of the four chromatids are involved in crossing over at any one point, the amount of recombination is 1/2 the frequency of crossing over. One map unit (cM) equals 1% recombination. Therefore, the map distance between the knobs is 10% or 10 cM.

8. The number of linkage groups is determined by the number of different chromosomes. The mouse has ($2n = 40$; $1n = 20$) 20 different chromosomes or 20 linkage groups.

10. The 34 map units is 34% recombination. Since recombination is 50% of the frequency of crossing over, we would expect to see 68% of the meioses with a single chiasma between the laciniate leaf and brown lint loci.

11. When the data are analyzed it is observed that yellow ($73 + 32 = 105$) and red ($77 + 28 = 105$) fruit segregate in a 1:1 ratio, and yellow ($77 + 32 = 109$) and white ($73 + 28 = 111$) flowers segregate in a 1:1 ratio. Therefore the deviation away from a 1:1:1:1 testcross ratio is not due to inviability of certain genotypes or factors that interfere with segregation of the genes, but is caused by linkage.

    (a) To calculate the recombination ratio between these genes first add the individuals to obtain a total ($77 + 73 + 28 + 32 = 210$). The two most abundant phenotypes (77 red fruit, yellow flowers; 73 yellow fruit, white flowers) are the parental types or nonrecombinants. The two less frequent phenotypes (28 red fruit, white flowers; 32 yellow fruit, yellow flowers) are the recombinants. The recombination frequency is calculated by dividing the number of recombinants ($28 + 32 = 60$) by the total number of progeny (210). This is 0.2857 or 28.57%.

    (b) The nonrecombinant chromosomes from the dihybrid parent were *R Wf* and *r wf*. Therefore, the genotype of the dihybrid was $\frac{R\ Wf}{r\ wf}$ and the linkage phase was coupling.

12. To work this problem it is important to consider the additive nature of the map distances. One way to start the problem is to consider the two genes that are farthest apart in terms of map recombination distance, i.e., L^0-Lg. = 50%. But 50% is the maximum amount of recombination and is not definitive by itself. Next consider *sxl - Lg* as a starting point. Add L^0 to the map. It will be 3 units to the left of *sxl*, so as to lengthen the distance between *Lg* and L^0. Bring in the additional markers, again trying to arrange them in an additive fashion.

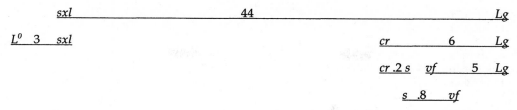

sxl	44	*Lg*
L⁰ 3 *sxl*	*cr* 6 *Lg*	
	cr .2 *s* *vf* 5 *Lg*	
	s .8 *vf*	

sxl _____ 44 _____ Lg

L^0 3 sxl cr 6 Lg

cr .2 s vf 5 Lg

s .8 vf

Not enough data are provided to assemble an unambiguous sequence. The map order is likely either

L^0 -sxl - cr - s - vf - Lg or L^0 - sxl -Lg - vf - s - cr

13. Since a chiasma occurs within the inversion loop in 54% of the meioses, this represents 54% crossing over within the inversion loop. Since the amount of recombination is 1/2 the frequency of crossing over, this equates to 27% recombination. The inversion loop, therefore, covers 27 cM.

14. Since the dihybrid is in the coupled phase, the recombinant chromosomes will be $\underline{Rd \quad v}$ and $\underline{rd \quad V}$. The loci are 10 cM apart. Therefore, each recombinant will occur in a frequency of 0.05. The parental types will make up 90% of the progeny. The $\underline{Rd \quad V}$ and $\underline{rd \quad v}$ gametes each will occur at a frequency of 0.45. The phenotypic frequency of the progeny from selfing the $\underline{Rd \quad V}$ plant can be determined by using a Punnett square.

	$\underline{Rd \quad V}$ (0.45)	$\underline{rd \quad v}$ (0.45)	$\underline{Rd \quad v}$ (0.05)	$\underline{rd \quad V}$ (0.05)
$\underline{Rd \quad V}$ (0.45)	*RdRd VV* (.2025)	*Rdrd Vv* (.2025)	*RdRd Vv* (.0225)	*Rdrd VV* (.0225)
$\underline{rd \quad v}$ (0.45)	*Rdrd Vv* (.2025)	*rdrd vv* (.2025)	*Rdrd vv* (.0225)	*rdrd Vv* (.0225)
$\underline{Rd \quad v}$ (0.05)	*RdRd Vv* (.0225)	*Rdrd vv* (.0225)	*RdRd vv* (.0025)	*Rdrd Vv* (.0025)
$\underline{rd \quad V}$ (0.05)	*Rdrd VV* (.0225)	*rdrd Vv* (.0225)	*Rdrd Vv* (.0025)	*rdrd VV* (.0025)

The phenotypes can then be summed as follows:

Red, nonvirescent (*Rd_ V_*): .2025 + .2025 + .0225 + .0225 + .2025 + .0225 + .0025 + .0225 + .0025 = .7025

Red, virescent (*Rd _ vv*): .0225 + .0225 + .0025 = .0475

green, nonvirescent (*rdrd V_*): .0225 + .0225 +.0025 = .0475

green, virescent (*rdrd vv*): .2025

15. To start solving this problem first calculate the frequency of the double recombinant gametes. These are expected to occur at a frequency of recombination between *rs* and *Dt* (0.05) X frequency of recombination between *Dt* and *cu* (0.10) X the coefficient of coincidence (0.9). The frequency of the two double recombinants should total 0.0045.

rs^+ *Dt* cu^+ = .00225

rs Dt^+ *cu* = .00225

The recombination between the rose and dichaete locus is .05. Since the map distance includes the double recombinants, we must subtract the frequency of the double recombinants (.0045) from .05. The frequency of these single recombinants totals .0455.

$$rs^+ \quad Dt \quad cu = .02275$$

$$rs \quad Dt^+ \ cu^+ = .02275$$

The recombination between the dichaete and curled loci is .10. Again, subtract the frequency of the double recombinants to obtain a frequency of the single recombinant gametes of .0955.

$$rs^+ \quad Dt^+ \quad cu = .04775$$

$$rs \quad Dt \quad cu^+ = .04775$$

The remaining gametes will be the nonrecombinants. These are obtained by subtracting the total number of recombinants from one. NR = (1 - [.0045 + .0455 + .0955]) = .8545.

$$rs^+ \quad Dt^+ \quad cu^+ = .42725$$

$$rs \quad Dt \quad cu = .42725$$

16. (a) When you look at the data it is seen that the eight phenotypic classes are distributed in two groups of four. This suggests that two genes are linked and a third gene is segregating independently. The key to solving the problem is to detect the linked genes. It may be easier to convert the phenotypes to allelic symbols.

B	c^h	S	35
b	c^{ch}	S	36
B	c^h	s	35
b	c^{ch}	s	37
B	c^{ch}	S	13
b	c^h	S	14
B	c^{ch}	s	13
b	c^h	s	17

The linked genes are B and c. S segregates independently. This is apparent if the data are rearranged as follows:

B	c^h	S	35	b	c^{ch}	S	36	B	c^{ch}	S	13	b	c^h	S	14

$$\underline{B \quad c^h} \ S \ 35 \qquad \underline{b \quad c^{ch}} \ S \ 36 \qquad \underline{B \quad c^{ch}} \ S \ 13 \qquad \underline{b \quad c^h} \ S \ 14$$

$$\underline{B \quad c^h} \ s \ 35 \qquad \underline{b \quad c^{ch}} \ s \ 37 \qquad \underline{B \quad c^{ch}} \ s \ 13 \qquad \underline{b \quad c^h} \ s \ 17$$

(b) The nonrecombinants are $\underline{B \quad c^h}$ and $\underline{b \quad c^{ch}}$ and the recombinants are $\underline{B \quad c^{ch}}$ and $\underline{b \quad c^h}$. The recombinants total 57. There are 198 total progeny. The amount of recombination = 57/198 = .2879. The map distance between these linked genes is 28.79 cM.

17. (a) This problem involves sex-linked genes. This is apparent from the trihybrid female X wild-type male producing all wild-type female progeny, but various phenotypic combinations among male progeny.

(b) To determine gene order requires recognizing nonrecombinant and double recombinant phenotypes among the male progeny. The 2,000 male progeny are distributed in eight phenotypes, the two most frequent ($y^+ \ ct^+ \ w$ and $y \ ct \ w^+$) representing the parental or nonrecombinant chromosomes, and the two least frequent ($y \ ct \ w$ and $y^+ \ ct^+ \ w^+$) the double recombinant chromosomes that came from the trihybrid female. The genotype of the trihybrid parent in terms of the linkage phases is:

$$\frac{y^+ \quad ct^+ \quad w}{y \quad ct \quad w^+}$$

A simple way to determine the order of the genes is to look at the genotypes (in regard to the chromosome received from the trihybrid) of the double recombinant phenotypes among the progeny.

$$y \quad ct \quad w$$

and

$$y^+ \quad ct^+ \quad w^+$$

Since a double crossover switches the middle gene pair, look for the gene pair that would have switched in the trihybrid designated above to generate the double recombinants. This gene pair is w/w^+. Therefore the white locus is positioned between yellow and cut. The correct gene order and genotype of the trihybrid female follows:

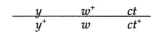

$$\frac{y \qquad w^+ \qquad ct}{y^+ \qquad w \qquad ct^+}$$

(c) The map distances are determined by first summing the phenotypes representing single recombinants between the loci. Remember that double recombinants result from recombination in both regions simultaneously and must be included. The single recombinants between the yellow and white loci are $y\ w\ ct^+$ (16) and $y^+\ w^+\ ct$ (15), and the single recombinants between the white and cut loci are $y\ w^+\ ct^+$ (201) and $y^+w\ ct$ (209).

$$\frac{y \qquad w^+ \qquad ct}{y^+ \qquad w \qquad ct^+}$$

15	201
16	209
3	3
1	1
35	414

The map distance between the yellow and white loci is (35/2,000)(100) = 1.75 cM.

The map distance between the white and cut loci is (414/2,000)(100) = 20.7 cM.

(d) The coefficient of coincidence (CC) is calculated by the formula

$$CC = \frac{\text{observed number of double recombinants}}{\text{expected number of double recombinants}} = \frac{.002}{(.207)(.0175)} = \frac{.002}{.0036} = .56$$

Interference = 1 - CC = 1 - .56 = .44

18. To solve this problem it is beneficial to substitute allelic symbols for the phenotypes. Also remember that in a testcross the phenotypes of the progeny indicate the alleles that came from the testcross parent (trihybrid in this case). It follows that the following chromosomes in the progeny were from the trihybrid parent:

j^+	hl	lf$^+$	142
j	hl$^+$	lf	139
j^+	hl$^+$	lf	9
j	hl	lf$^+$	10
j^+	hl$^+$	lf$^+$	96
j	hl	lf	103
j	hl$^+$	lf$^+$	1
j^+	hl	lf	0
			500

One genotype did not occur. This was one of the double recombinants and so it was put in the table with a frequency of zero. The two most frequent phenotypes represent the parental types. These can be put together to reconstruct the genotype of the trihybrid as indicated below:

$$\frac{j^+ \quad hl \quad lf^+}{j \quad hl^+ \quad lf}$$

The gene order can be immediately determined by comparing the trihybrid indicated above with the genotype of the double recombinants. The double recombinants, the least frequent phenotypes are:

$$\underline{\quad j \quad hl^+ \quad lf^+ \quad}$$

and

$$\underline{\quad j^+ \quad hl \quad lf \quad}$$

The gene pair that was switched to produce the genotypes of the double recombinants from the trihybrid was lf^+ / lf. Therefore, the leafy gene must be the middle gene.

(a) The correct sequence of genes must be $j \quad lf \quad hl$.

(b) The correct linkage phase is $\dfrac{j^+ \quad lf^+ \quad hl}{j \quad lf \quad hl^+}$.

(c) The single recombinants between the jointless and leafy loci are $j \quad lf^+ \quad hl$ (10) and $j^+ \quad lf \quad hl^+$ (9). The double recombinants (1 and 0) also are added to make the total number of recombinants between these loci 20. The frequency of recombination in this region is $20 / 500 = .04$ or 4 cM.

The single recombinants between the leafy and the hairless loci are $j^+ \quad lf^+ \quad hl^+$ (96) and $j \quad lf \quad hl$ (103). The double recombinants (1 and 0) also are added to make the total number of recombinants between these loci 200. The frequency of recombination in this region is $200 / 500 = .40$ or 40 cM.

(d) The $CC = \dfrac{.002}{(.40)(.04)} = .125$.   Interference $= 1 - CC = 1 - .125 = .875$.

19. (a) Since III-3 and III-4 already have one son with DMD, it means that III-3 is heterozygous for the DMD allele. Since the DMD gene is X-linked recessive, III-3 and III-4 have a 50% chance that if they parent another son he will have DMD.

(b) The father of III-3 did not have DMD, but III-3's mother had a brother with DMD. Therefore, III-3 inherited the chromosome carrying the DMD from her mother. The X-chromosome of III-3 carrying the recessive colorblind allele came from her father who was colorblind. Since the DMD and the colorblind loci are 10 cM apart, we would expect the one recombinant carrying both the DMD and colorblind alleles to occur with a frequency of 0.05. Therefore, if III-3 and III-4 have another son, the probability that he will be both colorblind and DMD is 5%.

8

The Genetics of Microorganisms

MICROORGANISMS IN GENETICS

THE GENETICS OF VIRUSUS
    Bacteriophages T4 and Lambda
    Mapping Genes in Bacteriophage
    Bacteriophage T4: A Linear Chromosome and a Circular Genetic Map

THE GENETICS OF BACTERIA
    Mutant Genes in Bacteria
    Unidirectional Gene Transfer in Bacteria

MECHANISMS OF GENETIC EXCHANGE IN BACTERIA
    Transformation
    Conjugation
    Transduction
    Plasmids and Episomes
    F' Factors and Sexduction
    Using Partial Diploids to Map Closely Linked Genes

THE EVOLUTIONARY SIGNIFICANCE OF GENETIC EXCHANGE IN BACTERIA

THE GENETICS OF FUNGI
    Detecting Linkage and Mapping Genes in Yeast
    Mapping Centromeres Using Ordered Tetrad Data
    Constructing Linkage Maps using Ordered Tetrad Data

A MILESTONE IN GENETICS: CONJUGATION IN *ESCHERICHIA COLI*

IMPORTANT CONCEPTS

  A. The analysis of microorganisms has allowed researchers to dissect the chemical nature of genes and their products
    1. Microorganisms have several advantages over higher plants and animals.
      a. They are small, reproduce quickly and can rapidly produce very large populations.
      b. They can be grown on biochemically defined culture media.
      c. They have relatively simple structure and physiology.
      d. Genetic variability is easy to detect in microorganisms.
  B. A wealth of information is known about the genomes of many microorganisms.
    1. Complete nucleotide sequences are known for many viruses, bacteria, protozoa, and fungi. that are providing detailed information about the genetic control of metabolism.
  C. Bacteriophages, e.g., T4, are viruses that infect bacteria.
    1. The T4 virus consists of a proteinaceous head that houses its DNA and a tail that functions in injecting the phage DNA into the bacterium.
      a. T4 contains 150 characterized genes and its genome is about 168,800 base pairs.
      b. Bacteriophage T4 has a life cycle of about 25 minutes.
      c. The phage attaches to the host cell and injects its DNA into the cell.
      d. The host cell's metabolic machinery is used for production of progeny phage that are released by lysis of the cell.
      e. The phage DNA contains hydroxymethylcytosine instead of cytosine and this serves to protect it from the host cell's DNA degrading enzymes and phage-encoded nucleases.
  D. Bacteriophage lambda (λ) is smaller than T4 but its life cycle is more complex.

1. The lambda genome is 48,502 nucleotide pairs and contains about 50 genes.
2. Lambda may reproduce by either a lytic cycle or lysogenic pathway.
   a. The lytic cycle results in lysis of the bacterium.
   b. In the lysogenic pathway the λ chromosome is inserted into the DNA of the bacterium.
      (1) The integrated λ chromosome is called a prophage.
      (2) Integration of the λ chromosome into the bacterial chromosome is site-specific.
   c. Occasionally, the λ chromosome excises from the bacterial genome and enter the lytic pathway.
      (1) Anomalous excision results in bacteria DNA being excised with phage DNA. The virus can, therefore, transfer bacterial DNA sequences from one bacterium to another.
E. The mechanism of recombination in phage is different from that in sexually reproducing organisms, but the strategy of constructing a phage genetic map is similar.
   1. Phage genes are mapped by simultaneously infecting bacterial cells with genetically different phage and then analyzing the progeny for recombination events.
F. The phage T4 has a circular genetic map but a physically linear chromosome.
   1. This paradox is explained by the terminally redundant, circularly permutated T4 chromosome.
G. The genetic information of most bacteria is stored in a single chromosome carrying a few thousand genes and a variable number of plasmids and episomes.
H. Bacteria do not reproduce sexually, but rather have developed parasexual ways to generate new gene combinations. Gene transfer in bacteria is unidirectional, i.e., from donor cells to recipient cells. There are three main processes of parasexual recombination: transformation, conjugation, and transduction.
   1. Transformation is the genetic alteration of bacteria brought about by the incorporation of foreign DNA into the bacterial cells.
      a. Linkage relationships can be determined in transformation experiments by determining the frequencies of co-transformation.
   2. Conjugation is a unidirectional transfer of genetic information from a donor cell to a recipient cell via a conjugation tube.
      a. A plasmid called the F factor mediates this transfer.
      b. Cells that carry the F factor are F^+ and those lacking it are F^-.
      c. If the F factor is integrated into the chromosome, the cell is Hfr (high frequency of recombination).
         (1) Hfr cells transfer chromosomal DNA from the donor to F^- recipients.
         (2) Sometimes, when the F factor exits the chromosome, it carries with it a piece of the host chromosome. This F factor is called an F' factor.
      d. Conjugation can be used to map genes.
         (1) Conjugation mapping is based on the time of donor gene transfer to recipient cells.
   3. Transduction is the transfer of DNA and recombination in bacteria mediated by bacterial viruses called bacteriophages.
      a. In generalized transduction, any segment of the bacterial genome may be transferred from one bacterium to another by a phage.
      b. In specialized transduction, only the specific region of the bacterial chromosome bracketing the prophage insertion site is transferred from one bacterium to another.
I. Distinguishing between transduction, transformation, and conjugation involves determining whether cell contact is required and whether the process is sensitive to DNase.
   1. This is experimentally tested by employing a U-tube experiment.
J. Parasexual mechanisms enhance the ability of bacteria to adapt to environmental changes, which makes it more difficult for humans to control them with antibiotics.
K. *Saccharomyces cerevisiae* (yeast) and *Neurospora crassa* (bread mold) are two fungi that are model systems for genetic analysis.
   1. Linkage can be readily detected by tetrad analysis.
      a. In two-point crosses in yeast, for example, linkage is detected when parental ditype tetrads are more frequent than nonparental ditype tetrads.
      b. The ordered tetrads of *Neurospora* make it possible to map centromeres relative to genes.
      c. The centromere-to-gene distance is one-half the frequency of the second division segregation tetrads, because only half of the spores of a tetrad arrangement resulting

from crossing over are recombinant.

IMPORTANT TERMS

In the space allotted, concisely define each term.

bacteriophage:

lytic phage:

lysogenic pathway:

prophage:

plaque:

rapid lysis (*r*):

host range:

terminally redundant:

circularly permutated:

concatamers:

headful mechanism:

parasexual processes:

prototrophs:

auxotrophs:

transformation:

conjugation:

transduction:

competent:

heteroduplex:

F factor:

autonomous state:

integrated state:

F+ cell:

Hfr cell:

selective medium:

generalized transduction:

specialized transduction:

transducing particles:

plasmid:

episome:

sexduction:

ascus:

ascospore:

tetrad analysis:

unordered tetrads:

ordered tetrads:

parental ditype:

nonparental ditype:

tetratype:

first division segregation:

second division segregation:

IMPORTANT NAMES

In the space allotted, concisely state the major contribution made by the individual.

Andre Lwoff:

Alfred Hershey and Max Delbruck:

Emory Ellis:

Seymour Benzer:

Frederick Griffith:

Richard Sia and Martin Dawson:

Oswald Avery, Colin MacLeod, and Maclyn McCarty:

Joshua Lederberg and Edward Tatum:

Norton Zinder and Joshua Lederberg:

Francois Jacob and Elie Wollman:

Mary Houlahan, George Beadle, and Hermione Calhoun:

TESTING YOUR KNOWLEDGE

In this section, fill in the blanks, answer the questions and solve the problems in the space allotted. Problems noted with an * are solved in the Approaches to Problem Solving section at the end of the chapter.

1. A virus that infects a bacterial cell is called a _____.

2. What are the two types of bacteriophage mutations that are easiest to study genetically?

3. Name two examples of conditional mutations of bacteriophages.

4. A clear area in a confluent lawn of bacterial cells that results from lysis or killing of contiguous cells by several cycles of bacteriophage growth is called a _____.

5. Why do occassional plaques occur when *rII* mutants are plated on a lawn of *E. coli* K12 (λ)?

6. Why do plaques never occur when *rII* deletion mutants are plated on a lawn of *E. coli* K12 (λ)?

7. The DNA of bacteriophage T4 may be protected from destruction by enzymes of the bacteria because it has _____ instead of cytosine in its DNA.

*8. Strain B *E. coli* cells were simultaneously infected with two *rII* mutants, *r1* and *r2*. Identically diluted aliquots of the progeny phage were plated on (1) *E. coli* strain B to detect the total number of phage progeny, and on (2) *E. coli* strain K12(λ) to detect the number of wild-type phage. The plaques on strain B numbered 10,000 and those on K12(λ) numbered 10. What is the frequency of recombination between these mutants?

*9. A series of *rII* mutants were crossed and the following frequencies of wild-type recombinants were recovered:

$r1 \times r2 \rightarrow$ .011

$r1 \times r3 \rightarrow$ .005

$r1 \times r4 \rightarrow$ .003

$r2 \times r3 \rightarrow$ .006

$r2 \times r4 \rightarrow$ .008

$r3 \times r4 \rightarrow$ .002

Indicate the order of the mutants and the map distances between them.

*10. Strain B *E. coli* cells were coinfected with *rII* mutants *x* and *y*. Progeny phage plated in a dilution series on both *E. coli* strains B and K12(λ) gave the following results:

Bacterial lawn	Dilution	Number of Plaques
Strain B	10^{-8}	6
Strain K12(λ)	10^{-4}	4

(a) What is the recombination frequency between the sites of mutants *x* and *y*?

(b) If one map unit is 1 kb pairs, how many nucleotides apart are these mutants?

*11. Hershey and coworkers demonstrated that two-point mapping in bacteriophage T2 could be done using rapid lysis mutants (r^+ = small plaque, r = large plaque) and host range mutants (h^+ infects only *E. coli* strain B, h infects strains B and B/2). Phage of the genotype hr^+ and h^+r were used to coinfect *E. coli* strain B. The progeny phages were collected and plated on a mixed culture of *E. coli* B and B/2. From the following results calculate the frequency of recombination between the h and r genes.

Plaque morphology	Frequency
small, turbid	49
small, clear	425
large, turbid	413
large, clear	45

The turbid plaques result from only the lysis of strain B bacteria of the mixture of B and B/2.

12. When a bacteriophage genome has integrated into the host bacterium's DNA, it is called a (an) _____.

13. The uptake of exogenous DNA from the culture medium and its incorporation by recombination into the DNA of the bacterium is called _____.

14. A bacteriophage that lyses the host bacterium is called a _____.

15. The incorporation of bacterial genes into an F' factor and their subsequent transfer by conjugation to an F⁻ cell is called _____.

16. A male bacterial cell, with the F factor integrated into its DNA is called a (an) _____.

17. An organism such as a bacterium that requires for growth a specific compound added to the minimal medium is called _____.

18. The ability of a bacterial cell to incorporate exogenous DNA and become genetically transformed is called _____.

19. A partial diploid in bacteria produced by the process of partial genetic exchange, such as sexduction, is called a (an) _____.

20. The bacterial plasmid that confers the ability to function as a genetic donor in conjugation is called a (an) _____.

21. A phage that invades but does not destroy (lyse) the host bacterium is called a _____.

22. The life cycle of a bacteriophage which leads to the production of progeny phages, and the lysis of the host bacterial cell is called the _____ pathway.

23. The phenomenon when a bacteriophage particle carries a random or near random segment of bacterial DNA from one bacterium to another is called _____.

24. The life-style of a bacteriophage which leads to integration via site-specific recombination into the bacterial chromosome is called the _____ pathway.

25. The phenomenon when a bacteriophage particle transduces only genetic markers located in one small region of the bacterial chromosome is called _____.

*26. In interrupted mating experiments involving *E. coli*, five different Hfr strains donate the following markers in the order shown.

<u>Order of entry of markers</u>

Hfr strain	1st	2nd	3rd	4th	5th
A	arg	xyl	metC	lysA	aroD
B	purE	metD	arg	xyl	metC
C	trpA	purE	metD	arg	xyl
D	xyl	arg	metD	purE	trpA
E	metC	lysA	aroD	trpA	purE

On a generalized map, indicate the order of these markers on the circular chromosome, the position where the sex factor integrated in each Hfr strain, and the direction of transfer mediated by each integrated F factor.

*27. In *E. coli*, the genes *ton*A (resistance to phages) and *pan* (auxotropic for pantothenate) are closely linked and map to position 3 on the time map. Another gene *aziC* (azide resistance) maps one time unit from these genes. Hfr bacteria of the genotype *azi⁺ tonA⁺ pan⁺* were mated with an F⁻ strain of the genotype *azi tonA pan* for a time interval sufficient for the entry of the genes. The number of colonies scored of the different recombinant classes follows:

Genotype	Number
$aziC^+ ton^+ pan^+$	833
$aziC^+ ton\ pan$	200
$aziC^+ ton\ pan^+$	45
$aziC^+ ton^+ pan$	2

What is the linkage order and recombination frequency between the genes?

*28. In *E. coli*, a series of three-point transduction tests were performed to map the order of mutant sites in the *leu* gene. The unselected, linked marker was the *thr* gene. In each cross, *leu⁺* recombinants were selected and then scored for *thr⁺* or *thr⁻*. From the following table of results, deduce the linear order of the *leu⁻* mutant.

Cross		*thr* allele in	
Donor markers	Recipient markers	wild-type recombinants	% thr⁺
$thr^+ leu^1$	$thr^- leu^2$	60 thr^+:300 thr^-	17
$thr^+ leu^3$	$thr^- leu^2$	185 thr^+:175 thr^-	51
$thr^+ leu^2$	$thr^- leu^1$	350 thr^+:349 thr^-	50
$thr^+ leu^2$	$thr^- leu^3$	50 thr^+:250 thr^-	17

29. In fungi, what is the designation for a tetrad of spores that contains four different types, e.g., *AB, Ab, aB, ab*?

30. In fungi, a tetrad containing two kinds of spores, e.g., 2 *AB* and 2 *ab*, is called a _____

*31. A new spore color mutant, maroon (*m*) is discovered in *Neurospora*. A maroon-spored fungus is crossed to a normal gray spore color (*m*⁺) fungus. The segregation patterns of the spores in individual asci are listed below:

Number	Genotype of spore pairs			
39	m	m	m^+	m^+
40	m^+	m^+	m	m
9	m^+	m	m^+	m
8	m	m^+	m	m^+
7	m^+	m	m	m^+
6	m	m^+	m^+	m

What is the distance in cM between the centromere and the gene determining maroon and gray spores?

*32. The following data are from a cross of a wild-type *Neurospora* with a strain that has purple spores (*p*) and shiny spore coats (*s*). The following ordered tetrads of sexual spores resulted:

	Spore pairs			Number
1	2	3	4	of asci
$p +$	$p +$	$+ s$	$+ s$	15
$p\ s$	$p\ s$	$+ +$	$+ +$	99
$p\ s$	$p +$	$+ s$	$+ +$	64
$p\ s$	$+ s$	$p +$	$+ +$	3
$p\ s$	$+ +$	$p\ s$	$+ +$	17
$p +$	$+ s$	$p +$	$+ s$	2
$p\ s$	$+ +$	$+ s$	$p +$	4
				204

(a) What is the map distance between purple and shiny?

(b) What is the distance between purple and the centromere?

(c) What is the map distance between shiny and the centromere?

THOUGHT CHALLENGING EXERCISE

In the beef production business, antibiotics are often used as a supplement in cattle feed, even though no cattle are sick with bacterial infections. Also, well-meaning physicians tend to readily provide broad spectrum antibiotics to relieve potential discomfort or suffering due to bacterial or suspected bacterial infections. In light of your newly acquired knowledge of bacterial reproduction, comment on the positive and potentially negative aspects of the use of antibiotics for disease prevention.

SUMMARY OF KEY POINTS

Their small size, short generation time, and simple structures have made microorganisms valuable model systems for genetic studies.

Many basic concepts of genetics were first deduced from studies of microorganisms.

Viruses are obligate parasites that can reproduce only by infecting living host cells.

Bacteriophages are viruses that infect bacteria.

Bacteriophage T4 is a lytic phage that infects *E. coli*, reproduces, and lyses the host cell.

Bacteriophage lambda (λ) can enter a lytic pathway, like T4, or it can enter a lysogenic pathway, during which its chromosome is inserted into the chromosome of the bacterium.

In its integrated state the λ chromosome is called a prophage, and its lytic genes are kept turned off.

Phage T4 has a circular genetic map, but a linear chromosome. The circular map is produced because T4 chromosomes are terminally redundant and circularly permutated.

Bacteria usually contain one main chromosome.

Wild-type bacteria are prototrophs; they can synthesize everything they need to grow and reproduce given an energy source and some inorganic molecules.

Auxotrophic mutant bacteria require additional metabolites for growth.

Gene transfer in bacteria is unidirectional; genes from a donor cell are transferred to a recipient cell, with no transfer from recipient to donor.

Three parasexual processes – transformation, conjugation, and transduction – occur in bacteria. These processes can be distinguished by two criteria; whether the gene transfer is inhibited by deoxyribonuclease and whether it requires cell contact.

Transformation involves the uptake of free DNA by bacteria.

Conjugation occurs when a donor cell makes contact with a recipient cell and then transfers DNA to the recipient cell.

Transduction occurs when a virus carries bacterial genes from a donor cell to a recipient cell.

Plasmids are self-replicating extrachromosomal genetic elements.

Episomes can replicate autonomously or as integrated components of bacterial chromosomes.

F-factors that contain chromosomal genes (F' factors) are transferred to F⁻ cells by sexduction.

Closely linked genes can be mapped in bacteria by three-factor crosses.

Parasexual recombination produce new combinations of genes in bacteria.

Parasexual mechanisms enhance the ability of bacteria to adapt to changes in the environment.

The genotypes of all four products of meiosis – the tetrad – can be analyzed in Ascomycetes such as the yeast *Saccharomyces cerevisiae* and the bread mold *Neurospora crassa*.

The unordered tetrads of yeast can be used to detect linkage and map genes.

The ordered tetrads of Neu*rospora crassa* can be used to map the positions of both centromeres and genes on chromosomes.

ANSWERS TO QUESTIONS AND PROBLEMS

1) bacteriophage **2)** plaque morphology; conditional **3)** host range; temperature sensitive **4)** plaque **5)** *r* → *r*⁺ back mutations occurred **6)** once part of a gene is lost, it generally can't be regained **7)** glucosylated hydroxymethylcytosine **8 - 11)** see *Approaches to Problem Solving* section that follows **12)** prophage **13)** transformation **14)** virulent **15)** sexduction **16)** Hfr **17)** auxotrophic **18)** competence **19)** merozygote **20)** F factor **21)** temperate **22)** lytic **23)** generalized transduction **24)** lysogenic **25)** specialized transduction **26 to 28)** see *Approaches to Problem Solving* section at the end of the chapter. **29)** tetratype **30)** ditype **31, 32)** see *Approaches to Problem Solving* section below.

Ideas concerning thought stimulating exercise:

Cattle fed with antibiotic-supplemented feed generally grow faster. However, the prolonged use of antibiotics may create an environment that selects for those bacteria, nonpathogenic and pathogenic, that carry antibiotic-resistance genes. The use of antibiotics in treatment of human bacterial diseases prevents much suffering and in many cases is essential for complete healing or even survival. The extensive use of antibiotics in medicine does create problems in that it results in selection for antibiotic-resistant bacteria. If the resistance genes end up in pathogenic bacteria, their control by antibiotics is more difficult.

APPROACHES TO PROBLEM SOLVING

8. Since the aliquots were of the same dilution they should contain equal numbers of progeny phage. Therefore, we can calculate the number of recombinants [number of plaques on strain K12(λ)] divided by the total number of phage in the aliquot (number of plaques on strain B). Remember, both r mutants and *r*⁺ phage infect and lyse strain B, whereas only wild-type (*r1*⁺ *r2*⁺) will infect and lyse strain K12(λ). Also note that only the *r1*⁺ *r2*⁺ recombinants are detected. The double mutant *r1 r2* does not infect and lyse strain K12(λ). Since only half of the recombinants are detected we use the formula:

$$\text{Recombination (R)} = \frac{2[\text{number plaques on strain K12}(\lambda)]}{\text{number of plaques on strain B}} = \frac{2(10)}{10,000} = 0.002$$

9. The wild-type phage that infect *E. coli* strain K12(λ) represent only one-half of the recombinants. Therefore, double the detected recombination frequencies to estimate the total frequency of recombination. Since one map unit equals 1% recombination, multiply the recombination frequencies by 100.

$$
\begin{array}{lll}
r1 \times r2 & \rightarrow .011 \times 2 = .022 & \text{map distance} = 2.2 \\
r1 \times r3 & \rightarrow .005 \times 2 = .010 & \text{map distance} = 1.0 \\
r1 \times r4 & \rightarrow .003 \times 2 = .006 & \text{map distance} = 0.6 \\
r2 \times r3 & \rightarrow .006 \times 2 = .012 & \text{map distance} = 1.2 \\
r2 \times r4 & \rightarrow .008 \times 2 = .016 & \text{map distance} = 1.6 \\
r3 \times r4 & \rightarrow .002 \times 2 = .004 & \text{map distance} = 0.4 \\
\end{array}
$$

A map is constructed by assigning mutants a sequence that is additive and compatible with the individual map distances. The two mutants displaying the largest map distance are often at the opposite ends of the map, in this case *r1* and *r2*.

r1 _____ 2.2 _____ r2

Next find the compatible order of the other mutants.

r1 _____ 1.0 _____ r3

r1 __ 0.6 __ r4   r3 _____ 1.2 _____ r2

r4 __ 0.4 __ r3

r4 _____ 1.6 _____ r2

The map is   r1 _____ r4 _____ r3 _____ r2
                        0.6   0.4   1.2

10. The problem is approached the same as problem 8, except that the dilution of the aliquot is different for the phage plated on strain B compared with that plated on *E. coli* strain K12(λ). Since the dilution factor is 10,000-fold, the number of plaques on strain B must be multiplied by 10,000 (10,000 × 6 = 60,000). Four recombinants were detected out of 60,000 plaques. Remember, only one-half of the recombinants are detectable.

(a) Recombination (R) $= \dfrac{2[\text{number plaques on strain K12}(\lambda)]}{\text{number of plaques on strain B}} = \dfrac{2(4)}{60,000} = 0.00013$ (map distance = 0.013).

(b) If one map unit is 1 kb pairs, then (.013 × 1000 = 13) is the number of nucleotides that separate these mutants.

11. To approach this problem note that all the genotypes, including both recombinants, can be detected on the mixed lawn of *E. coli* strain B and strain B/2. The genotype of each plaque is indicated as follows:

Plaque morphology	Frequency	Genotype
small, turbid	49	$r^+ \ h^+$
small, clear	425	$r^+ \ h$
large, turbid	413	$r \ h^+$
large, clear	45	$r \ h$
total =	932	

The recombinants r^+h^+ and $r\,h$ comprise 94 of the total of 932 plaques. The frequency of recombination is therefore 0.101.

26. To approach this problem, the data on the order of the genes can be assembled either directly on a circle or on an overlapping linear map that subsequently can be presented as a circle. Using the latter approach we obtain the following overlapping linkages. The integrated F factors are indicated in bold face and the arrow indicates the direction of transfer during conjugation.

$$\leftarrow \mathbf{F^E} \text{ - } metC \text{ - } lysA \text{ - } aroD \text{ - } tryA \text{ - } purE \text{ -}$$
$$\leftarrow \mathbf{F^A} \text{- } arg \text{ - } xyl \text{ - } metC \text{ - } lysA \text{ - } aroD\text{-}$$
$$\leftarrow \mathbf{F^B} \text{-} purE \text{ - } metD \text{ - } arg \text{ - } xyl \text{ - } metC \text{ -}$$
$$\leftarrow \mathbf{F^C} \text{ -} trpA \text{ -} purE \text{ - } metD \text{ - } arg \text{ - } xyl \text{ -}$$
$$\text{ - } tryA \text{ -} purE \text{ - } metD \text{ - } arg \text{ - } xyl \text{ - } \mathbf{F^D} \rightarrow$$

The linear linkage maps above can be presented in circular form as follows.

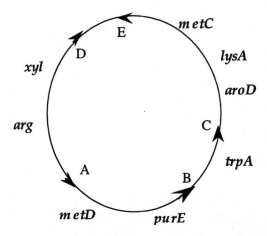

27. In this problem it is observed that all the progeny scored are recombinants for $aziC^+$, which is a marker in this Hfr strain that enters the F^- cell after ton^+ and pan^+. The amount of recombination is measured relative to the total number of $aziC^+$ colonies. All viable recombinants require even numbers of crossovers.

The $aziC^+\,ton^+\,pan^+$ recombinants resulted from crossovers flanking these three genes. These do not represent crossing over between $aziC$ and ton or between ton and pan.

The $aziC^+\,ton\,pan$ recombinants resulted from two crossovers, one before $aziC^+$ and one in the region between $aziC^+$ and the $ton\,pan$; $\underline{aziC^+\,ton^+\,pan^+}$.
$aziC\;ton\;pan$

The $aziC^+\,ton\,pan^+$ recombinants allow the order of the genes to be determined. This is done by determining which order requires the least number of crossover to give $aziC^+\,ton\,pan^+$. If the order is $aziC\,ton\,pan$, four crossovers, $\underline{aziC^+\,ton^+\,pan^+}$, would be needed to generate the $aziC^+\,ton\,pan^+$
$aziC\;ton\;pan$
recombinant. However, if the order was $aziC\,pan\,ton$, only two crossovers, $\underline{aziC^+\,pan^+\,ton^+}$, would
$aziC\;pan\;ton$
be required to generate the recombinant. The gene order is, therefore, $aziC\,pan\,ton$.

The $aziC^+\,ton^+\,pan$ recombinant requires four crossovers to occur, $\underline{aziC^+\,pan^+\,ton^+}$. This is the
$aziC\;\mathbf{pan}\;ton$
rarest recombinant class which is compatible with our determined gene order.

The total recombination between *azi*C and *pan* is 200 + 2/833 +200 + 45 +2 = 202/1080 = 0.1870 = 18.70 map units.

The total recombination between *pan* and *ton* is 45 + 2/1080 = 0.04352 = 4.35 map units.

28. To determine the order of the mutants in the *leu* gene, compare the number of crossovers it takes to produce the wild-type recombinants relative to thr^+ and thr for the two possible gene orders. The first cross is as follows:

$$\frac{thr^+ \quad leu_1 \quad leu_2^+}{thr \quad leu_1^+ \quad leu_2} \qquad \text{or} \qquad \frac{thr^+ \quad leu_2^+ \quad leu_1}{thr \quad leu_2 \quad leu_1^+}$$

If the first order was the correct one, it would take four crossovers to produce wild-type recombinants for thr^+ and two for thr^-. Therefore, we would expect the frequencies of thr^+ to be much less than thr^- among the wild-type recombinants. This is compatible with the observed 17% of the wild-type recombinants that are thr^+. If the second gene order is correct, then it would take two crossovers to produce either thr^+ or thr^- alleles among the wild-type recombinants and they would occur in more equal proportions. The gene order is thr - leu_1 - leu_2 .

The second cross is depicted as follows:

$$\frac{thr^+ \quad leu_3 \quad leu_2^+}{thr^- \quad leu_3^+ \quad leu_2} \qquad \text{or} \qquad \frac{thr^+ \quad leu_2^+ \quad leu_3}{thr^- \quad leu_2 \quad leu_3^+}$$

The correct gene order will produce about equal numbers of thr^+ (51%) and thr^- alleles among the wild-type recombinants. The first gene order is not correct because it would take four crossovers to produce thr^+ wild-type recombinants, and only two crossovers to produce thr^- wild-type recombinants. The second gene order requires only two crossovers to produce either thr^+ or thr^- wild-type recombinants. We would then expect thr^- and thr^+ recombinants to occur in about equal frequency among the wild-type recombinants, as observed. Therefore, the correct gene order is thr - leu_2 - leu_3.

The third cross is as follows:

$$\frac{thr^+ \quad leu_2 \quad leu_1^+}{thr^- \quad leu_2^+ \quad leu_1} \qquad \text{or} \qquad \frac{thr^+ \quad leu_1^+ \quad leu_2}{thr^- \quad leu_1 \quad leu_2^+}$$

The second gene order requires only two crossovers to produce either thr^+ or thr^- among the wild-type recombinants. This is compatible with the observed frequency of thr^+ (50%) among the wild-type recombinants. The gene order is thr -leu_1 - leu_2.

The fourth cross is as follows:

$$\frac{thr^+ \quad leu_2 \quad leu_3^+}{thr^- \quad leu_2^+ \quad leu_3} \qquad \text{or} \qquad \frac{thr^+ \quad leu_3^+ \quad leu_2}{thr^- \quad leu_3 \quad leu_2^+}$$

If the first order was correct we would expect to see far fewer thr^+ (require four crossovers) than thr^- (require two crossovers) among the wild-type recombinants. Since only 17% of the recombinants are thr^+, this gene order appears to be correct. If the second gene order was correct, we would expect to see about equal numbers of thr^+ and thr^- among the wild-type recombinants, since each type would require only two crossovers. The correct gene order is thr - leu_2 - leu_3.

Combining the results from the four crosses we conclude that the order of the mutants of the *leu* gene is thr - leu_1 - leu_2 - leu_3.

31. The first step in solving this problem is to designate the genotypic patterns of the spore pairs as 1st division (no crossover between the gene and centromere) or 2nd division (crossover between the gene and the centromere) segregation.

Number	Genotype of spore pairs				Division of segregation
39	*m*	*m*	*m+*	*m+*	1st
40	*m+*	*m+*	*m*	*m*	1st
9	*m+*	*m*	*m+*	*m*	2nd
8	*m*	*m+*	*m*	*m+*	2nd
7	*m+*	*m*	*m*	*m+*	2nd
6	*m*	*m+*	*m+*	*m*	2nd

The total number of asci scored was 109, thirty of which resulted from crossing over between the gene and the centromere. Since only two of the four chromatids are involved in a single crossover, The frequency of recombination is 1/2 the frequency of second division segregation:

Recombination = $\dfrac{0.5 \text{ (freq. of asci showing 2nd division segregation)}}{\text{total number of asci scored}} = \dfrac{0.5(30)}{109} = 0.1376 = 13.76$ cM.

32. This cross was between *p s* and *+ +* strains. The recombinants will be *p +* and *+ s*. The tetrads are classified as parental ditype (PD), nonparental ditype (NPD), or tetratype (T). They can also be classified as 1st or 2nd division segregation patterns with respect to each marker.

	Spore pairs			Number of asci	Type of ascus	Segregation for *p*	Segregation for *s*
1	2	3	4				
p +	*p +*	*+ s*	*+ s*	15	NPD	1st division	1st division
p s	*p s*	*+ +*	*+ +*	99	PD	1st division	1st division
p s	*p +*	*+ s*	*+ +*	64	T	1st division	2nd division
p s	*+ s*	*p +*	*+ +*	3	T	2nd division	1st division
p s	*+ +*	*p s*	*+ +*	17	PD	2nd division	2nd division
p +	*+ s*	*p +*	*+ s*	2	NPD	2nd division	2nd division
p s	*+ +*	*+ s*	*p +*	4	T	2nd division	2nd division
				204			

(a) We analyze the data and conclude that the genes are linked because the NPD (15 + 2 = 17) are much less numerous than the PD tetrads (99 + 17 =116). The standard mapping formula can be used to estimate the distance between *p* and *s*.

$$\dfrac{(1/2)\ T + NPD}{\text{total}} = \dfrac{(1/2)\ (71) + 17}{204} = .257 = 25.7 \text{ cM}$$

(b) The distance of *p* from the centromere is calculated as:

$$\dfrac{(1/2)\ (\text{second division segregation})}{\text{total asci scored}} = \dfrac{(1/2)(26)}{204} = 0.064 = 6.4 \text{ cM}$$

(c) The distance of *s* from the centromere is calculated as:

$$\dfrac{(1/2)\ (\text{second division segregation})}{\text{total asci scored}} = \dfrac{(1/2)(87)}{204} = 0.213 = 21.3 \text{ cM}$$

The genes are linked on opposite arms of the chromosome as follows:

p — 6.4 — centromere — 21.3 — *s*

9

DNA and the Molecular
Structure of Chromosomes

IMPORTANT CONCEPTS

    A. The genetic information of all living organisms, except certain viruses, is stored in DNA. RNA is the genetic material of some viruses. Some naked RNAs (viroids) and proteinaceous structures (prions) act as transmissible infectious agents.

        1. DNA is a polynucleotide that usually has a double helical structure in which the two strands are held together by hydrogen bonding between the two bases of each nucleotide pair and hydrophobic bonding in the core of stacked base pairs.

            a. The base pairing is specific: adenine always pairs with thymine, and guanine always pairs with cytosine.

            b. The two strands of a DNA double helix are complementary: once the sequence of bases in one strand is established, the sequence of bases in the other (complementary) strand is fixed.

            c. The two complementary strands of a DNA double helix have opposite chemical polarity, one 3' to 5' and the other 5' to 3'.

    B. Chromosomes of prokaryotic organisms are highly compacted *in vivo*.

        1. The DNA molecules are separated into many negatively supercoiled loops by RNA-containing cross-links.

C. Each eukaryotic chromosome contains a single giant molecule of DNA that extends from one end of the chromosome through the centromere to the other end of the chromosome.
   1. The DNA of interphase chromatin is held in a negatively supercoiled configuration by association with histones.
      a. Interphase chromosomes are characterized by 10-nm-diameter fibers.
      b. Isolated interphase chromatin has a beads-on-a-string structure.
         (1) Each bead or nucleosome is an ellipsoid-like structure about 11 nm in diameter and 6 nm high.
         (2) Each contains a highly conserved core consisting of a 146-nucleotide-pair-long segment of DNA wrapped around an octomer of histones.
         (3) Each octomer contains two molecules each of the histones H2a, H2b, H3, and H4.
   2. During mitosis and meiosis, the chromatin is further condensed, probably by a second level of coiling, into a 30-nm chromatin fiber.
      a. At metaphase, the coiled or folded 30-nm chromatin fiber is in turn coiled or folded into the highly condensed structure that can be seen with the light microscope; this third level of condensation involves a chromosomal scaffold that is composed of nonhistone chromosomal proteins.
   3. The centromere (spindle fiber-attachment regions) and telomeres (ends) of the chromosomes have unique structures.
   4. Eukaryotic genomes contain unique or single-copy sequences (present 1 to 10 times per genome), moderately repetitive DNA sequences (those repeated 10 to 10^5 times), and highly repetitive DNA sequences (repeated more than 10^5 times).
      a. Single-copy DNA sequences make up about 30% to 85% of the genomes of eukaryotes, whereas they represent almost all of the sequences in prokaryotes.
         (1) Most of the genes that encode proteins are present in the unique fraction of the genome.
      b. Moderately repetitive sequences make up from as little as 5% to as much as 80% of the genomes of eukaryotes.
         (1) The most prevelent of the moderately repetitive sequences are transposable genetic elements, or inactive elements derived from transposable elements.
         (2) Much of the dispersed moderately repetitive DNA in the human genome consists of families of transposable elements belonging to four classes: LINEs (long interspersed nuclear elements), SINEs (short interspersed nuclear elements), transposable elements containing long terminal repeats (LTRs) and DNA transposons.
      c. Highly repetitive sequences are often located in centromeric or telomeric regions of the chromosomes.

IMPORTANT TERMS

In the space allotted, concisely define each term.

replication:

gene expression:

mutation:

deoxyribonucleic acid (DNA):

ribonucleic acid (RNA):

deoxyribonuclease (DNase):

ribonuclease (RNase):

protease:

viroids:

prions:

nucleotide:

adenine:

guanine:

thymine:

cytosine:

uracil:

purines:

pyrimidines:

double-helix:

complementarity:

antiparallel (opposite chemical polarity):

B-DNA:

A-DNA:

Z-DNA:

negative supercoiling:

folded genome:

domains:

chromatin:

histones (H1, H2a, H2b, H3, and H4):

nonhistone chromosomal proteins:

protamines:

nucleosome:

multineme:

unineme:

lampbrush chromosomes:

pulse-field gel electrophoresis:

autoradiography:

viscoelastometry:

nucleosome:

linkers:

nucleosome core:

chromatin fibers (10-nm and 30-nm diameter):

scaffold:

telomeres:

repetitive DNA:

satellite bands:

satellite DNA:

denaturation:

renaturation:

unique (single-copy) DNA sequences:

moderately repetitive DNA sequences:

 highly repetitive DNA sequences:

transposable genetic elements:

in situ hybridization:

FISH:

LINEs:

SINEs:

LTRs:

DNA transposons:

IMPORTANT NAMES

In the space allotted, concisely state the major contribution made by the individual or group.

Johann Freidrich Miescher:

Frederick Griffith:

Oswald Avery, Colin Macleod, and Maclyn McCarty:

Alfred Hershey and Martha Chase:

Heinz Fraenkel-Conrat:

Theodore Diener:

Stanley Prusiner:

James Watson:

Francis Crick:

Erwin Chargaff:

Maurice Wilkins:

Rosalind Franklin:

Ruth Kavenoff, Lynn Klotz, and Bruno Zimm:

TESTING YOUR KNOWLEDGE

In this section, answer the questions, fill in the blanks, and solve the problems in the space allotted. Problems noted with an * are solved in the Approaches to Problem Solving section at the end of the chapter.

1. The basic building block of a nucleic acid is called a _____.

2. The proteins that are rich in basic amino acids and function in the initial packaging of DNA in eukaryotes are called _____.

3. The 11 nm by 6 nm particle that consists of an octomer of histones around which is coiled a 146-basepair-long sequence of DNA is called a _____.

4. If a double-stranded DNA molecule contains 15% guanine, it must contain _____ % thymine.

5. An enzyme that catalyzes the destruction of a DNA molecule by breaking phosphodiester bonds is called _____.

6. The two pyrimidines found in DNA are _____ and _____.

7. The purines found in DNA are _____ and _____.

8. If a double-stranded DNA molecule contains 12% cytosine, then it must contain _____ % guanine.

9. The individuals who first showed that DNA was the genetic material using transformation were _____, _____, and _____.

10. The pyrimidines found in RNA are _____ and _____.

11. The persons who provided additional evidence that DNA is the genetic material using bacteriophage T2 were _____ and _____.

*12. A segment of DNA has the base-pair sequence   A T G G C A
                                                         T A C G G T

     How many hydrogen bonds occur in this six base-pair sequence?

*13. A nucleic acid with the base composition of 45% adenine, 45% cytosine, 5% thymine and 5% guanine is:
     (a) double-stranded DNA
     (b) double-stranded RNA
     (c) single-stranded DNA
     (d) single-stranded RNA

14. A nucleic acid with the base composition of 30% adenine, 20% cytosine, 30% thymine and 20% guanine is:

     (a) double-stranded DNA
     (b) double-stranded RNA
     (c) single-stranded DNA
     (d) single-stranded RNA

*15. A nucleic acid with a base composition of 22% adenine, 20% cytosine, 30% uracil and 28% guanine is:

     (a) double-stranded DNA
     (b) double-stranded RNA
     (c) single-stranded DNA
     (d) single-stranded RNA

16. The three scientists who shared a Nobel prize for discovering the structure of DNA were _____, _____, and _____.

17. Which one of the histones is not found in the nucleosome core particle?

18. Supercoils are introduced into DNA by enzymes called _____.

19. DNA that coils to the left and has 12 base-pairs per turn is called _____.

\

20. The interphase chromatin consists of:

   (a) supercoiled DNA molecules free of histones
   (b) 30-nm fibers
   (c) 10 nm-fibers and a scaffold of nonhistone proteins
   (d) 10 nm-fibers consisting of DNA and histones

*21. One strand of a DNA molecule has the following sequence 5' - ACGTATGA - 3'.
   The complementary strand must be:

   (a) 5'-TGCATACT-3'
   (b) 5'-TCATACGT-3'
   (c) 5'-UCAUACGT-3'
   (d) 5'-ACGTATGA-3'

*22. An important crop plant has a genome size of about 1 pg (1 X 10^9 base pairs).

   (a) How many nucleosomes would it have assuming each nucleosome and spacer DNA comprises 200 bp of DNA?

   (b) How many turns of the double helix are present in the DNA of this plant?

   (c) If this plant has its DNA equally partitioned among four chromosomes, how many centimeters of DNA would be in each chromosome?

THOUGHT CHALLENGING EXERCISE

It has been observed in eukaryotes, that: (1) there appears to be a great excess in DNA, much more than that needed to code for proteins (< 1% of the DNA in some species has a protein-coding function); (2) DNA varies greatly among species (even among closely related species); and (3) DNA content does not correlate positively with evolutionary advancement or genetic complexity. These phenomena led to what has been called the DNA C-value paradox. What do you think the role is of the huge excess in DNA found in most eukaryotes, including humans?

SUMMARY OF KEY POINTS

The genetic material must perform three essential functions: the genotypic function–replication, the phenotypic function–gene expression, and the evolutionary function–mutation.

The genetic information of most living organisms is stored in deoxyribonucleic acid (DNA).

In some viruses, the genetic information is present in ribonucleic acid (RNA).

Viroids and prions are infectious naked molecules of RNA and protein, respectively.

DNA usually exists as a double helix, with the two strands held together by hydrogen bonds between the complementary bases: adenine paired with thymine and guanine paired with cytosine.

The complementarity of the strands of a double helix makes DNA uniquely suited to store and transmit genetic information.

The two strands of a DNA double helix have opposite chemical polarity.

RNA usually exists as a single-stranded molecule containing uracil instead of thymine.

The functional DNA molecules in cells are negatively supercoiled.

The DNA molecules in prokaryotic and viral chromosomes are organized into negatively supercoiled domains.

Bacterial chromosomes contain circular molecules of DNA segregated into about 50 domains.

Each eukaryotic chromosome contains one giant molecule of DNA packaged into 10-nm ellipsoidal beads called nucleosomes.

The condensed chromosomes that are present in mitosis and meiosis are composed of 30-nm chromatin fibers.

At metaphase, the 30-nm fibers are segregated into domains by scaffolds composed of nonhistone chromosomal proteins.

The centromeres (spindle-fiber-attachment regions) and telomeres (termini) of chromosomes have unique structures that facilitate their functions.

Eukaryotic genomes contain repeated DNA sequences, with some sequences present a million times or more.

The DNA sequences in eukaryotic genomes are commonly grouped into three classes: (1) unique or single-copy sequences present in one to a few copies per genome, (2) moderately repetitive sequences present in 10 to 10^5 copies, and (3) highly repetitive sequences present in over 10^5 copies per genome.

Two families of moderately repetitive transposable elements make up 10 percent of the human genome and play an important role in its evolution.

ANSWERS TO QUESTIONS AND PROBLEMS

1) nucleotide 2) histones 3) nucleosome 4) 35 5) nuclease 6) thymine and cytosine 7) adenine and guanine 8) 12 9) Avery, Macleod, and McCarty 10) uracil and cytosine 11) Hershey and Chase 12) 15 13) c 14) a 15) d 16) Watson, Crick, and Wilkins 17) H1 18) topoisomerases 19) Z-DNA 20) d 21) b 22) a. 5×10^6 b. 1×10^8 c. 8.5

Ideas on thought challenging exercise:

One of the great enigmas of modern biology is the apparently excessive amount of DNA beyond that needed for coding functions, and the abundant variation in DNA content among organisms. Some molecular biologists view this excessive DNA as junk or parasitic. In this theory, variation in DNA content is considered to be without phenotypic effects and accumulates simply because some sequences have the capability to differentially replicate themselves and spread throughout the genome. However, extensive studies over the last three decades have shown that DNA content variation does have measurable and somewhat predictable effects on the phenotype. Studies of plants show that DNA content positively correlates with nuclear volume, meristematic cell volume, and the duration of the mitotic cycle. A variety of correlations have also been observed between DNA content and the environment in which a plant grows. The effects of DNA content on the cell, independent of any coding function, have been called the nucleotype by Michael Bennett. Other investigators have postulated that heterochromatin, that often contains repetitive DNA sequences, may have multiple functions in the nucleus including gene regulation, positions of chiasmata, and chromosome pairing.

APPROACHES TO PROBLEM SOLVING

12. There are two hydrogen bonds between adenine and thymine, and three hydrogen bonds between guanine and cytosine. Therefore, fifteen hydrogen bonds occur in the six base-pair sequence containing three A:T base pairs and three G:C base pairs.

13. In a double-stranded DNA molecule the number of purines (A + G) must equal the number of pyrimidines (C + T) and A = T and G = C. In this DNA there is 45% adenine and 5% thymine. Since A ≠ T, the DNA must be single-stranded.

15. This nucleic acid contains uracil instead of thymine; therefore, it must be RNA. Since A ≠ U, and C ≠ G, this RNA is single-stranded.

21. The strands of a DNA are held together by complementary base-pairs. The strands are also of opposite polarity. Therefore, the strand complementary to 5' -ACGTATGA -3' must be 3' - TGCATACT - 5'.

22. (a) Assume that each nucleosome and its spacer average 200 bp of DNA. To obtain the number of nucleosomes in 1 pg DNA, simply divide 1×10^9 base pairs by 2×10^2 bp per nucleosome = 5×10^6 nucleosomes.

    (b) The DNA helix makes a complete turn every 10 base-pairs. Therefore, 1×10^9 bp/10 bp per turn of the double helix = 1×10^8 turns.

    (c) The base-pairs of the DNA helix are spaced 3.4 Å or 3.4×10^{-10} meters apart. Multiply (1×10^9 bp)(3.4×10^{-10} meters per bp) to obtain the length of the DNA containing 1×10^9 bp. This calculates to be 0.34 meters or 34 cm. If the DNA is partitioned equally among four chromosomes, there must be 8.5 cm DNA per chromosome.

10

Replication of DNA and Chromosomes

IMPORTANT CONCEPTS

A. Replication of DNA is semiconservative in both prokaryotes and eukaryotes.
   1. During replication the two complementary strands unwind, and each single strand serves as a template directing the synthesis of a new complementary strand.
      a. The net result of replication is two progeny DNA molecules identical to the parental double helix.
   2. Replication often begins at unique origins and proceeds bidirectionally from these origins.
      a. Prokaryotic chromosomes usually contain a single primary origin of replication.
   3. DNA replication is catalyzed by enzymes called DNA polymerases.
      a. All known DNA polymerases, both prokaryotic and eukaryotic, have an absolute requirement for a 3'-hydroxyl primer.
      b. DNA polymerases catalyze covalent extension only in the 5' → 3' direction. Thus, synthesis of the strand growing in the 3' → 5' direction is discontinuous.
      c. Short segments called Okazaki fragments are synthesized in the 5' → 3' direction and then are covalently joined by DNA ligase.
      d. No known polymerases can catalyze the initiation of new DNA chains.

(1) Nascent DNA chains are initiated by short RNA primers synthesized by DNA primase.

B. Chromosome replication in *E. coli* is accomplished by a complex replication apparatus or replisome composed of many different proteins and enzymes.
1. DNA topoisomerases produce transient breaks in DNA that function as swivels during the unwinding of the complementary strands of a parental double-helix.
2. DNA helicase uses energy from ATP to unwind the complementary strands of a parental DNA molecule.
3. Single-strand DNA-binding protein coats the DNA single strands and keeps them in an extended configuration for replication.
4. The DNA polymerase III holoenzyme, which catalyzes semiconservative DNA replication in *E. coli*, is a multimeric protein containing at least 20 polypeptides.
   a. The 3' → 5' exonuclease activity built into prokaryotic DNA polymerases provides a proofreading function that is essential for accurate DNA replication.
5. Negative supercoils are introduced into and removed from *E. coli* DNA by DNA gyrase.

C. Eukaryotic DNA replication, with a few unique aspects, is basically the same as that in prokaryotes.
1. RNA primers and Okazaki fragments are shorter in eukaryotes than in prokaryotes.
2. Eukaryotic chromosomes contain multiple origins of replication.
3. Eukaryotic replication requires three different DNA polymerases.
   a. DNA polymerase α exists in a stable complex with DNA primase and extends the growing RNA-DNA polynucleotide chain to about 30 nucleotides.
   b. DNA polymerase δ extends the RNA-DNA primer chains and synthesizes most of the chromosomal DNA.
   c. DNA polymerase ∈is required for *in vivo* DNA replication but its exact function (s) has not been deciphered.

D. DNA replication and histone synthesis are tightly coupled in eukaryotes, and the progeny DNA molecules are rapidly packaged into nucleosomes.

E. Telomerase, an enzyme with a built-in RNA template, adds the unique termini or telomere sequences to eukaryotic chromosomes, therefore, preventing the ends of chromosomes from becoming shorter during each replication.

IMPORTANT TERMS

In the space allotted, concisely define each term.

semiconservative replication:

replication fork:

origin of replication:

replication bubble:

autonomously replicating sequences (ARS elements):

DNA polymerase I:

primer DNA:

template DNA:

nuclease (endonuclease, exonuclease):

5' → 3' exonuclease activity:

3' → 5' exonuclease activity:

DNA polymerases II, III, IV, and V:

holoenzyme :

proofreading:

continuous synthesis:

discontinuous synthesis:

leading strand:

lagging strand:

Okazaki fragments:

DNA ligase:

RNA primer:

DNA primase:

DNA helicases:

single-stranded DNA-binding proteins (SSB proteins):

DNA topoisomerases:

DNA gyrase:

oriC:

prepriming proteins:

DnaA protein:

DnaC protein:

primosome:

replisome:

rolling-circle replication:

replicon:

DNA polymerase α:

DNA polymerase δ:

telomerase:

progerias:

IMPORTANT NAMES

In the space allotted, concisely state the major contribution made by the individual or group.

Matthew Meselson and Franklin Stahl:

J. Herbert Taylor, Philip Woods, and Walter Hughes:

John Cairns:

Arthur Kornberg:

Paula DeLucia:

Joel Huberman and Arthur Riggs:

TESTING YOUR KNOWLEDGE

*In this section, answer the questions, fill in the blanks, and solve the problems in the space allotted. Problems noted with an * are solved in the Approaches to Problem Solving section at the end of the chapter.*

1. The enzyme that catalyzes the formation of a covalent bond between the 5' and 3' ends of newly synthesized single strands of DNA during DNA replication is called _____.

2. The enzyme that replicates most of the DNA in *E. coli* that also has a proofreading function is

    _____.

3. During DNA replication of *E. coli,* the enzyme that removes RNA primers and replaces them with DNA is called _____.

4. The relatively short, single-stranded DNA fragments that are synthesized during discontinuous DNA replication and joined together to form a continuous strand are called _____.

5. The enzyme that uncoils DNA into single strands prior to DNA replication is called _____.

6. The synthesis of the leading strand of DNA proceeds from _____ → _____ direction and the lagging strand is synthesized in the _____ → _____ direction.

7. Short RNA primers are synthesized during DNA replication by an enzyme called _____.

8. The enzyme that forms a phosphodiester bond between adjacent nucleotides, but does not extend the strand is called _____.

9. The sequence of DNA in *E. coli* that functions in the origin of DNA replication is called _____.

10. During DNA synthesis, the extended single strands of DNA are maintained by _____.

11. Enzymes that catalyze temporary single-strand breaks in DNA molecules during uncoiling, and use covalent linkages to themselves to hold on to the cleaved molecules are called _____.

12. What are the first three proteins necessary in *E. coli* for the initial steps in formation of the replication fork in the *oriC* region?

13. In *E. coli*, the initiation of DNA synthesis on the lagging strand is carried out by a protein complex containing DNA primase and DNA helicase called the _____.

14. The enzyme that removes supercoils from DNA, one at a time, by producing transient single-strand breaks is called _____.

15. The complete replication apparatus moving along the DNA molecule at the replication fork is called the _____.

16. The enzyme that removes or introduces supercoils from DNA, two at a time, by producing transient double-strand breaks is called _____.

17. An enzyme complex that plays a key role in triggering the onset of DNA synthesis in eukaryotes is called _____.

18. In eukaryotes, the DNA polymerase complex that also has primase activity, but is otherwise most similar to DNA polymerase III of prokaryotes is called _____.

19. In eukaryotes, the DNA polymerase that apparently catalyzes the replication of the leading strand is called _____.

20. The enzyme that has a built-in RNA strand template that initiates DNA synthesis at the end of chromosomes is called _____.

21. In humans, a type of disorder that is characterized by premature aging is called _____.

*22. A culture of bacteria has been grown for many generations in a medium containing N^{15}. Bacteria are removed and grown on a culture containing only normal nitrogen, i.e., N^{14}.

    (a) Diagram the gradient density profiles that would be generated by the three possible types of replication after one and two generations of growth on the normal culture medium.

    (b) After one generation of growth in the normal culture medium can you differentiate between a semiconservative, conservative, and dispersive form of DNA replication? Which mode of replication does the experimentally observed density gradient profile support and why?

THOUGHT CHALLENGING EXERCISE

Design an experiment to measure the length of the cell cycle in eukaryotes.

SUMMARY OF KEY POINTS

DNA replicates by a semiconservative mechanism: as the two complementary strands of a parental double helix unwind and separate, each serves as a template for the synthesis of a new complementary strand.

The hydrogen-bonding potentials of the bases in the template strands specify complementary base sequences in the nascent strands.

Replication is initiated at unique origins and usually proceeds bidirectionally from each origin.

DNA synthesis is catalyzed by enzymes called DNA polymerases.

All DNA polymerases require a primer strand, which is extended, and a template strand, which is copied.

All DNA polymerases have an absolute requirement for a free 3'-OH on the primer strand, and all DNA synthesis occurs in the 5' to 3' direction.

The 3' → 5' exonuclease activities of DNA polymerases proofread nascent strands as they are synthesized, removing any mispaired nucleotides at the 3' termini of primer strands.

DNA replication is complex, requiring the participation of a large number of proteins.

DNA synthesis is continuous on the progeny strand that is being extended in the overall 5' → 3' direction, but is discontinuous on the strand growing in the overall 3' → 5' direction.

New DNA chains are initiated by short RNA primers synthesized by DNA primase.

The enzymes and DNA-binding proteins involved in replication assemble into a replisome at each replication fork and act in concert as the fork moves along the parental DNA molecule.

The large DNA molecules in eukaryotic chromosomes replicate bidirectionally from multiple origins.

Two or three DNA polymerases, (α, δ, and/or ϵ) are present at each replication fork in eukaryotes.

Telomeres, the unique sequences at the ends of chromosomes, are added to chromosomes by a unique enzyme called telomerase.

ANSWERS TO QUESTIONS AND PROBLEMS

1) DNA ligase 2) DNA polymerase III 3) DNA polymerase I 4) Okazaki fragments
5) DNA helicase 6) 5' → 3'; 5' → 3' 7) DNA primase 8) DNA ligase 9) *OriC*
10) single-stranded DNA-binding proteins 11) topoisomerases 12) DnaA; DnaB; DnaC
13) primosome 14) topoisomerase I 15) replisome 16) topoisomerase II 17) start kinases 18) DNA polymerase α 19) DNA polymerase δ 20) telomerase 21) progerias 22) see *Approaches to Problem Solving* section at the end of the chapter.

Ideas concerning thought challenging exercise:

Either root tips of a plant, or a cell culture of an animal could be treated with colchicine for a short period of time to inhibit spindle fiber formation that will result in tetraploid cells. After treatment, the tissue could be transferred to growth conditions lacking colchicine. Cells could be sampled every couple of hours over a period of two or three days. Slides prepared from the samples could be microscopically analyzed for diploid and tetraploid metaphases. The proportion of tetraploid metaphases plotted over time should yield cyclic peaks as the tetraploid nuclei pass through the cell cycles. The interval between the peaks is a measure of the length of the mitotic or cell cycle. A similar experiment could be carried out by pulse-labeling the cells with tritiated thymidine. The metaphases on microscope slides with incorporated radioactive thymidine could be detected by a procedure called autoradiography.

APPROACHES TO PROBLEM SOLVING

22. (a) The following density gradient profiles would be generated by the three possible types of replication following one generation of growth in normal medium after removing from the N^{15}-containing medium.

		semiconservative	conservative	dispersive
1st generation	light		1/2 _____	
	intermediate	all _____		all _____
	heavy		1/2 _____	

The following density gradient profiles would be generated by the three possible types of replication following two generations growth in normal medium after removing from the N^{15}-containing medium.

		semiconservative	conservative	dispersive
2nd generation	light	1/2 _____	3/4 _____	
	intermediate	1/2 _____		all _____
	heavy		1/4 _____	

(b) Semiconservative and dispersive modes of replication would result in a single intermediate band after one generation. Therefore, you can't differentiate between these in the first generation. If a conservative mode of replication occurred there would be two bands, one heavy and one light, in equal proportions after the first generation. Therefore, if conservative replication occurred, it would be apparent in the first generation. A distinct centrifugation profile should be apparent for each mode of replication in the second generation. A dispersive mode would have one band halfway between the intermediate and the light band. The conservative mode would result in two bands; one at the heavy position and another about three times as intense at the light position. Semiconservative replication would result in two bands of equal intensity; one at the intermediate position and one at the light position. Data consistent with the semiconservative model of replication are experimentally observed.

11

Transcription and RNA Processing

IMPORTANT CONCEPTS

A. The pathway of information flow by which a gene exerts its effect on the phenotype of an organism is often very complex. The first two steps in this pathway are transcription and translation.
   1. Transcription is the production of an RNA complementary to the template strand of the DNA of a gene.
   2.   Five types of RNA molecules play essential roles in gene expression.
      a. Messenger RNA (mRNA) functions as an intermediary in protein synthesis, using one of the two strands in the gene as a template.
         (1) Messenger RNA is translated, according to the specifications of the genetic code, into the sequence of amino acids in the polypeptide gene-product.
      b. Transfer RNAs (tRNAs) are small RNA molecules that function as adaptors between amino acids and the codons in the mRNA during translation.
      c. Ribosomal RNAs (rRNAs) are structural and catalytic components of the ribosomes.
      d. Small nuclear RNAs (snRNAs) are structural components of the spliceosomes, the nuclear structures that excise introns from nuclear genes.

e.   Micro RNAs (miRNAs), short single-stranded RNAs cleaved from small hairpin-shaped precursors, block the expression of complementary or partially complementary mRNAs.
3.   Transcription is catalyzed by a complex, multimeric enzyme called RNA polymerase.
4.   Eukaryotes have three RNA polymerase complexes.
   a.   RNA polymerase I is located in the nucleolus and catalyzes the synthesis of large ribosomal RNAs.
   b.   RNA polymerase II transcribes the nuclear genes into mRNAs that encode protein products.
   c.   RNA polymerase III catalyzes the synthesis of tRNAs, most of the snRNAs, and the 5S ribosomal RNA.

B.   In eukaryotes, the primary transcripts usually undergo several types of processing.
1.   Processing frequently includes the excision of segments of primary transcripts and the addition of 5' 7-methyl guanosine caps and 3' poly(A) tails.
2.   In some cases, especially in mitochondria of trypanosomes and plants, the nucleotide sequences of primary transcripts are altered by RNA editing processes.

C.   Most eukaryotic genes contain noncoding sequences called introns that separate the coding sequences called exons.
1.   The entire nucleotide-pair sequence of each split gene, including both intron and exon sequences, is transcribed to produce a pre-mRNA molecule.
2.   The intron sequences are removed from the primary transcript or pre-mRNA by splicing processes to produce the mature, functional RNA molecule.
   a.   Three distinct splicing processes are known.
      (1)   Introns in many pre-tRNA molecules are excised by simple splicing nucleases and ligases.
      (2)   Introns in certain rRNA precursors and some primary transcripts of genes in mitochondria and chloroplasts are excised autocatalytically (self-catalyzed with no protein involved).
      (3)   Introns of nuclear pre-mRNAs are excised by complex ribonucleoproteins called spliceosomes.
3.   The intron sequences of nuclear genes are highly divergent.
   a.   In primary transcripts of nuclear genes, the only completely conserved sequences of different introns are the dinucleotide sequences at the ends of introns, i.e., 5' exon-GT-intron-AG-exon 3'.
   b.   There is only one short, poorly conserved sequence, the TACTAAC box, located about 30 nucleotides upstream from the 3' splice site of introns.
      (1)   The adenine base at position six in the TACTAAC box is completely conserved and plays a key role in the splicing reaction.

IMPORTANT TERMS

In the space allotted, concisely define each term.

metabolic pathway:

messenger RNA (mRNA):

amino acids:

translation:

genetic code:

codons:

ribosomes:

messenger RNA (mRNA):

primary transcript:

pre-mRNAs:

transfer RNA (tRNA):

ribosomal RNA (rRNA):

small nuclear RNAs (snRNAs):

micro RNAs (miRNAs):

anticodon:

ribonucleoside triphosphates:

DNA template strand:

DNA nontemplate strand:

RNA polymerases:

promoters:

transcription bubble:

transcription unit:

RNA polymerase holoenzyme:

sigma (δ) factor:

promoter sites:

termination signal:

heterogeneous nuclear RNA (hnRNA):

5′ cap:

3′ poly(A) tail:

RNA polymerase I, II, and III:

TATA box:

CAAT box:

GC box:

octomer box:

basal transcription factors:

poly(A) polymerase:

RNA editing:

guide RNAs:

introns:

exons:

TACTAAC box:

splicing endonuclease:

splicing ligase:

self-splicing (autocatalytic):

spliceosomes:

snRNPs:

IMPORTANT NAMES

In the space allotted, concisely state the major contribution made by each individual or group.

Oscar Miller and Barbara Hamlako:

Philip Sharp:

Richard Robert:

Richard Flavell:

Philip Leder:

Pierre Chambon:

Bert O'Malley:

TESTING YOUR KNOWLEDGE

1. Noncoding intervening sequences in eukaryotic genes that interrupt the coding sequences are called _____.

2. A specific DNA sequence to which the RNA polymerase binds to start transcription is called a _____.

3. The molecule that transmits information, stored in the DNA of the nucleus, to the ribosomes in the cytoplasm where it is read is called _____.

4. The molecule that caps the 5' end of the pre-messenger RNA is _____.

5. The 3' end of a pre-mRNA is modified by the addition of a _____ tail.

6. The locally unwound segment of DNA which is produced by RNA polymerase is called a _____.

7. The protein in *E. coli* whose function is to recognize and bind RNA polymerase to the promoter site in DNA is called a _____.

8. In eukaryotes, the population of primary transcripts is called _____.

9. The conserved part of a promoter of eukaryotic genes that is centered at about thirty bases upstream of the transcription initiation site is called the _____.

10. The conserved part of the promoter of a eukaryotic gene that is located about eighty bases upstream of the transcription initiation site is called the _____.

11. The sequences of eukaryotic genes that code for amino acid sequences or RNA sequences of the processed gene product are called _____.

12. The enzyme that adds 3' poly(A) tails to pre-mRNAs is called _____.

13. The complex ribonucleoprotein structures that carry out pre-mRNA splicing are called
_____.

14. The eukaryotic enzyme complex that catalyzes the transcription of protein-coding genes
is called _____.

15. A disease that results from mutations that alter the splicing of human beta hemoglobin pre-mRNA
is called _____.

16. The small ribonucleoproteins that are found in spliceosomes are called _____.

*17. The chicken ovalbumin gene has seven introns interrupting the coding sequences or exons. How
many R-loops will be observed if DNA of the gene and the mRNA coded by it are mixed under
conditions that allow the single-stranded mRNA to hybridize and displace one single strand of the
DNA double helix?

*18. How many single-stranded loops will be observed if the ovalbumin mRNA is hybridized to single-
stranded DNA of the ovalbumin gene?

*19. The following is a partial sequence of a gene in *E. coli* and sequences upstream from the coding
sequence:

    5'- A A G C T A T A A T C G T A C G A C C T A A G G A G G T A A G C G -3'
    3'- T T C G A T A T T A G C A T G C T G G A T T C C T C C A T T C G C -5'

Indicate the RNA that is transcribed by RNA polymerase for the portion of the gene shown.

*20. The following is a partial sequence of a gene in humans with some of the upstream sequences:

    5'- T A T A A A A C G T G C A T C G A C G T A C G A T A T C G C C G T A C C G C T G C A C -3'
    3'- A T A T T T T G C A C G T A G C T G C A T G C T A T A G C G G C A T G G C G A C G T G -5'

Indicate the RNA that is transcribed by RNA polymerase II from the portion of the gene shown.

THOUGHT CHALLENGING EXERCISE

It has been observed that bacteria transformed with cloned eukaryotic genes often produce a nonfunctional eukaryotic gene product. What aspects of the flow of genetic information in prokaryotes compared to eukaryotes might explain this phenomenon? How might this problem be prevented so that a bacterial cell culture could be used to produce a functional eukaryotic protein?

SUMMARY OF KEY POINTS

Metabolism occurs by sequences of enzyme-catalyzed reactions, with each enzyme specified by one or more genes.

The central dogma of molecular biology is that genetic information flows from DNA to DNA during chromosome replication, from DNA to RNA during transcription, and from RNA to protein during translation.

Transcription involves the synthesis of an RNA transcript complementary to one strand of DNA of a gene.

Translation is the conversion of information stored in the sequence of nucleotides in the RNA transcript into the sequence of amino acids in the polypeptide gene product, according to the specifications of the genetic code.

In eukaryotes, genes are present in the nucleus, whereas polypeptides are synthesized in the cytoplasm.

Messenger RNA molecules function as intermediaries that carry genetic information from DNA to the ribosomes, where proteins are synthesized.

RNA synthesis, catalyzed by RNA polymerases, is similar to DNA synthesis in many respects.

RNA synthesis occurs within a localized region of strand separation, and only one strand of DNA functions as a template for RNA synthesis.

RNA synthesis occurs in three stages: (1) initiation, (2) elongation, and (3) termination.

RNA polymerases–the enzymes that catalyze transcription–are complex, multimeric proteins.

The covalent extension of RNA chains occurs within locally unwound segments of DNA.

Chain elongation stops when RNA polymerase encounters a transcription-termination signal.

Transcription, translation, and degradation of mRNA molecules often occur simultaneously in prokaryotes.

Three different RNA polymerases are present in eukaryotes, and each polymerase transcribes a distinct set of genes.

Eukaryotic gene transcripts usually undergo three major modifications: (1) the addition of 7-methyl guanosine caps to the 5' termini, (2) the addition of poly(A) tails to the 3' ends, and (3) the excision of intron sequences.

The information content of some eukaryotic transcripts is altered by RNA editing, which changes the nucleotide sequences of transcripts prior to their translation.

Most, but not all, eukaryotic genes are split into coding sequences called exons and noncoding sequences called introns.

Some genes contain very large introns; others harbor large numbers of small introns.

The biological significance of introns is still open to debate.

Noncoding intron sequences are excised from RNA transcripts in the nucleus prior to their transport to the cytoplasm.

Introns in tRNA precursors are removed by the concerted action of a splicing endonuclease and ligase, whereas introns in some rRNA precursors are spliced out autocatalytically—with no protein involved.

The introns in nuclear pre-mRNAs are excised on complex ribonucleoprotein structures called spliceosomes.

The intron excision process must be precise, with accuracy to the nucleotide level, to assure that codons in exons distal to introns are read correctly during translation.

ANSWERS TO QUESTIONS AND PROBLEMS

1) introns 2) promoter 3) messenger RNA (mRNA) 4) 5', 7-methyl guanosine 5) poly (A)
6) transcription bubble 7) sigma factor 8) heterogeneous nuclear RNA (hnRNA) 9) *TATA box* Hogness box) 10) *CAAT box* 11) exons 12) poly(A) polymerase 13) spliceosome 14) RNA polymerase II 15) β -thalessemia 16) snRNPs 17 - 20) see *Approaches to Problem Solving* section at the end of the chapter.

Some ideas concerning the thought challenging exercise:

Eukaryotic genes generally have introns interrupting the coding sequences, whereas prokaryotes do not. Therefore, if a eukaryotic gene containing introns was put into a bacterial cell, the mechanism to process out the introns would be lacking. The whole transcript including the sequences complementary to the introns would be translated. This would result in a nonfunctional protein. One way to avoid this problem is to isolate mRNA, which has had the introns removed. RNA can be converted to DNA by an enzyme called reverse transcriptase. The resulting DNA sequences, called cDNAs, then can be combined with a bacterial promoter and inserted into a bacterial cell. The transcription and translation by bacterial systems produces a protein with the correct linear order of amino acids.

APPROACHES TO PROBLEM SOLVING

17. R-loops result when purified mRNA transcripts are annealed with double-stranded DNA of the corresponding gene under conditions that favor DNA-RNA duplex formation. The RNA strand displaces the homologous DNA strand. The single-stranded DNA loops are visualized using electron microscopy. The R-loops correspond to the nontranscribed strand of the segment of double-stranded DNA from which the ovalbumin pre-mRNA was transcribed. The introns occur as double-

stranded and separate the R-loops. Therefore, there will always be one more R-loop than there are introns. The ovalbumin gene has seven introns and, therefore, would display eight R-loops.

18. If mRNA is hybridized to the DNA of the gene coding it, homologous pairing will occur. Since the mRNA does not have introns, there will be pairing between the mRNA and DNA, except for the introns. The DNA sequences comprising the introns will appear as single-stranded loops. Since the ovalbumin gene has seven introns, one will observe seven single-stranded DNA loops.

19. To approach this problem, one first needs to determine where transcription will initiate. In prokaryotes, transcription initiates at a site 10 base-pairs from the center of the *TATA box*. The sense strand of DNA is the strand opposite from which the *TATA box* is read. The location of the *TATA* box, and the start point of initiation of transcription are indicated in bold face below:

5'- A A G C **T A T A A T** C G T A C G A C C T A A G G A G G T A A G C G - 3'
3'- T T C G A T A T T A G C A T G C **T** G G A T T C C T C C A T T C G C - 5'
                                 5' - A C C U A A G G A G G U A A G C G - 3' Transcribed RNA

20. The *TATA box* in eukaryotes is centered about 30 bp upstream of the point of transcription initiation. By analyzing the sequence of nucleotides in the portion of the gene indicated, the *TATA* box and the start point of transcription are found and indicated in bold face.

5'-**TA T A A A** A C G T G C A T C G A C G T A C G A T A T C G C C G T A C C G C T G C A C - 3'
3'-AT A T T T T G C A C G T A G C T G C A T G C T A T A G C G G C A T G G C G A C G T G - 5'
                                                              5' A C C G G U G C A C - 3'
The base sequence of the mRNA transcribed from the DNA is shown complementary to the sense strand of the DNA double-helix.

12

Translation and the Genetic Code

IMPORTANT CONCEPTS

A. Genes control growth and development largely by acting through protein intermediaries.
1. Proteins play structural roles in cells and catalyze an enormous array of metabolic reactions.
    a. Proteins are composed of up to twenty different amino acids with highly varied chemical properties.
    b. Their three-dimensional structures provide proteins with enormous functional versatility.
    c. The primary structure, amino acid sequence, of proteins is determined by specific sequences of nucleotide-pairs in the genes that encode them.
2. The process of protein synthesis (translation) involves the translation of the genetic information stored in the nucleotide sequences of mRNA molecules into sequences of amino acids in polypeptide chains.
    a. Translation is complex and takes place on ribosomes in the cytoplasm.
        (1) Each ribosome consists of two subunits, which together contain three to four different RNA molecules and 50 to 80 different proteins. The exact composition varies from species to species.
    b. Translation requires the participation of 40 to 60 small RNA molecules called transfer RNAs (tRNAs).
        (1) Each tRNA is activated for protein synthesis by the covalent attachment of a specific amino acid.
        (2) The attachment of an amino acid to a tRNA is catalyzed by an enzyme called aminoacyl-tRNA synthetase.
        (3) There is at least one tRNA and one aminoacyl-tRNA synthetase for each of the twenty amino acids that are commonly found in proteins.

        c.   Translation requires several soluble proteins called initiation, elongation, and termination factors.

        d.   Translation involves enzymatic activity by peptidyl transferase.

           (1) Peptidyl transferase activity resides in the 23S rRNA molecule of the large subunit of the ribosome, rather than a ribosomal protein.

B. The genetic code has several important characteristics.

    1. The genetic code is triplet, with a sequence of three nucleotides making up each codon.

    2. All 64 possible triplet codons are used; 61 specify amino acids, two of which are used for polypeptide chain initiation, and three function in terminating translation.

    3. The code is degenerate, i.e., several different codons often specify the same amino acid.

        a.   The base-pairing between the third (3′) base of the codon and the first (5′) base of the anticodon does not follow strict base-pairing rules; instead, there is wobble at this site, permitting base-pairs to form other than the usual A:U and G:C. Therefore, the anticodon of a single tRNA may recognize one, two, or three different codons.

    4. The genetic code is nonoverlapping. Each nucleotide in an mRNA belongs to just one codon.

    5. The genetic code is nearly universal, i.e., the codons have the same meaning, with minor exceptions, in all species.

C. Mutations that affect nucleotides of a gene may affect the amino acid sequence of the encoded polypeptide.

    1. Mutations that produce chain-termination triplets within genes are called nonsense mutations.

        a.   Some suppressor mutations alter the anticodons of tRNAs so that the mutant tRNAs recognize chain-termination codons and insert amino acids in response to their presence in mRNA molecules. The resulting polypeptide will be functional only if the amino acid inserted by the suppressor mutation does not significantly alter the protein's chemical properties.

    2. A single base change in a triplet codon so that it specifies a different amino acid is called a missense codon.

    3. A gain or loss of one or a few base pairs in the protein coding region of a gene may alter the reading frame and the sequence of encoded amino acids.

IMPORTANT TERMS

In the space allotted, concisely define each term.

translation:

amino acid:

free amino group:

free carboxyl group:

side group (R):

peptide:

peptide bonds:

primary structure:

secondary structure:

tertiary structure:

quaternary structure:

chaperones:

α helices:

β sheets:

ionic bonds:

hydrogen bonds:

hydrophobic interactions:

Van der Waals interactions:

inteins:

aminoacyl-tRNA synthetase:

signal sequence:

nucleolar organizer regions:

A or aminoacyl site:

P or peptidyl site:

E or exit site:

initiation factors:

methionyl-tRNA$_f^{Met}$ (tRNA$_f^{Met}$):

Shine-Dalgarno sequence:

Kozak's Rules:

elongation factor Tu (EF-Tu):

EF-Tu·GTP:

elongation factor Ts (EF-Ts):

peptidyl transferase:

elongation factor G (EF-G):

chain termination codons:

release factors (Rfs):

triplet code:

suppressor mutations:

reading frame:

degenerate code:

nearly universal code:

wobble hypothesis:

nonsense mutations:

missense mutations:

IMPORTANT NAMES

In the space allotted, concisely state the major contribution made by each individual or group.

Robert Holley:

Francois Chapeville and Gunter von Ehrenstein:

Masayasu Nomura:

Francis Crick:

Marilyn Kozak:

Marshall Nirenberg:

Ghobind Khorana:

Oscar Miller:

Harris Bernstein:

TESTING YOUR KNOWLEDGE

In this section, answer the questions, fill in the blanks, and solve the problems in the space allotted. Problems noted with an * are solved in the Approaches to Problem Solving section at the end of the chapter.

The following mRNA codons are useful in answering some of the questions and solving problems of this section and in subsequent chapters.

Triplet nucleotide sequence of mRNA codons

UUU	Phe	UCU	Ser	UAU	Tyr	UGU	Cys
UUC	Phe	UCC	Ser	UAC	Tyr	UGC	Cys
UUA	Leu	UCA	Ser	UAA	Stop	UGA	Stop
UUG	Leu	UCG	Ser	UAG	Stop	UGG	Trp
CUU	Leu	CCU	Pro	CAU	His	CGU	Arg
CUC	Leu	CCC	Pro	CAC	His	CGC	Arg
CUA	Leu	CCA	Pro	CAA	Gln	CGA	Arg
CUG	Leu	CCG	Pro	CAG	Gln	CGG	Arg
AUU	Ileu	ACU	Thr	AAU	Asn	AGU	Ser
AUC	Ileu	ACC	Thr	AAC	Asn	AGC	Ser
AUA	Ileu	ACA	Thr	AAA	Lys	AGA	Arg
AUG	Met	ACG	Thr	AAG	Lys	AGG	Arg
GUU	Val	GCU	Ala	GAU	Asp	GGU	Gly
GUC	Val	GCC	Ala	GAC	Asp	GGC	Gly
GUA	Val	GCA	Ala	GAA	Glu	GGA	Gly
GUG	Val	GCG	Ala	GAG	Glu	GGG	Gly

1. The energy for activating amino acids, i.e., attaching them to their respective tRNA, comes from

   _____.

2. A triplet of nucleotides in a tRNA that recognizes, by complementary base pairing, the triplets of nucleotides in the mRNA is called an _____.

3. The scientist most responsible for breaking the genetic code was _____.

4. A single base-pair change in a gene that results in a single amino acid substitution in the corresponding protein is called a _____ mutation.

5. The primary energy source for protein synthesis is _____.

6. A complex structure composed of protein and RNA molecules that functions as the site of amino acid polymerization is called a _____.

7. The person who was awarded a Nobel Prize for research on the structure of tRNA was _____.

8. A single base-pair mutation that results in premature termination of translation of the corresponding mRNA is called a _____ mutation.

9. The phenomenon that one tRNA for a particular amino acid can recognize more than one codon for the amino acid can be explained by Crick's _____ hypothesis.

10. A base sequence in an mRNA, upstream from the point of initiation of translation, that hydrogen bonds to a sequence in the rRNA of the small ribosomal subunit is called the _____.

11. Which end (5' or 3') of an mRNA is synthesized first?

12. Which end of a polypeptide is the oldest (amino terminal or carboxyl terminal)?

13. If the anticodon in a tRNA is 3' - CCU - 5', what is the complementary sequence in the gene coding for this tRNA?

14. Which group of an amino acid covalently bonds to the tRNA?

*15. What is the minimum number of aa-tRNAs necessary to recognize all the codons for the amino acid arginine? Hint: Consider the base pairing between the 5' base of the anticodon and the 3' base of the codon as depicted in Table 12.2.

*16. Three of the genetic codons are nonsense and function in termination of translation. Like any other codons, these can change because of mutational base-pair substitutions in the DNA. Considering only one base-pair change at a time, indicate which amino acids could be inserted in a protein by mutation of the nonsense codon UAG.

*17. A protein was sequenced in several mutant individuals. It was found that these mutants resulted in single amino acid substitutions at position number eight from the amino terminal end of the polypeptide. Arrows indicate the sequence in which the mutants were recovered. Assuming single nucleotide changes, indicate the most likely codon for each of the amino acids shown below.

Wild-type		mutant 1		mutant 2		mutant 3		mutant 4
Thr ____	→	Ala ____	→ Val ____		→ Met ____		→ Leu ____	

*18. A fragment of a polypeptide, Met-Leu-Ala-Gly, is encoded by the following sequence of DNA:

Strand A   - G T T A C A A C C G G C C A -
Strand B   - C A A T G T T G G C C G G T -

(a) Which strand is the template strand?

(b) Indicate the 3′ and 5′ ends of the template strand of the DNA.

(c) Which polypeptide bond in the above polypeptide was formed first?

(d) Which amino acid is at the amino terminal end of the polypeptide?

*19. The amino acid sequence of a protein encoded by two mutant alleles was compared to the amino acid sequence of the protein encoded by the normal gene.

	1	2	3	4	5	6	7	8
Normal	Met	Lys	Tyr	Ser	Glu	Trp	Val	Val
Mutant I	Met	Lys	Tyr					
Mutant II	Met	Lys	Tyr	Arg	Ser	Gly	Trp	

(a) Indicate the type of change at the mRNA level that resulted from mutant number I and the specific codon affected.

(b) Indicate the type of change at the mRNA level that resulted from mutant number II, the initial codon affected, and which possible codons were utilized for each amino acid of the normal protein.

*20. The following is a sequence of nucleotides in a DNA double helix that codes for a short polypeptide. The messenger RNA encoded by this DNA has both translational initiation and termination codons.

    Strand A -   A A A T C A A T A G T T A G A A C C C A T C T T G
    Strand B -   T T T A G T T A T C A A T C T T G G G T A G A A C

(a) Which strand is the template strand?

(b) What is the polarity of the strands of the DNA double helix?

(c) What is the base sequence of the mRNA coded by this DNA?

(d) What is the amino acid sequence of the polypeptide encoded by this mRNA?

THOUGHT CHALLENGING EXERCISE

It was suggested over one hundred years ago that nuclein (DNA) may be the genetic material. However, it took until the late 1950s before it was generally accepted that genes were sequences of polynucleotide in the DNA. Why do you think that it took over fifty years after the rediscovery of Mendel's laws in 1900 before much interest was shown by biochemists in the nucleic acid component of chromatin?

SUMMARY OF KEY POINTS

Most genes exert their effect(s) on the phenotype of an organism through proteins, which are large macromolecules composed of polypeptides.

Each polypeptide is a chain-like polymer assembled from different amino acids.

The amino acid sequence of each polypeptide is specified by the nucleotide sequence of a gene.

The vast functional diversity of proteins results in part from their complex three-dimensional structures.

Genetic information carried in the sequence of nucleotides in mRNA molecules is translated into sequences of amino acids in polypeptide gene products by intricate macromolecular machines called ribosomes.

The translation process is complex, requiring the participation of many different RNA and protein molecules.

Transfer RNA molecules serve as adaptors, mediating the interaction between amino acids and codons in mRNA.

The process of translation involves the initiation, elongation, and termination of polypeptide chains and is governed by the specifications of the genetic code.

Each of the 20 amino acids in proteins is specified by one or more nucleotide triplets in mRNA.

Of the 64 possible triplets, given the four bases in mRNA, 61 specify amino acids and 3 signal chain termination.

The code is nonoverlapping, with each nucleotide part of a single codon, degenerate, with most amino acids specified by two or four codons, and ordered, with similar amino acids specified by related codons.

The genetic code is nearly universal; with minor exceptions, the 64 triplets have the same meaning in all organisms.

The wobble hypothesis explains how a single tRNA can respond to two or more degenerate codons.

Some suppressor mutations alter the anticodons of tRNAs so that the mutant tRNAs recognize chain-termination codons and insert amino acids in response to their presence in mRNA molecules.

Comparisons of the nucleotide sequences of genes with the amino acid sequences of their polypeptide products have verified the codon assignments deduced from *in vitro* studies.

ANSWERS TO QUESTIONS AND PROBLEMS

1) ATP 2) anticodon 3) Nirenberg 4) missense 5) GTP 6) ribosome 7) Holley 8) nonsense 9) wobble 10) Shine-Dalgarno sequence 11) 5' 12) amino terminal 13) 5' GGA 3' 14) carboxyl 15) 3 16) tyrosine, serine, leucine, trytophan, lysine, glutamic acid, glutamine 17) ACG, GCG, GUG, AUG, UUG or CUG 18) a. strand A; b. 3'- GTTACAACCGGCCA - 5'; c. Met-Leu; d. Met 19) see Approaches to Problem Solving section 20. a. strand A; b. Strand A, 5' → 3'; Strand B, 3' → 5'; c. 5'-AUGGGUUCUAACUAUUGAUUU-3'; d. fMet-Gly-Ser-Asn-Tyr.

Ideas concerning the thought challenging exercise:

Advancements in science often come very rapidly and many top scientists often do their research in areas that are most exciting and receiving attention at a given time. During the first half of the 20th century, exciting advances were being made in the discipline of biochemistry. Much was being learned about proteins and their functions as enzymes. This was an era where much of the attention was centered around proteins. The chemical components of DNA, i.e., nucleotides, had been characterized in the 1920s and it was known that nucleic acids were polymers of nucleotides. However, it was thought that nucleic acids were simply a polymer consisting of a repeated tetrameric nucleotide unit, e.g., A - T - G - C, and hence were too simple to be the genetic material. The fallacious idea that protein was likely the genetic material did not die easily. This view could still be found in the literature of the late 1950s and early 1960s, long after conclusive evidence had been presented that DNA was the genetic material.

APPROACHES TO PROBLEM SOLVING

15. To solve this problem, first list the six codons for arginine.

CGU
CGC
CGA
CGG
AGA
AGG

Next, recall that, due to wobble, there does not have to be a different tRNA for each codon of an amino acid. Therefore, a tRNA with 3' - UCU - 5' anticodon can recognize both the 5' - AGA - 3' and 5'- AGG - 3'. The tRNA with the anticodon GCI can recognize the codons CGA, CGU, and CGC. It requires a third tRNA with the anticodon GCU or GCC to recognize the codon CGG.

16. The approach to working this problem is straightforward. Simply list all the codons that can result from single base changes of UAG and then check their coding properties from the table of codons.

UAC	tyrosine
UAU	tyrosine
UAA	nonsense
UCG	serine
UUG	leucine
UGG	tryptophan
AAG	lysine
CAG	glutamine
GAG	glutamic acid

17. Since all mutations are assumed to have occurred by single base-pair changes in the DNA, first look at the possible codons for each of the amino acids substituted at position number eight. All of these can be accounted for by changes at the first or second position of the codon. Any problem with the degeneracy is minimized because methionine has only one codon, AUG. Therefore, we know that the third base in the codons must be G.

Thr <u>ACG</u> --> Ala <u>GCG</u> --> Val <u>GUG</u> --> Met <u>AUG</u> --> Leu <u>UUG</u> or <u>CUG</u>

18. (a) The template strand can be determined by looking for the strand of DNA that is complementary for the codons for the Met-Leu-Ala-Gly. The codon for Met is 5' - AUG - 3'. The complementary strand of DNA must contain a 3'-TAC-5' sequence. TAC starts with the third base from the left on strand A. The transcription of strand A, 3' - GTTACAACCGGCCA - 5', results in an mRNA with the reading frame sequence of 5' - CA**AUGUUGGCCGGU**- 3' which translates as fMet.Leu.Ala.Gly.

(b) Since the messenger RNA is translated 5' → 3' and has reversed polarity of the sense strand of DNA, the sense strand of the DNA must be

3' - GTTACAACCGGCCA - 5'.

(c) Translation of the mRNA proceeds from 5' → 3'. Therefore, the peptide bond connecting methionine and leucine was the first formed.

(d) Since polymerization of amino acids occurs by formation of a peptide bond between the carboxyl group of the growing polypeptide chain and the amino group of the incoming amino acid, methionine must occupy the amino terminal end of the above polypeptide.

19. (a) When you look at the product of mutant I, it is noticed that it is only a partial polypeptide. This suggests that a nonsense mutation may account for it. A single base substitution in the DNA could change a codon for serine to a nonsense codon, i.e., UCA → UGA, UCA → UAA, or UCG → UAG.

(b) Mutant II has the first three amino acids in its polypeptide identical to that of the normal. The rest of the amino acids are different, suggesting that a single addition or deletion of a base has affected the reading frame. To determine what has happened and to deduce as much as possible about the codons utilized to produce the normal, start by inserting possible codons for the amino acids of the normal and mutant protein. When this is done we don't have to designate any changes in the first three codons. The mutant and normal differ, starting at the fourth position. A deletion of the **U** base of the serine codon UCG or the **U/C** base of the tyrosine codon shifts the reading frame one base to the right, so that the next codon is CGG for Arg.

Normal	Met	Lys	Tyr	Ser	Glu	Trp	Val	Val
	AUG	AAG	UAU	UCA	GAA	UGG	GUG	GU_
		A	C	G	G			

	AUG	AAG	UAU	CG G	AGU	GG G	UG G
		A	C				
Mutant II	Met	Lys	Tyr	Arg	Ser	Gly	Trp

The next codon is either AAU or AGU. Since the next amino acid in the mutant protein is serine, the codon must be AGU. The next codon is GGG for glycine, followed by UGG for tryptophan. The assignments of the frameshifted codons in many cases dictate the codons used for the normal protein. The codon assignments for the normal mRNA are Met (AUG) - Lys (AAG or AAA) - Tyr (UAU or UAC) - Ser (UCG) - Glu (GAA) - Trp (UGG) - Val (GUG) - Val (GUU, GUC, GUA, or GUC).

20. The key to solving this problem is that both translational initiation (AUG) and termination sequences (UAA, UAG, or UGA) are present in the mRNA. Generally, you can determine the template strand of the DNA by looking for an open reading frame flanked by sequences complementary to initiation (3'-TAC-5') and termination (3'-ATT-5', 3'-ATC-5' or 3'-ACT-5') codons. These are found in Strand A of the DNA:

Strand A-   A A A T C A A T A G T T A G A A C C C A T C T T G-
Strand B-   T T T A G T T A T C A A T C T T G G G T A G A A C-

(a) Strand A is the template strand.

(b) The polarity of the strands is as follows:

Strand A   5' - A A A T C A A T A G T T A G A A C C C A T C T T G - 3'
Strand B   3' - T T T A G T T A T C A A T C T T G G G T A G A A C - 5'

(c) The translated mRNA sequence coded by the template strand has the following nucleotide sequence:

5'- AUGGGUUCUAACUAUUGAUUU-3'

(d) By reading the triplet codons starting with the initiation codon AUG and terminating with the nonsense codon UGA, we get the following polypeptide:

fMet-Gly-Ser-Asn-Tyr-

13

Mutation, DNA Repair, and Recombination

MUTATION: SOURCE OF THE GENETIC VARIABILITY REQUIRED FOR EVOLUTION

MUTATION: BASIC FEATURES OF THE PROCESS
    Mutation: Somatic or Germinal
    Mutation: Spontaneous or Induced
    Mutation: Usually a Random, Nonadaptive Process
    Mutation: A Reversible Process

MUTATION: PHENOTYPIC EFFECTS
    Mutations with Phenotypic Effects: Usually Deleterious and Recessive
    Effects of Mutations in Human Globin Genes
    Mutations in Humans: Blocks in Metabolic Pathways
    Conditional Lethal Mutations: Powerful Tools for Genetic Studies

THE MOLECULAR BASIS OF MUTATION
    Induced Mutations
    Mutations Induced by Chemicals
    Mutations Induced by Radiation
    Mutations Induced by Transposable Genetic Elements
    Expanding Trinucleotide Repeats and Inherited Human Diseases

SCREENING CHEMICALS FOR MUTAGENICITY: THE AMES TEST

DNA REPAIR MECHANISMS
    Light-Dependent Repair
    Excision Repair
    Other DNA Repair Mechanisms

INHERITED HUMAN DISEASES WITH DEFECTS IN DNA REPAIR

DNA RECOMBINATION MECHANISMS
    Recombination: Cleavage and Rejoining of DNA Molecules
    Gene Conversion: DNA Repair Synthesis Associated with Recombination

A MILESTONE IN GENETICS: MULLER DEMONSTRATES THAT X RAYS ARE MUTAGENIC

IMPORTANT CONCEPTS

    A. Mutations are heritable changes in the genetic material.
        1. Mutations that involve changes at specific sites of a gene are called point mutations.
        2. Forward mutations are those that change a wild-type allele to an allele that results in an abnormal phenotype. A mutation of a mutant allele to the wild-type allele is a back mutation.
        3. The induction of mutations in stationary-phase cells, induced by environmental stress, is widespread among bacteria species and contributes to their ability to adapt to diverse environments.
           a. Stationary-phase mutagenesis is caused by the induction of error-prone DNA repair processes.

      b.   Randomly occurring adapted mutations are produced along with other non-adaptive mutations.

    4.   Mutations may occur spontaneously or may be induced by agents that interact with DNA and RNA.

      a.   Irradiation can induce mutation.

         (1) Examples of mutagenic ionizing radiation are X rays, gamma rays, and neutrons.

         (2) Ultraviolet light is nonionizing. It is maximally absorbed by DNA at a wavelength of 254 nm, the wavelength at which it is most mutagenic. Thymine dimerization is the major mutagenic effect of UV.

      b.   Many chemicals that react with DNA and RNA are potent mutagenic agents.

         (1) Base analogs have structures similar to normal bases, are incorporated into DNA during replication, and cause transition mutations (purine substituting for purine, and pyrimidine substituting for pyrimidine) through base mispairing due to tautomeric shifts.

         (2) Nitrous acid deaminates adenine to hypoxanthine, which pairs with cytosine during replication, and cytosine to uracil, which pairs with adenine during replication. It causes transition mutations.

         (3) Acridine dyes such as proflavin and acridine orange cause frameshift mutations because they intercalate between stacked base pairs in DNA and cause addition or deletion of one or a few base pairs during replication.

         (4) Alkylating agents, exemplified by ethyl methanesulfonate (EMS), add alkyl groups to bases in the DNA. Alkylating agents cause transversions, transitions, frameshifts, and chromosomal breaks which may lead to aberrations.

         (5) Hydroxylamine (NH_2OH) reacts specifically to hydroxylate cytosine to hydroxylaminocytosine, which pairs with adenine during DNA replication. Therefore it causes only GC -> AT transitions.

    5.   Mutations provide the raw material for evolution. Thus, some level of mutation is needed to allow organisms to adapt to changes in the environment.

    7.   Most mutations are detrimental to the organism in which they reside. Mutation rates that are too high would be disadvantageous to a species, except possibly in a rapidly changing environment.

    8.   The potential benefits of the use of X-rays, nuclear reactors, and other types of ionizing radiation must be carefully weighed against the known and estimated potential risks. Similar precautions must be taken to prevent the continued pollution of our environment with mutagenic and carcinogenic chemicals. These risk estimates and precautions must take into account the potential harm to future generations of living organisms, including humans, keeping in mind the increased frequencies of deleterious mutations that may result.

B.   The integrity of the genetic information of living organisms is protected by a battery of repair mechanisms, each functioning to correct a specific type of defect in DNA.

    1.   *E. coli* has at least five distinct mechanisms for the repair of DNA defects.

      a.   Photoreactivation is catalyzed by photolyase and repairs thymine dimers by removing cross-linkages.

      b.   Excision repair involves recognition of a damaged region by an enzyme complex that excises bases of the defective single strand, fills in the gap with DNA polymerase activity, and seals the break left by DNA polymerase by DNA ligase.

         (1) Base excision repair systems remove abnormal or chemically modified bases from DNA.

         (2) Nucleotide excision repair pathways remove larger defects like thymine dimers.

      c.   The mismatch repair pathway provides a backup to replicative proofreading by correcting mismatched bases. This system uses a difference in methylation state to excise mismatched nucleotides in the newly synthesized strand. This requires a GATC-specific endonuclease, exonucleases, DNA polymerase, and DNA ligase.

      d.   Postreplicative repair is a recombination-dependent repair process.

      e.   An error-prone SOS repair system operates when the DNA has been heavily damaged.

         (1) During the SOS response, a whole battery of DNA repair, recombination, and replication proteins are synthesized. This allows DNA replication to proceed across damaged segments in template strands and subsequently results in an increased frequency of replication errors and mutations.

C. Individuals with inherited disorders in DNA repair pathways suffer from developmental abnormalities and health problems, often including susceptibility to certain types of cancer.

    1. An example of a human inherited disease with defects in DNA repair is *Xeroderma pigmentosum.*

        a. Individuals with *Xeroderma pigmentosum* are extremely sensitive to sunlight, which results in a high frequency of skin cancer.

D. Recombination of genes located on homologous chromosomes is produced by breakage and rejoining of the resident DNA molecules.

    1. Recombination involves the breakage of DNA molecules in individual chromatids and the exchange of parts.

        a. The breakage and reunion process usually is associated with a small amount of DNA repair synthesis.

    2. When crosses are performed using genetic markers that are far apart, recombination is usually reciprocal. However, when markers are closely linked, a type of nonreciprocal recombination called gene conversion is often observed.

        a. Gene conversion results from the repair of nucleotide-pair mismatches that are produced during the recombination process.

IMPORTANT TERMS

In the space allotted, concisely define each term.

mutation:

point mutations:

germinal mutations:

somatic mutations:

spontaneous mutation:

induced mutation:

mutagens:

replicate plating:

forward mutation:

reversion (reverse mutation):

back mutation:

adaptive mutation:

stationary-phase mutagenesis:

suppressor mutation:

isoalleles:

null alleles:

recessive lethals:

neutral mutation:

auxotrophs:

conditional lethal mutation:

auxotrophs:

tautomeric shift:

transition:

transversion:

frameshift mutation:

base analogs:

nitrous acid (HNO_2):

acridine dyes:

alkylating agents:

hydroxylating agents:

ionizing radiation:

nonionizing radiation (ultraviolet light):

roentgen (r):

transposons:

trinucleotide repeats:

carcinogens:

light-dependent repair (photoreactivation):

DNA photolyase:

base excision repair:

nucleotide excision repair:

excinuclease:

mismatch repair:

error-prone repair:

GATC-specific endonuclease:

postreplication repair:

SOS response:

recombination deficient mutants:

RecBCD complex:

RecA protein:

single-strand assimilation:

chi forms:

gene conversion:

heteroduplexes:

CLB chromosome:

IMPORTANT NAMES

In the space allotted, concisely state the contribution made by each individual or group.

Jean Lamarck:

Joshua and Esther Lederberg:

Charlotte Auerbach:

Bruce Ames:

Robin Holliday:

Herman J. Muller:

TESTING YOUR KNOWLEDGE

*In this section, answer the questions, fill in the blanks, and solve the problems in the space allotted. Problems noted with an * are solved in the Approaches to Problem Solving section at the end of the chapter.*

1. A single base pair change in a gene that results in a single amino acid substitution in the corresponding protein is called a _____ mutation.

2. Nitrous acid is mutagenic because it deaminates _____ to _____,   and _____ to _____.

3. Which disease results if a human has a genetic block between phenyalanine and tyrosine?

4. Which type of mutation may result if a single base pair is inserted or deleted in the DNA double helix?

5. Which disease results if a human cannot metabolize the brain ganglioside GM2?

6. Humans who lack homogentisic acid oxidase activity have a disease called _____.

7. If a human cannot repair thymine dimers, which disease will occur?

8. A point mutation that results in premature termination of translation of the corresponding mRNA is called a _____ mutation.

9. Sickle cell hemoglobin differs from hemoglobin A in that it has _____ instead of _____ at the sixth position in the beta polypeptide.

10. A mutation of a DNA strand in which a purine replaces a pyrimidine and vice versas is called a _____.

11. The chemical mutagen, hydroxylamine (H_2NOH), reacts with cytosine to form _____.

12. An enzyme catalyzed, light-independent, process of repair of thymine dimers in DNA that involves removal of a dimer and some flanking nucleotides, synthesis of a new segment of DNA complementary to the undamaged strand, and DNA ligase is called _____.

13. An example of a mutagenic base analog is _____.

14. An example of an intercalating agent, that inserts itself between adjacent base pairs and causes base pair addition or deletions during DNA replication is _____.

15. The enzyme required for light-dependent repair of thymine dimers is called _____.

16. A mutation of a DNA strand in which a different purine replaces a purine, and a pyrimidine replaces a pyrimidine is called a _____.

17. Sickle cell hemoglobin differs from hemoglobin C only in that it has _____ instead of _____ at the sixth position in the beta polypeptide.

*18. A culture of *Bacillus* was treated with nitrous acid and several mutants were recovered. From one of these mutants, wild-type mutations revertants (back mutations) could be obtained by treating with 5-bromouracil, but not with hydroxylamine.

  (a) Illustrate, using a DNA base pair, the pathway by which the original mutant was generated.

  (b) Indicate the steps by which bromouracil may have produced the back mutation.

*19. The base-pair sequence in a region of a gene coding for a polypeptide is:

        Strand A   3′ - T T T A C C A C C C C T A C C A G A A C A C T T - 5′
        Strand B   5′ - A A A T G G T G G G G A T G G T C T T G T G A A - 3′

  (a) If this gene codes for a polypeptide that contains three tryptophans, which strand is transcribed?

  (b) What is the nucleotide sequence of the mRNA transcribed from this DNA?

  (c) What is the polypeptide sequence of the protein translated from this mRNA?

(d) Hydroxylamine affects the bold-faced base pair and causes a mutation during DNA replication. What is the amino acid sequence of the polypeptide coded by the mutant gene?

*20. The following is an eight amino acid sequence from the amino terminal end of a wild-type enzyme of *E. coli*.

fMet - Trp - Cys - Trp - Gln - Leu - Asn - Trp - (aa)$_n$

After treatment of the bacteria culture with an acridine, a mutant is isolated that lacks the enzyme activity. Only a partial polypeptide consisting of five amino acids is detected as the gene product in the mutant strain. The amino acid sequence is:

fMet - Gly - Ala - Gly - Asn

A culture of the mutant bacteria is grown in the presence of an acridine and a revertant with partial enzyme activity is recovered. The mutant had a normal sized polypeptide chain, differing from the wild-type in that it had different amino acids at positions two, three, four and five. The amino acid sequence of the partial revertent was:

fMet - Gly - Ala - Gly - Asn - Leu - Asn - Trp - (aa)$_n$

Determine, as much as possible, the nucleotide sequence of the mRNA coding for the polypeptide of the wild-type, mutant and partial revertent strains.

THOUGHT CHALLENGING EXERCISE

Mutation can affect the genome in many different ways. For now, let's consider mutations that (1) result in changes of the coding sequences of genes, and (2) result in changes of nucleotide sequences of introns. Which type of mutation do you feel would have the most evolutionary potential, and why?

SUMMARY OF KEY POINTS

Mutations are heritable changes in the genetic material that provide the raw material for evolution.

Mutations occur in both germ-line and somatic cells, but only germ-line mutations are transmitted to progeny.
Mutations can occur spontaneously or be induced by mutagenic agents in the environment.

Mutation usually is a nonadaptive process in which an environmental stress simply selects organisms with preexisting, randomly occurring mutations.

Adaptive, or stationary-phase, mutagenesis occurs in bacteria that have been exposed to an environmental stress such as starvation.

Restoration of the wild-type phenotype in a mutant organism can result from either back mutation or a suppressor mutation.

The effects of mutations on the phenotypes of living organisms range from undetectable to lethal changes.

Most mutations exert their effects on phenotype by altering the amino acid sequences of polypeptides, the primary gene products.

The mutant polypeptides, in turn, cause blocks in metabolic pathways.

Conditional lethal mutations provide powerful tools with which to dissect biological processes.

Mutations are induced by chemicals, ionizing irradiation, ultraviolet light, and endogenous transposable genetic elements.

Point mutations are of three types: (1) transitions—purine for purine and pyrimidine for pyrimidine substitutions; (2) transversions—purine for pyrimidine and pyrimidine for purine substitutions; and (3) frameshift mutations—additions or deletions of one or two nucleotide pairs, which alter the reading frame of the gene distal to the site of the mutation.

Several inherited human diseases are caused by expanded trinucleotide repeats.

Bruce Ames and coworkers developed an inexpensive and sensitive method for testing the mutagenicity of chemicals with auxotropic mutants of *Salmonella*.

Multiple DNA repair systems have evolved to safeguard the integrity of genetic information in living organisms.

Each repair pathway corrects a certain type of damage in DNA.

The importance of DNA repair pathways is documented convincingly by inherited human disorders that result from defects in DNA repair.

Certain types of cancer are also associated with defects in DNA repair pathways.

Crossing over involves the breakage of homologous DNA molecules and the rejoining of parts in new combinations.

When genetic markers are closely linked, nonreciprocal recombination, or gene conversion, often occurs, yielding 3:1 ratios of the segregating alleles.

Gene conversion results from DNA repair synthesis that occurs during the recombination process.

ANSWERS TO QUESTIONS AND PROBLEMS

1) missense **2)** adenine to hypoxanthine and cytosine to uracil **3)** phenylketonuria (PKU)
4) frameshift **5)** Tay-Sach **6)** alkaptonuria **7)** *Xeroderma pigmentosum* **8)** nonsense **9)** valine,
glutamic acid **10)** transversion **11)** hydroxylaminocytosine **12)** nucleotide excision repair
13) 5-bromouracil, 2-aminopurine **14)** acridine orange, proflavin **15)** photolyase **16)** transition **17)**
valine, lysine **18-20)** see *Approaches to Problem Solving* section.

Some ideas concerning the thought challenging exercise:

The amino acid sequence of proteins is the result of millions of years of evolution. Therefore mutations
that affect the coding regions of genes are likely to be detrimental or neutral in their effects, rather than
improving upon the function of the protein. There is a lot more to learn about the nature of introns.
However, we do have examples of mutations of introns that affect the rate of processing of pre-mRNAs
and reduce the amount of mRNA and protein produced. Mutations that affect the activity, or timing of
the activity, of genes are considered to be much more likely to be of evolutionary significance. It is likely
that some mutations of introns may affect splicing choices and result in the production of a different but
related protein. A mutation resulting in the production of a novel protein would more likely to be of
evolutionary significance than a mutation that affects the amino acid sequence of an existing protein.

APPROACHES TO PROBLEM SOLVING

18. A key to working this problem is to know the action of the chemical mutagens. The transition
    mutation induced by nitrous acid could be reverted by treatment with 5-bromouracil, but not with
    hydroxylamine. Therefore, the original mutant involved a **CG -> TA** transition. Hydroxylamine
    reacts specifically with cytosine and can not cause transitions of A-T base pairs.
    (a) The original mutation occured as follows:

    CG ---> nitrous acid → UG - DNA replication → UA   DNA replication → **TA**
              deaminates   uracil pairs with   adenine pairs
              cytosine to   adenine   with thymine
              uracil

    (b) 5-bromouracil led to the back mutation as follows:

    TA---> 5-BU (B) present during → BA --5-BU shifts to its → BG --DNA replication → **CG**
              DNA replications gets   enol tautomer during   guanine pairs
              incorporated in its keto   the next replication   with cytosine.
              tautomer in place   and pairs with guanine.
              of thymine.

19. (a) To solve this problem, first pick the DNA strand that has three triplets, ACC, that are
    complementary to the tryptophan codon, UGG. That is strand A.

    (b) Next transcribe strand A of the DNA to obtain the mRNA sequence:

    5'- A A A U G G U G G G G A U G G U C U U G U G A A -3'

    (c) Translate this mRNA in the reading frame that results in three tryptophans in the resulting
    polypeptide

    5' - A A A U G G U G G G G A U G G U C U U G U G A A - 3'

        Lys   Trp   Trp   Gly   Trp   Ser   Cys   Glu

    (d) Hydroxylamine reacts specifically with cytosine to form hydroxlyaminocytosine, which in turn
    pairs with adenine during DNA replication to cause CG → TA transitions. Therefore, if the bold-
    faced base pair is affected, the **C** will be substituted with **T** in the mutant.
                           G   **A**

This will result in the following nucleotide sequence of the affected mRNA with just a single base change indicated by bold-facing:

5'-AAAUGGUGG**AGA**UGGUCUUGUGAA-3'

The new codon AGA, resulting from the single base change, codes for Arg. Therefore, the amino acid sequence coded by the mutant gene is Lys Trp Trp **Arg** Trp Ser Cys Glu.

20. The key to solving this problem is knowing that acridines cause single base-pair additions or deletions in the DNA double helix. When these occur in the coding region of a gene they result in frameshift mutations.

The next step is to indicate all possible codons for the amino acids of the wild type protein as follows:

fMet -	Trp -	Cys -	Trp -	Gln -	Leu -	Asn -	Trp - (aa)$_n$
AUG	UGG	UGU	UGG	CAA	UUA	AAU	UGG
		UGC		CAG	UUG	AAC	
				CU_			

Compare the amino acid sequence of the mutant with that of the wild-type. Find the first amino acid at which they vary and see whether it is a single base addition or deletion that has caused the change in amino acid sequence. The deletion of the U in the first Trp codon in the mRNA changes the reading frame to generate the codon GGU for Gly.

fMet - Gly - Ala - Gly - Asn

AUGU<u>GG</u>UGC

The codons for the amino acids of the mutant are determined by comparing to the possible wild-type codons and realizing that the reading frame has been shifted one base to the right. The nucleotide sequence of the mutant mRNA is therefore determined to be:

fMet -	Gly -	Ala -	Gly -	Asn		
AUGU<u>GGU</u>	GCU	GGC	AAU	UGA	AUUGG
		C	A		C

The nonsense codon generated downstream from the initial point of the frameshift causes premature termination of translation of the mutant mRNA and only a polypeptide that is five amino acids long. The nucleotide sequence of the wild-type mRNA is:

fMet -	Trp -	Cys -	Trp -	Gln -	Leu -	Asn -	Trp - (aa)$_n$
AUG	UGG	UGC	UGG	CAA	UUA	AAU	UGG
				C G		C	

Only amino acids two through five differ between the partial reverent and the wild-type protein. The mutation that partially restored enzymatic activity was induced by an acridine. An addition of a base pair (AT in this case) in the gene and a U in the mRNA restored the correct reading frame.

fMet -	**Gly -**	**Ala -**	**Gly -**	**Asn -**	Leu -	Asn -	Trp - (aa)$_n$
AUG	GGU	GCU	GGC	AA U	UUA	AAU	UGG
				C	G	C	

14

Definitions of the Gene

IMPORTANT CONCEPTS

    A. The definition of a gene changed as more genetic knowledge accumulated.
        1. The gene was first defined as the unit of genetic material controlling one phenotypic characteristic or one trait.
        2. Subsequently, the gene was more precisely defined as the unit of genetic material encoding one polypeptide (or one RNA molecule in the case of genes not coding for proteins).
        3. Prior to 1941, the gene was also believed not to be subdivisible by mutation or recombination.
        4. Many mutable sites separable by recombination exist within each gene.
            a. The unit of genetic material not subdivisible by mutation or recombination is now known to be the single nucleotide pair.
    B. The gene is operationally defined by the complementation test.
        1. A cis-trans test is performed by constructing cis heterozygotes and trans heterozygotes and determining whether they have mutant or wild-type phenotypes.
            a. Cis heterozygotes carry the two mutations on one chromosome and the corresponding wild-type alleles on the homologous chromosome.

(1) The *cis* heterozygotes must exhibit the wild-type phenotype or the results of the *trans* test will not be informative.

    b. *Trans* heterozygotes carry the two mutations on different homologous chromosomes.

        (1) The complementation test, the *trans* part of the *cis-trans* test, is used to determine whether two mutations are in the same gene or in different genes. If a *trans* heterozygote has the mutant phenotype, the two mutations are in the same gene. If a *trans* heterozygote has the wild-type phenotype, the two mutations are in different genes.

  2. The utility of the complementation test is limited in some cases by intragenic complementation, polar effects of chain-termination mutations in operons, and intergenic noncomplementation resulting from the interaction between gene products.

    a. Intragenic complementation frequently occurs when the active form of the protein gene product is a multimer consisting of two or more polypeptide products of a given gene.

    b. Delimiting genes by complementation tests should be done whenever possible using null mutations (mutations resulting in no gene product, a partial gene product, or an otherwise totally nonfunctional gene product) to minimize the possibility of confounding effects of intragenic complementation.

C. The *rII* region in T4 contains two genes, *A* and *B*.

  1. Over 2400 mutants in these two genes were analyzed by Benzer using pairwise crosses and deletions.

    a. The mutants mapped to about 300 different sites.

    b. This was one of the most detailed genetic analyses of a gene ever undertaken.

D. The bacteriophage φX174 codes for more proteins than its DNA could linearly encode.

    a. This paradox was resolved when it was discovered that φX174 has genes contained within genes. Sequences of nucleotides are read in different frames so that a single nucleotide may be part of more than one gene.

E. Eukaryotic genes frequently are mosaics of introns and exons.

  1. In some cases, the exons of a gene are spliced together in different combinations so that a family of protein isoforms results. In these cases, the gene can be defined as a DNA sequence that is a single unit of transcription and encodes a set of closely related polypeptides.

F. In the case of antibody synthesis in vertebrates, functional genes are assembled from gene segments during the development of the antibody-producing cells of the immune system. Thus, the definition of the gene must remain flexible enough to encompass several variations on the one gene-one polypeptide concept.

IMPORTANT TERMS

In the space allotted, concisely define each term.

alkaptonuria:

one mutant gene-one metabolic block:

one gene-one enzyme:

one gene-one polypeptide:

position effect:

cis-trans position effect:

coupling (*cis*) configuration:

cis heterozygote:

repulsion (*trans*) configuration:

trans heterozygote:

complementation test (*trans* test):

cis-trans test:

cistron:

homoalleles:

heteroalleles:

intragenic complementation:

deletion mapping:

hot spots:

overlapping genes:

genes-within-genes:

protein isoforms:

gene segments:

antibodies:

IMPORTANT NAMES

In the space allotted, concisely state the major contribution made by the individual or group.

George W. Beadle and Edward L. Tatum:

Clarence Oliver:

Edward B. Lewis:

Sir Archibald Garrod:

Seymour Benzer:

Eric Lander:

Melvin M. Green:

Charles Yanofsky:

Sydney Brenner:

TESTING YOUR KNOWLEDGE

*In this section, fill in the blanks, answer the questions, and solve the problems in the space allotted. Problems noted with an * are solved in the Approaches to Problem Solving section at the end of the chapter.*

1. The *one gene-one enzyme* hypothesis was proposed by _____ and _____.

2. What are the instances in which the results of complementation tests **cannot** be used to unambiguously delimit genes?

3. The interaction of *trans* coded gene products is called _____.

4. If a *trans*-heterozygote has the mutant phenotype, are two mutations of the same or different genes?

5. If a *trans*-heterozygote has the wild-type phenotype, are the two mutants in the same or different genes?

6. The term *cistron* is synonymous with the term _____.

7. Who first proposed the concept of *one mutant gene-one metabolic block*?

8. The three basic tools of genetics are _____, _____, and _____.

9. Mutations of the same site that do not undergo recombination are called _____ alleles.

10. Two mutations that do not complement are of the same gene and are called _____ alleles.

11. Mutations that are both structurally and functionally allelic are called _____.

12. A mutation that not only results in a defective product of the gene in which it is located, but also interferes with the expression of one or more adjacent alleles is called a _____ mutation.

*13. In barley, single gene mutations occur that result in hood-like structures on the awns. Two recessive mutants with similar phenotypes are *subjacent-hooded* and *calcaroides*. A cross between these two mutants produced hybrids with hood-like structures on the awn. Are these mutants of the same or different genes? Justify your answer.

*14. Bacteriophage T4 that are mutant for the rapid lysis gene *rII* will enter but not lyse *E. coli* strain K12(λ). Wild-type T4 will infect and lyse strain K12(λ) bacteria. Benzer performed complementation tests between numerous pairs of *rII* mutants. The results of several complementation tests, done by simultaneously infecting K12(λ) bacteria with pairs of *rII* mutants are presented in the following table ("+" represents complementation and "o" no complementation):

Mutant:	r1	r2	r3	r4	r5	r6	r7	r8	r9
r9	o	+	+	o	+	+	o	o	o
r8	o	+	+	o	+	+	o	o	
r7	o	+	+	o	+	+	o		
r6	+	o	o	+	o	o			
r5	+	o	o	+	o				
r4	o	+	+	o					
r3	+	o	o						
r2	+	o							
r1	o								

(a) From these data, the nine *rII* mutants can be assigned to how many genes?

(b) Which mutants are assignable to the same gene?

*15. Seven different mutants of *Salmonella typhimurium* were isolated which were unable to grow in the absence of tryptophan (trp). These were examined in all possible *cis* and *trans* heterozygotes (partial diploids). All of the *cis* heterozygotes were able to grow in the absence of tryptophan. Some of the *trans* heterozygotes grew in the absence of tryptophan; some did not. The experimental results, using (+) to indicate growth and (o) to indicate no growth, are given in the following table. How many genes do these seven mutations represent? Which mutations are in the same gene(s)?

Mutant	Growth of *Trans* heterozygotes (minus tryptophan)						
	1	2	3	4	5	6	7
7	+	+	+	+	+	+	o
6	+	+	+	+	o	o	
5	+	+	+	+	o		
4	+	+	o	o			
3	+	+	o				
2	o	o					
1	o						

*16. Srb and Horowitz isolated several *Neurospora* mutants that were auxotrophic for arginine, i.e., they could not produce their own arginine and to grow required the addition of arginine to the culture media. The mutants were individually tested for their ability to grow in the presence of compounds suspected to be intermediate in the biochemical pathway leading to arginine. Mutant 1 grew if ornithine, citrulline or arginine was added to the media. Mutant 2 grew if citrulline or arginine, but not ornithine was added to the media. Mutant 3 grew if arginine was added to the media, but did not grow if either citrulline or ornithine was added.

(a) From analysis of the above data propose a metabolic pathway leading to arginine and indicate which step is controlled by each gene.

(b) Which precursor would be expected to accumulate in mutant 2?

*17. The following illustrates a set of Benzer A cistron deletion mutants with the boundary of each mutant numbered:

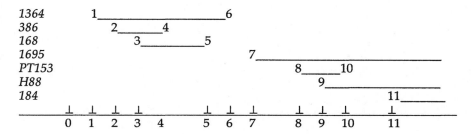

Four *rII* point mutations of the bacteriophage T4 were crossed with the seven deletion mutants where (+) indicates the appearance of r^+ recombinants and (-) indicates the lack of recombinants among the progeny.

				Deletion Mutant			
Mutant #	1364	386	168	1695	PT153	H88	184
r1	+	+	+	-	+	+	+
r2	-	+	+	+	+	+	+
r3	+	+	+	-	-	+	+
r4	-	+	-	+	+	+	+

Indicate the region, e.g., 0 - 1, 1 - 2, 2 - 3, etc., of the A cistron where each of the mutants occur.

THOUGHT CHALLENGING EXERCISE

Albinism in humans is a recessive genetic disorder resulting from the lack of the melanin pigment that is derived from the amino acid tyrosine. Generally, all children of parents who are both albino are also albino. However, some families where both parents are albino have only normally pigmented children. What does this tell us about the genetic nature of albinism and about the derivation of melanin?

SUMMARY OF KEY POINTS

The concept of the gene has undergone many refinements since the rediscovery of Mendel's work in 1900.

Most genes encode one polypeptide and can be operationally defined by the complementation test.

The existence of a basic genetic element—the gene—that controlled a specific phenotypic trait was established by Mendel's work in 1866.

During the last century, the concept of the gene has evolved from the unit that can mutate to cause a specific block in metabolism, to the unit specifying one enzyme, to the sequence of nucleotide pairs in DNA encoding one polypeptide chain or one RNA molecule.

The concept of the gene has evolved from a bead on a string, not divisible by recombination or mutation, to a sequence of nucleotide pairs in DNA encoding one polypeptide chain.

The unit of genetic material not divisible by recombination or mutation is the single nucleotide pair.

The complementation or *trans* test provides an operational definition of the gene; it is used to determine whether mutations are in the same gene or different genes.

Intragenic complementation may occur when a protein is a multimer containing at least two copies of one gene product.

Benzer mapped over 2400 mutations to 308 distinct sites in the two genes at the *rII* locus of T4.

Benzer's results showed that the base pair was the smallest unit altered by mutation and that recombination occurred between adjacent base pairs.

Overlapping deletions were used to speed up the mapping of large numbers of mutations.

Benzer's results solidified the concept of a gene as a unit of function that is divisible by mutation and recombination.

The chromosome of bacteriophage ϕX174 contains overlapping genes and genes-within-genes composed of DNA sequences specifying mRNAs that are translated in different reading frames.

The transcripts of some genes undergo alternate pathways of splicing to produce mRNAs with different exons joined together.

Translation of these mRNAs commonly produces closely related polypeptides called protein isoforms.

Other genes, such as those encoding antibody chains, are assembled from gene segments during development by regulated processes of somatic recombination.

ANSWERS TO QUESTIONS AND PROBLEMS

1) Beadle and Tatum **2)** dominant and codominant mutations, mutations that exhibit intragenic complementation, polar mutations, and *cis* acting genes that don't code for diffusible products **3)** complementation **4)** same **5)** different **6)** gene **7)** Garrod **8)** mutation, recombination, complementation **9)** structural **10)** functional **11)** homoalleles **12)** polar **13-17)** see *Approaches to Problem Solving* section at the end of the chapter.

Some ideas concerning thought challenging exercise:

The albino phenotype is caused by a recessive mutant. If there was just one gene that could mutate to cause the albino phenotype, then we would expect all progeny from parents who were both albino to have only albino children. The normal children born to albino parents indicates complementation.

Therefore, two different genes must be involved, with one parent homozygous at one locus and the other parent homozygous at a second locus. The fact that there are two different recessive mutant genes that can result in albinism indicates that there must be at least two steps in the biochemical pathway leading from tyrosine to melanin.

APPROACHES TO PROBLEM SOLVING

13. This is an example of using a complementation test in a plant. The hybrid from the cross between *subjacent-hooded* and *calcaroides* displayed the hood-like structure on the awns. Therefore the two independently derived mutants must be of the same gene, i.e., they are allelic.

14. (a) To approach this problem remember that mutants of the same gene do not generally complement, whereas mutants of different genes do. The r mutants 1, 4, 7, 8, and 9 don't complement each other, and therefore are of the same gene. The r mutants 2, 3, 5, and 6 don't complement each other, but do complement 1, 4, 7, 8, and 9. These mutants can be assigned to two genes. (b) Mutants *r1*, *r4*, *r7*, *r8*, and *r9* are assigned to one gene. Mutants *r2*, *r3*, *r5*, and *r6* are assigned to a second gene.

15. The approach to this problem is the same as the previous problem. Simply see which mutants complement (mutants of different genes) and which ones do not (mutants of the same gene). To double check your assignments, make sure that all mutants that you have assigned to one gene don't complement, but that they do complement all mutants assigned to different genes. Mutant 7 complements all other mutants. Therefore, 7 is in one gene. Mutants 5 and 6 don't complement each other, but complement all other mutants listed in the table. This tells us that mutants 5 and 6 are of a second gene. Mutant 3 and 4 don't complement but complement all the other mutants. Mutants 3 and 4 must belong to a third gene. Mutants 1 and 2 don't complement, but complement all other mutant genes. Mutants 1 and 2 must be of a fourth gene.

16. The data presented allows a biochemical pathway and the step affected by each gene to be deduced.

    (a) Since all mutants were screened for the inability to grow without the addition of arginine, they must be unable to carry out one or more steps in the production of arginine. Mutant 3 grows only if arginine is added. Therefore, it cannot convert a precursor to arginine.

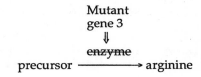

The wild-type allele of mutant 3 must code for the enzyme that catalyzes this step.

Mutants 2 grows if citrulline or arginine is added. Therefore, citrulline must be the immediate precursor to arginine, and mutant 2 blocks the pathway at the step that in normal strains converts a precursor to citrulline.

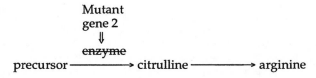

Mutant 1 grows if ornithine, citrulline, or arginine is added to the medium. It must affect a step in the pathway before these three compounds are produced.

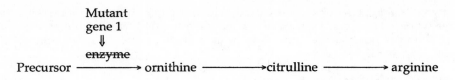

Mutant
gene 1
⇓
~~enzyme~~

Precursor ⟶ ornithine ⟶ citrulline ⟶ arginine

The wild-type allele of mutant gene 1 must code for the enzyme that converts a precursor to ornithine.

The biochemical pathway and the step controlled by each gene are indicated below:

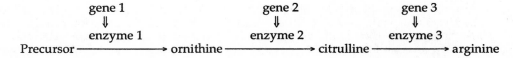

gene 1   gene 2   gene 3
⇓   ⇓   ⇓
enzyme 1   enzyme 2   enzyme 3

Precursor ⟶ ornithine ⟶ citrulline ⟶ arginine

(b) In mutant 2, ornithine is not converted to citrulline. Therefore, ornithine would accumulate.

17. To assign each mutant to a region requires an analysis of the recovery or lack of recovery of wild-type recombinants when the mutants are crossed with different deletion mutants. Wild-type recombination indicates that the mutant is not in the area of the gene missing in the deletion mutant. Lack of recombinants indicates that the deletion covers the site corresponding to the position of the point mutation.

Mutant *r1* shows lack of recombination only with deletion mutant *1695*. This limits the location of the point mutation to the area covered by *1695*. The fact that it shows recombination with deletion mutants *PT153*, *H88*, and *184*, limits the location of mutant *r1* to region 7 - 8.

Mutant *r2* does not show recombination with deletion mutant *1364*, but does with all others. This limits the location of the point mutation *r2* to the area covered by *1364*. It must reside in either region 1 - 2 or region 5 - 6, because regions 1 - 2 and 5 - 6 are the only areas covered by *1364* that do not overlap the other deletion mutations.

Mutant *r3* shows recombination with all deletion mutants except *1695* and *PT153*. This indicates that *r3* is located within the area covered by deletions *1695* and *PT153*, or 8 - 10. Because the *r3* mutant shows recombination with *H88*, it can be more narrowly defined to region 8 - 9.

Mutant *r4* does not recombine with deletion mutants *1364* and *168*, but recombines with all others. This restricts it somewhere to the overlapping region of deletion mutants *1364* and *168* or region 3 - 5. Because it recombines with deletion mutant *368*, it must be located outside *368* or in region 4 - 5.

15

The Techniques of Molecular Genetics

IMPORTANT CONCEPTS

A. Molecular technologies have been developed that allow the construction of recombinant DNA molecules that contain sequences from unrelated species, and the cloning of these molecules in appropriate host cells.

  1. The ability to clone and sequence any gene or DNA sequence of interest depends on a class of enzymes called restriction endonucleases.

    a. Restriction endonucleases, discovered by Hamilton Smith and David Nathans in 1970, cut DNA molecules at specific sequences called restriction sites.

      (1) Restriction sites are commonly palindromes, i.e., nucleotide-pair sequences that read the same forward and backward ($5' \rightarrow 3'$) from a center of symmetry.

    b. Many restriction endonucleases produce DNA fragments (restriction fragments) with complementary single-stranded ends, but some produce DNA fragments with blunt ends.

    c. Two restriction fragments, regardless of source, with the same single-stranded ends can be covalently fused to produce a recombinant DNA molecule.

      (1) The first recombinant DNA molecules were produced in Paul Berg's laboratory at Stanford University in 1972.

  2. Various applications of recombinant DNA technology require not only the construction of

recombinant DNA molecules, but also the amplification or cloning of these molecules in cloning vectors.

    a. A cloning vector has three essential components: an origin of replication; a dominant selectable marker gene, usually a gene that confers drug resistance to the host cell; and at least one unique restriction endonuclease cleavage site.

        (1) Stanley Cohen and his colleagues at Stanford University inserted an *Eco*RI restriction fragment from one DNA molecule into a cleaved, unique *Eco*RI restriction site of a circular, self-replicating DNA molecule. This recombinant DNA molecule, when transformed into *E. coli* cells exhibited normal autonomous replication, thus being the first cloning vector.

        (2) More versatile plasmids containing several unique restriction enzyme cleavage sites and genes conferring resistance to more than one antibiotic were soon developed.

        (3) Many of the cloning vectors used today are second-and third-generation derivatives of the plasmid pBR322, the first widely used cloning vector.

    b. Cloning vectors have been developed to clone larger DNA fragments.

        (1) The chromosome of bacteriophage lambda, with the central one-third of its DNA deleted, can be used as a vector to accommodate inserts of 10 kb to 15 kb.

        (2) Cosmids are recombinant DNA molecules of plasmid and lambda chromosomes that combine the plasmid's ability to replicate autonomously in *E. coli* cells and the *in vitro* packaging capability (35 to 45 kb) of the λ chromosome.

        (3) Shuttle vectors have been developed that contain both *E. coli* and yeast *S. cerevisiae* origins of replication and, therefore, can replicate in both cells of *E. coli* and yeast.

        (4) Yeast artificial chromosomes (YACs), bacterial artificial chromosomes (BACs), and P1 artificial chromosomes (PACs) are cloning vectors that were developed to clone DNA fragments several hundred kb in length. Because of their simplicity and relative ease of construction, BAC and PAC vectors have largely replaced YAC vectors in the study of large eukaryotic genomes.

  3. The polymerase chain reaction (PCR), devised by Kary Mullis, is an extremely powerful procedure that allows the amplification of a selected DNA sequence in the genome a million-fold or more without the use of living cells during the process.

    a. PCR technologies provide shortcuts for many cloning and sequencing applications.

        (1) Amplication of small available amounts of DNA by PCR allows amplified genes and DNA sequences to be structurally analyzed.

        (2) PCR is used for prenatal diagnosis of inherited human diseases where limited amounts of fetal DNA are available.

        (3) PCR DNA fingerprinting is used in forensic cases involving the identification of individuals from very small tissue samples.

B. The first step in cloning a gene or DNA sequence of a given organism usually involves the construction of a library for screening cloned DNA sequences.

  1. Genomic DNA libraries are usually prepared by isolating total genomic DNA of an organism, digesting the DNA with a restriction endonuclease, and ligating the restriction fragments into an appropriate cloning vector.

  2. Libraries can also be produced from complementary DNAs (cDNAs).

    a. Complementary DNAs are double-stranded DNA molecules produced from mRNAs by an enzyme called reverse transcriptase.

  3. Screening eukaryotic DNA libraries for a specific gene or other DNA sequence of interest requires the identification of a single DNA sequence among thousands to millions of other such sequences.

    a. The most powerful screening procedure is genetic selection: searching for a DNA sequence in the library that can restore the wild-type phenotype to a mutant organism.

    b. The first eukaryotic DNA sequences to be cloned were genes that are highly expressed in specialized cells, which facilitate the purification of mRNA and the production of cDNA.

        (1) Radioactive ovalbumin and hemoglobin cDNAs were used to screen genomic libraries by *in situ* colony or plaque hybridization.

C. Various techniques are available to prepare cloned DNA and RNA sequences for *in vitro* manipulation.

  1. Phasmid cloning vectors (e.g., pUC118 and pUC119), contain components from both phage

chromosomes and plasmids. After addition of a helper phage, they produce single-stranded DNA replicates and package them into phage coats.
   a.   Single-stranded DNA molecules are used in DNA sequencing and *in vitro* mutagenesis protocols.
   2. Oligonucleotide-directed site-specific mutagenesis allows scientists to change nucleotide sequences in a specific manner.
D. The development of recombinant DNA techniques has spawned a whole array of new approaches to analyze genes and gene products.
   1. Gel electrophoresis provides a powerful tool for the separation of macromolecules with different sizes and charges.
   2. In 1975, E. M. Southern reported an important new procedure subsequently nickname "Southern blotting", that allowed investigators to identify the location of genes and other DNA sequences on restriction fragments separated by electrophoresis.
      a.   The essential feature of Southern blotting is the transfer of the DNA molecules, separated by gel electrophoresis, to nylon membranes followed by the hybridization of a radioactive DNA probe of interest. The probe will hybridize only with the immobilized DNA on the membrane that contains a nucleotide sequence complementary to the sequence of the probe.
   3. RNA molecules separated by gel electrophoresis also can be transferred to nylon membranes and hybridized with radioactive probes. This is called northern blotting.
   4. Western blotting involves transfer of proteins, separated on polyacrylamide electrophoresis, to a nylon membrane followed by the detection of individual proteins by using specific antibodies.
E. The structure of genes and chromosomes can be analyzed by an array of molecular techniques.
   1. Physical maps of DNA molecules can be made that are based on restriction enzyme cleavage sites.
   2. The nucleotide pair sequence of a gene or DNA sequence can be determined by either of two sequencing procedures developed in the period from 1974 to 1977.
      a.   The Maxam and Gilbert procedure uses four different chemical reactions to cleave DNA chains at As, Gs, Cs, or Ts.
      b.   The second approach, developed by Fred Sanger and colleagues, uses an enzymatic procedure and specific chain-terminators to generate four populations of fragments that terminate at As, Gs, Cs, and Ts, respectively.

IMPORTANT TERMS

In the space allotted, concisely define each term.

recombinant DNA technologies:

cloning:

cloning vectors:

restriction endonuclease:

recombinant DNA molecules:

restriction sites:

*Eco*RI:

methylation:

palindromes:

DNA ligase:

restriction fragments:

amplification:

clones:

dominant selectable marker:

polylinker (polycloning) site:

unique endonuclease cleavage site:

cosmids:

196 Chapter 15

shuttle vectors:

yeast artificial chromosome (YAC):

bacterial artificial chromosome (BAC):

P1 artificial chromosomes (PAC):

polymerase chain reaction (PCR):

Taq polymerase:

PCR DNA fingerprinting:

genomic DNA library:

complementary DNA (cDNA):

cDNA library:

terminal transferase:

genetic selection:

complementary screening:

in situ colony (plaque) hybridization:

physical containment:

biological containment:

phagemids:

helper phage:

pUC118 and pUC119:

directional cloning:

site-specific mutagenesis:

oligonucleotide-directed site-specific mutagenesis:

Southern blots:

northern blots:

western blotting:

physical maps:

restriction maps:

2′, 3′-dideoxyribonucleoside triphosphates:

IMPORTANT NAMES

In the space allotted, concisely state the major contribution of the individual or group.

Hamilton Smith and Daniel Nathans:

Werner Arber:

Paul Berg:

Stanley Cohen:

Kary Mullis:

E. M. Southern:

Allan Maxam and Walter Gilbert:

Fred Sanger:

TESTING YOUR KNOWLEDGE

In this section, answer the questions, fill in the blanks, and solve the problems in the space allotted. Problems noted with an * are solved in the Approaches to Problem Solving section at the end of the chapter.

1. An enzyme that cleaves DNA at a specific sequence in which the bases read the same on complementary strands ($5' \rightarrow 3'$ and $3' \rightarrow 5'$) from a point of symmetry is called a (an) _____ _____.

2. A double-stranded DNA complementary to an mRNA that is produced by reverse transcriptase is called a (an) _____.

3. Which cloning vector would be the best for producing a genomic DNA library of mean insert size of about 1000 bp?

4. Which cloning vectors are best suited for producing a genomic DNA library of average insert size of about 200 Kbp?

5. Which cloning vector is best suited for cloning cDNAs?

6. The ability to insert a foreign DNA in a predetermined orientation into the polycloning region of pUC118-pUC119 vector system is called _____.

7. Three individuals who shared the 1986 Nobel Prize in Physiology and Medicine for experiments that led to, and the discovery of, restriction endonucleases were _____, _____, and _____.

8. A cloning vector that can replicate in both bacteria and yeast is called a (an) _____.

9. The individual who was a co-recipient of the 1980 Nobel Prize in Chemistry for his pioneering research on recombinant DNA molecules was _____.

10. The analysis of the functioning of a gene after changing one nucleotide at a time *in vitro* is called _____.

11. The individual who received a Nobel Prize in 1993 for inventing the polymerase chain reaction was _____.

12. The procedure by which proteins are transferred from electrophoretic gels to membranes and detected with antibodies is called _____.

13. A hybrid cloning vector used for the biological purification of single-stranded DNA from the double-stranded DNA insert is called _____.

14. The procedure by which DNA fragments, separated by electrophoresis, are transferred from the gel to a nylon membrane, is called _____.

*15. Which of the following RNA sequences is complementary to a palindrome sequence in the encoding DNA?
    (a) AUCUUCUA   (d) AUUACGCC
    (b) CCUAUAGG   (e) none of these
    (c) AAAAGGGG

*16. A geneticist is producing a genomic BAC library of a plant that has a genome size of 10^9 bp. The average insert size in the library is 100 Kbp. How many recombinant clones will have to be constructed in order to be 99% sure that the library will have a specific DNA sequence represented?

*17. A molecular geneticist determined the order of nucleotides in a ten base-pair region of a gene by the Sanger dideoxyribonucleoside method. The fragments in the four syntheses were electrophoresed, and visualized by autoradiography. From an analysis of the autoradiogram diagrammed below, determine the $5' \rightarrow 3'$ sequence of nucleotides in the strand sequenced.

lane 1 ddA	lane 2 ddT	lane 3 ddG	lane 4 ddC	Direction of migration
				↓

*18. A geneticist was developing a restriction map of a 10,000 bp linear DNA sequence. The DNA fragment was treated individually and in combination with the restriction enzymes EcoRI, BamHI, and HindIII. The enzyme reactions were allowed to go to completion and the DNA fragments were separated and sized by agarose gel electrophoresis. Produce a restriction map from the following fragment sizes detected from the various digestions.

	EcoRI	BamHI	HindIII	EcoRI + BamHI	EcoRI + HindIII	BamHI + HindIII	EcoRI + BamHI + HindIII
	500	2000	1500	500	500	500	500
	3500	8000	3500	1500	1000	1500	1000
	6000		5000	2000	2500	3000	2000
				6000	5000	5000	5000
total length	10000	10000	10000	10000	9000	10000	8500

19. A portion of homologous DNA from a normal and mutant jellybean plant was sequenced by the dideoxy chain-termination method. The gel analysis of fragments from normal and mutant DNA is shown below.

Gel analysis of fragments

	Normal				Mutant			
Origin −	ddA	ddT	ddG	ddC	ddA	ddT	ddG	ddC

(gel band pattern shown, running from Origin (−) at top to (+) at bottom)

(a) What is the sequence of the normal single-stranded DNA that served as a template for generating this pattern?

(b) What is the sequence of the mutant single-stranded DNA that served as a template for generating this pattern?

(c) Which of the DNA sequences, normal or mutant, may be recognized by a restriction endonuclease?

20. Which primers would you choose to amplify the region (bold font) of the following DNA sequence?

```
5'-ATGCTAATGTATTAAGCTATAATGTACGACCTAAGGAGGTAAGCGCATGTAATACTGA-3'
3'-TACGATTACATAATTCGATATTACATGCTGCTGGATTCCTCCATTCGCGTAGATTATGACT-5'
```

Primer 1:   5'-ATGCTAATGTATT-3'
Primer 2:   3'-CATGTAATACTGA-5'
Primer 3:   5'-TACGATTACATAA-3'
Primer 4:   3'-GTAGATTATGACT-5'

THOUGHT CHALLENGING EXERCISE

Some of the greatest discoveries in science that later had tremendous practical applications have come from purely basic research. Using examples, discuss scientific discoveries that fall into this category.

SUMMARY OF KEY POINTS

The discovery of restriction endonucleases—enzymes that recognize and cleave DNA in a sequence-specific manner—allowed scientists to produce recombinant DNA molecules *in vitro*.

DNA sequences can be inserted into small, self-replicating DNA molecules called cloning vectors and amplified by replication *in vivo* after being introduced into living cells by transformation.

A variety of cloning vectors have been constructed, each with advantages for certain research purposes.

The polymerase chain reaction (PCR) can be used to amplify specific DNA sequences *in vitro*.

DNA libraries can be constructed that contain complete sets of genomic DNA sequences or DNA copies (cDNAs) of mRNAs in an organism.

Specific genes or other DNA sequences can be isolated from DNA libraries by genetic complementation or by hybridization to labeled nucleic acid probes containing sequences of known function.

Phagemids provide a powerful tool for purifying the single DNA strands of cloned DNA sequences.

The nucleotide sequences of cloned genes can be changed as desired by *in vitro* site-specific mutagenesis.

DNA restriction fragments and other small DNA molecules can be separated by agarose or acrylamide gel electrophoresis and transferred to nylon membranes to produce DNA gel blots called Southern blots.

The DNAs on Southern blots can be hybridized to labeled DNA probes to detect sequences of interest by autoradiography.

When RNA molecules are by separated by gel electrophoresis and transferred to membranes for analysis, the resulting RNA gel blots are called northern blots.

When proteins are transferred from gels to membranes and detected with antibodies, the products are called western blots.

Detailed physical maps of DNA molecules can be prepared by identifying the sites that are cleaved by various restriction endonucleases.

The nucleotide sequences of DNA molecules provide the ultimate physical maps of genes and chromosomes.

ANSWERS TO QUESTIONS AND PROBLEMS

1) restriction endonuclease 2) complementary DNA or cDNA 3) A plasmid such as pBR322, pUC118, or pUC119 4) BACs and PACs 5) Plasmid-based vectors such as pBR322 6) forced cloning 7) Hamilton Smith, Daniel Nathans and Werner Arber 8) shuttle vector 9) Paul Berg 10) site-specific mutagenesis 11) Kary Mullis 12) western blotting 13) phagemids 14) Southern blotting 15-18, 20)

see *Approaches to Problem Solving* section **19) a.** 3'-AGAATTCG-5' **b.** 3'-AGATTTCG-5' **c.** the normal jellybean plant has the palindrome 5'-**AGAATTCG**-3' in its DNA that is not found in the mutant DNA sequence.   3'-**TCTTAAGC**-5'

Ideas on thought challenging exercise:

Most of the scientific discoveries presented in this chapter have made possible the array of molecular techniques now available for cloning and manipulating DNA. Yet, most were the rewards of excellent basic research. A case in point is the discovery of restriction enzymes. Their discovery came not because the scientists set out to find such enzymes, but rather were discovered during basic research on bacteria and bacteriophages.

APPROACHES TO PROBLEM SOLVING

15. To approach this problem, first indicate the sense DNA strand (complementary to the RNA) and the nonsense strand of the double helix for the indicated RNAs. When this is done, it is apparent that (b) contains a palindrome.

RNA	C C UA UA G G
DNA	**GGA**T AT C C
DNA	C C T **AT A**GG

16. To solve this problem, use the formula in your text $N = \ln (1 - P)/\ln (1 - f)$, P = probability, and f = fraction of genome present in a single clone. Thus, for this problem P = 0.99 and $f = 10^5/10^9 = 0.0001$.

$$N = \ln [1 - 0.99]/\ln [1 - 0.0001] = 46{,}049 \text{ clones}$$

17. The smaller fragments migrate faster in the gel. Since the polymerization stops with the addition of a dideoxyribonucleotide to the growing 3' end, position 10 is the 3' end and position 1 represents the 5' end of the polynucleotide being synthesized.

lane 1 ddA	lane 2 ddT	lane 3 ddG	lane 4 ddCC	Direction of migration	Base at each position in the polypeptide
____				10 ↓	A
		____		9	G
		____		8	G
			____	7	C
	____			6	T
____				5	A
____				4	A
		____		3	G
			____	2	C
			____	1	C

The 5' → 3' sequence of nucleotides in the strand sequenced is complementary to the strand synthesized and is, therefore, TCCGATTCGG

18. To approach this problem, first diagram the possible restriction map for *Eco*RI fragments. The two possibilities in kb pairs are:

or

Next the restriction fragment of *Bam*HI and of the codigests of *Bam*HI and *Eco*RI are analyzed. *Bam*HI has one cutting site that results in a 2kb fragment and an 8kb fragment. Superimposing these on the *Eco*RI map diagrammed above, and noting that restriction maps are additive, the restriction map is:

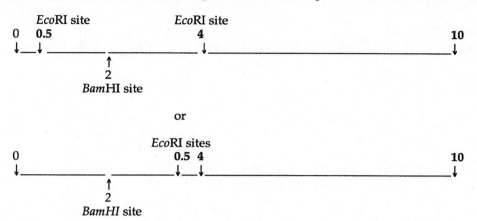

The next step is to analyze *Eco*RI and *Hind*III restriction sites. Again, this is done relative to *Eco*RI sites.

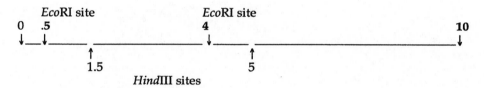

These data limit the position of the restriction site generating the 500 bp *Eco*RI fragment as indicated above. This is the only order that allows the 2500 bp *Eco*RI + *Hind*III fragment to be formed.

*Hind*II and *Bam*HI restriction sites considered together map as follows:

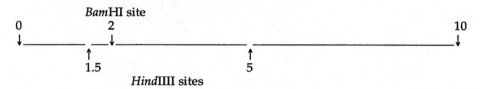

The final map pieces together as follows:

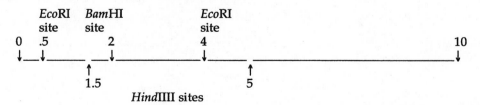

The reason the restriction fragments don't add up to 10000 in some of the digestions is that some of the restriction cleavages produce fragments of the same size.

20. DNA is always replicated from 5' to 3'. Therefore, PCR primers must be used that flank the DNA sequence targeted for amplification. The primer set that will amplify the region indicated in bold is primer 1 and primer 4. These will anneal to the single strands of the DNA as follows:

```
5'-ATGCTAATGTATTAAGCTATAATGTACGACCTAAGGAGGTAAGCGCATGTAATACTGA-3'
                                                    GTAGATTATGACT-5'
5'-ATGCTAATGTATT
3'-TACGATTACATAATTCGATATTAGCTGCTGGATTCCTCCATTCGCGTAGATTATGACT-5'
```

16

Genomics

IMPORTANT CONCEPTS

A. The development of recombinant DNA and gene cloning technologies has allowed for many new types of genetic analysis and manipulation.
1. Essentially any gene of any organism can be isolated, dissected, modified, and manipulated.
2. Synthetic, modified, or other foreign genes can be introduced into microorganisms, plants, and animals.
3. Recombinant DNA technologies have resulted in the explosive accumulation of new information in virtually every area of biology.

B. Genomics is a genetics subdiscipline that deals with mapping, sequencing and functions of entire genomes.
1. The genomics subdivision of mapping and sequencing genomes is called structural genomics.
2. The genomics subdivision that studies the function of genomes is called functional genomics.
3. Comparative genomics is the study of genome evolution.

C. Detailed genetic maps result from the correlation of genetic, cytogenetic, and physical maps of chromosomes.
1. Physical maps such as restriction maps are based on molecular distances (in base pairs) separating sites in the chromosomes.
   a. Physical maps often contain the locations of overlapping genomic clones (contigs) and unique nucleotide sequences called sequence-tagged sites (STSs).
2. Markers that are mapped both genetically and physically are called anchor markers.

3. Genetic maps are constructed from recombination frequencies, with 1 centimorgan (cM) equal to the distance that yields 1 percent recombination.
   a. Restriction fragment-length polymorphisms (RFLPs) are commonly used for increasing the density of markers on genetic maps.
      (1) In humans, the most useful polymorphisms are variable number tandem repeats (VNTRs).
      (2) Microsatellites are polymorphic tandem repeats (2 to 5 bp) that are extremely valuable in constructing high density maps.
4. Cytogenetic maps are based on the location of genes within, or near, microscopically-visible, morphological markers, e.g., chromosome bands.
   a. Cloned genes and DNA sequences can be positioned on the cytogenetic map by *in situ* hybridization.
D. Positional cloning of genes depends on the availability of detailed maps of the regions of the chromosomes where the genes reside.
   1. Positional cloning is accomplished by mapping the gene of interest, identifying a tightly-linked RFLP or other molecular marker, and then "walking" or "jumping" along the chromosome until the gene is reached.
E. The Human Genome Project was established with the goals of mapping all human genes and sequencing all 3×10^9 nucleotide pairs in the human genome.
   1. A nearly complete sequence of the human genome comprising 99 percent of the euchromatic DNA was released in October, 2004.
F. Functional genomics will accelerate progress toward understanding of programs of gene expression in development.
   1. Expressed sequences (ESTs) present in cDNA are complementary to the protein-coding content of the genome.
   2. Dot blot hybridization can be used to study gene expression.
      a. Gene-specific nucleotide sequences are applied to filters in specific patterns or arrays.
         (1) Radioactive or fluorescent RNA or cDNA preparations are hybridized to the DNA on the filters.
            (a) The amounts of RNA or cDNA hybridized to each probe on the dot blot can be quantified.
         (2) A high density micro dot blot containing thousands of hybridization probes on a single membrane or other solid support is called a gene chip.
            (a) Gene chips allow the study of a large number (potentially all) of the genes of an organism simultaneously.
   3. The green fluorescent protein (GFP) of the jellyfish provides a powerful new tool for studying gene expression at the protein level.
      a. Fusion genes encoding the protein of interest and GFP are introduced into cells by transformation.
         (1) GFP fluoresces bright green and thus can be used to study gene expression in living cells and to study protein localization and movement in cells over time.
G. Comparative genomics research has contributed to our understanding of genome evolution.
   1. Despite extensive chromosomal rearrangements, linkage relationships of blocks of unique DNA sequences and known genes are remarkably conserved among cereal grasses.
      a. Cereal grasses differ greatly in genome size and chromosome number, but have large chromosome segments containing genes with very similar linkage arrangements.
         (1) The chromosomes of all of the cereal grasses can be constructed by simply rearranging rice chromosome segments.
   2. Comparisons of the nucleotide sequences of the genomes of organisms have resulted in a better understanding of taxonomic relationships and of the changes responsible for the evolution of species from common ancestors.
   3. Bioinformatics is the science of gathering, manipulating, storing, retrieving, and classifying recorded information, especially DNA and protein sequences.
   4. The evolution of complex eukaryotes has involved a decrease in the proportion of the genome encoding proteins and a corresponding increase in the proportion of noncoding DNA.

5. The evolution of mammalian chromosomes involved three classes of conserved synteny; conservation of entire chromosomes; conservation of large segments of chromosomes; and the joining of segments of different chromosomes to produce new synteny.

IMPORTANT TERMS

In the space allotted, concisely define each term.

genomics:

structural genomics:

functional genomics:

comparative genomics:

transcriptome:

proteome:

proteomics:

physical maps:

anchor markers:

sequence-tagged sites (STSs):

expressed-sequence tags (ESTs):

restriction fragment-length polymorphism (RFLP):

variable number tandem repeats (VNTRs):

microsatellites:

contig:

positional cloning:

chromosome walks:

chromosome jumping:

Human Genome Project:

Human Genome Organization (HUGO):

radiation hybrid mapping:

dot blot hybridizations:

green fluorescent protein (GFP):

homologous:

paralogous:

chromosome painting:

IMPORTANT NAMES

In the space allotted, concisely state the major contribution made by the individual or group.

Kari Stefansson:

Thomas Roderick:

J. Craig Venter:

Francis Collins:

Walter Goad:

James Watson:

Graham Moore:

TESTING YOUR KNOWLEDGE

*In this section, answer the questions, fill in the blanks, and solve the problems in the space allotted. Problems noted with an * are solved in the Approaches to Problem Solving section at the end of the chapter.*

1. A cloning procedure that depends upon the availability of a detailed map of the region of the chromosome where the gene of interest resides is called _____.

2. A physical map constructed by ordering overlapping genomic DNA fragments is called a _____.

3. Markers that are mapped both genetically and physically are called _____.

4. In humans, a very useful class of RFLP markers involving short tandemly repeated sequences that vary in number is called _____.

5. Restriction maps of genomic clones may be computer analyzed and organized into overlapping sets of clones called _____.

6. A class of RFLPs consisting of polymorphic tandem repeats of sequences only two to five nucleotides long are called _____.

7. PCR amplified short unique DNA sequences that may be used as anchor sequences are called _____.

8. A procedure for visually determining the chromosomal position of a cloned gene or other DNA sequence is called _____.

9. DNA sequences obtained from a cDNA that may be used as probes for anchoring physical maps to RFLP maps or assembled as arrays on micro dot blots for studying gene expression are called _____.

10. Microarrays of oligonucleotides representing thousands of ESTs on a single silicon wafer of only a few square centimeters are called _____.

11. What properties of the green fluorescent protein (GTP) render it ideal to locate the location of proteins *in vivo*?

*12. A physical map is being constructed for a small chromosome of a fungus. Six BAC clones (I-VI) are available which cover the complete chromosome. These are probed with STSs (a, b, c, and d). The results are shown in the following table ("+" indicates that the probe hybridizes to the BAC DNA, "-" indicates no hybridization). The chromosome is also cut with a restriction enzyme that generates restriction fragments 1-5. The restriction fragments are also probed with the STSs generating the results also shown in the table. What can be deduced about the order of the BACs and of the restriction fragments?

	BACs						Restriction Fragments				
	I	II	III	IV	V	VI	1	2	3	4	5
STS a	-	+	-	+	+	-	-	-	-	+	-
STS b	-	-	+	+	+	-	-	-	-	-	+
STS c	-	-	-	-	-	+	-	-	+	-	-
STS d	+	-	+	-	-	+	-	+	-	-	-

THOUGHT CHALLENGING EXERCISE

What problems may be encountered when attempting to walk to genes in plants with large genome sizes and polyploid ancestry, e.g., *Triticum aestivum* (wheat) with a haploid DNA content of $> 1.7 \times 10^{10}$ bp?

SUMMARY OF KEY POINTS

Genomics is the subdivision of genetics devoted to the mapping, sequencing, and functional and comparative analyses of genomes.

Genetic maps of chromosomes are based on recombination frequencies between markers.

Cytogenetic maps are based on the location of markers within, or near, cytological features of chromosomes observed by microscopy.

Physical maps of chromosomes are based on distances in base pairs, kilobase pairs, or megabase pairs separating markers.

High-density maps that integrate genetic, cytological, and physical maps of chromosomes have been constructed for many chromosomes, including all of the human chromosomes.

Detailed genetic, cytogenetic, and physical maps of chromosomes permit researchers to isolate genes based on their location in the genome.

If a molecular marker such as a restriction fragment-length polymorphism (RFLP) maps close to a gene, the gene can usually be isolated by chromosome walks or chromosome jumps.

Researchers collaborating on the Human Genome Project have constructed detailed maps of all 24 human chromosomes.

Other participants of the Human Genome Project have determined the complete, or nearly complete, nucleotide sequences of the genomes of several important model organisms.

A nearly complete sequence of the euchromatic DNA in the human genome was released in October, 2004.

Once the complete sequence of a genome has been determined, scientists can study the temporal and spatial patterns of expression of all genes of the organism.

Microarrays of gene-specific hybridization probes on gene chips allow researchers to study the transcription of thousands of genes simultaneously.

Chimeric genes that contain the coding region of the green fluorescent protein of the jellyfish fused with the coding regions of the genes of experimental organisms can be used to study the localization of proteins in living cells.

Comparative genomics—comparing the nucleotide sequences of genomes—has provided new information about the relationships between various taxonomic groups.

Bioinformatics is the science of storing, comparing, and extracting information from biological systems, especially DNA and protein sequences.

The nucleotide sequences of diverse prokaryotic organisms have provided insights into their adaptation to unique environmental niches.

As eukaryotic organisms have increased in complexity, the proportion of their genomes that encodes proteins has decreased, and the function of most of the noncoding DNA is unknown.

Comparative genomics has revealed a remarkable conservation of synteny in related eukaryotic species, such as mammals and the cereal grasses.

ANSWERS TO QUESTIONS AND PROBLEMS

1) positional cloning **2)** contiguous clone map **3)** anchor markers **4)** VNTRs (variable number of tandem repeats) **5)** contigs **6)** microsatellites **7)** sequence-tagged sites (STSs) **8)** *in situ* hybridization **9)** expressed sequence tags (ESTs) **10)** gene chips **11)** small protein; requires only blue or UV light to fluoresce; mutants have been isolated that have different fluorescent intensities, emission wavelengths, and pH requirements **12)** see *Approaches to Problem Solving* section.

Some ideas concerning the thought challenging exercise:

Several problems may be encountered in attempting to walk to genes in plants with large genomes. Interspersed repetitive DNAs provide one potential road block. Another potential problem is that regions rich in repetitive DNA between genes may show very little recombination; therefore, a short map distance between genes may actually represent a large physical distance. This may render a walk unattainable. The presence of duplicate regions of linkage groups may result in a walk that jumps from one chromosome to another. Many plants are polyploid, or of polyploid ancestry, and therefore have duplication of whole genomes. Wheat is a hexaploid and therefore contains three related genomes.

APPROACHES TO PROBLEM SOLVING

12. The order of the BACs is determined by overlapping the regions of homology as shown below. In this case, these regions are detected by STS homologies. The BACs occur in the relative order VI, I, III, (IV/V), II. The order of BACs IV and V cannot be determined using the available data. The restriction fragment containing each STS is indicated in parentheses. The restriction fragments map in the relative order 3, 2, 5, 4. Restriction fragment 1 does not contain a sequence complementary to the STSs used.

				a (4)		II
			b(5)	a (4)		IV
I		d (2)	b(5)	a (4)		V
III		d(2)	b(5)			
VI	c (3)	d(2)				

17

Applications of Molecular Genetics

USE OF RECOMBINANT DNA TECHNOLOGY TO IDENTIFY HUMAN GENES
Huntington's Disease
Cystic Fibrosis

MOLECULAR DIAGNOSIS OF HUMAN DISEASES

HUMAN GENE THERAPY

DNA FINGERPRINTS
Paternity Tests
Forensic Applications

PRODUCTION OF EUKARYOTIC PROTEINS IN BACTERIA
Human Growth Hormone
Proteins With Industrial Applications

TRANSGENIC PLANTS AND ANIMALS
Transgenic Animals: Microinjection of DNA into Fertilized Eggs
Transgenic Plants: The Ti Plasmid of *Agrobacterium tumefaciens*

REVERSE GENETICS: DISSECTING BIOLOGICAL PROCESSES BY INHIBITING GENE EXPRESSION
Antisense RNA
T-DNA and Transposon Insertion
RNA Interference

A MILESTONE IN GENETICS: TRINUCLEOTIDE REPEATS AND HUMAN DISEASE

IMPORTANT CONCEPTS

A. RFLPs have been used to identify the genes responsible for several major human diseases by positional cloning, i.e., by cloning genes based on their chromosomal locations.
   1. Positional cloning of genes is greatly assisted by the preparation of physical maps of chromosomes, e.g., molecular maps prepared from the restriction maps of overlapping clones (contigs).
   2. The human genes responsible for Huntington's disease (HD)and cystic fibrosis were identified by positional cloning.
      a. The identification and characterization of these genes have provided new information about the molecular and cellular basis of these diseases, made DNA diagnosis of the defective genes possible, and raised hope for successful treatment of the diseases by somatic-cell gene therapy, i.e., providing mutant cells with functional copies of the defective gene.
      b. HD is a human genetic disorder associated with an unstable trinucleotide repeat.
B. Gene therapy involves adding a functional copy of a gene to the genome of an individual carrying defective copies of the gene.
   1. Several requirements must be fulfilled before a gene-therapy procedure will be approved.
      a. The gene must be cloned and well characterized, i.e., it must be available in pure form.
      b. An effective method must be available for delivering the gene into the target cells.
      c. The risk of gene therapy to the patient must have been carefully evaluated and shown to be minimal.
      d. The disease must not be treatable by other strategies.
      e. Data must be available from preliminary experiments with animal models or human cells

indicating that the proposed gene therapy should be effective.
2. The first attempts of human gene therapy occurred in 1990, when a four-year-old girl with adenosine deamininase-deficient severe combined immunodeficiency disease received her own transgenic white blood cells carrying functional ADA genes.
    a. A major problem encountered has been the silencing of the introduced ADA transgenes.
C. DNA fingerprinting provides valuable evidence in establishing individual identity, especially in paternity and forensic cases.
D. Transgenic microbes, plants, and animals are now being used to produce valuable proteins.
    1. Transgenic microbes are used in commercial production of specific gene products such as human insulin, human growth hormone (HGH), and proteins with industrial applications such as protease, amylase and rennin.
    2. Transgenic animals are routinely produced by microinjecting DNA into fertilized eggs.
    3. Transgenic plants are produced by using the Ti plasmid of *Agrobacterium tumefaciens* as a gene-transfer vector or by a procedure called microprojectile bombardment, which simply involves shooting DNA-coated tungsten particles into nuclei.
E. Biological processes can be dissected by a procedure called reverse genetics that utilizes known nucleotide sequences to inhibit the expression of specific genes.
    1. The expression of an endogenous gene can sometimes be shut off or reduced by the presence of antisense RNA.
        a. Antisense RNA is complementary to mRNA and blocks the translation of the mRNA (sense RNA).
            (1) The *FlavrSavr*™ tomato was the first genetically engineered plant approved for human consumption. This tomato was produced by antisense RNA technology to decrease the rate of expression of an endogenous gene encoding polygalacturonase, an enzyme that breaks down cell walls and causes softening in tomatoes as they ripen. This renders vine-ripened tomatoes too soft to survive the machine harvesting and handling required for marketing. *FlavrSavr*™ tomatoes remain firm during the ripening process, thus allowing them to remain on the vines longer before they are picked and shipped to market.
        b. The use of antisense RNAs for reverse genetics has been largely replaced by approaches using T-DNA and transposon insertional mutagenesis or RNA interference.
    2. Null ("knockout") mutations can be induced by insertional mutagenesis using either T-DNA or transposons.
    3. Reverse genetics can also be approached using RNA interference (RNAi).
        a. When cells are transformed with at least a portion of a gene in reverse orientation, or when a double-stranded RNA is injected, the expression of genes containing the same nucleotide sequence is inhibited by the RNA-inducing silencing complex (RISC).

IMPORTANT TERMS

In the space allotted, concisely define each term.

Tay-Sach's disease:

Huntington's disease (HD):

Cystic fibrosis (CF):

CpG islands:

Zoo blots:

CFTR protein (cystic fibrosis transmembrane conductance regulator):

gene therapy:

transgene:

transgenic:

somatic-cell gene therapy:

germline gene therapy:

adenosine deaminase-deficient severe combined immunodeficiency disease (ADA-SCID):

gene addition:

gene replacement:

DNA fingerprints:

human growth hormone (HGH):

G$_o$ generation:

microprojectile bombardment:

electroporation:

Agrobacterium tumefaciens-mediated transformation:

totipotency:

Ti plasmid:

T-DNA:

vir region:

chimeric selectable marker genes:

reverse genetics:

antisense RNA:

antisense gene:

insertional mutagenesis:

RNA interference (RNAi):

microRNAs

genetically modified crops (GM crops):

GM foods:

IMPORTANT NAMES

In the space allotted, concisely state the major contribution made by the individual or group.

James Gusella and Nancy Wexler:

Francis Collins and Lap-Chee Tsui:

Andrew Fire and Craig Mello:

TESTING YOUR KNOWLEDGE

*In this section, answer the questions, fill in the blanks, and solve the problems in the space allotted. Problems noted with an * are solved in the Approaches to Problem Solving section at the end of the chapter.*

1.   A cloning procedure that depends upon the availability of a detailed map of the region of the chromosome where the gene of interest resides is called _____.

2. Two human disease-causing mutant genes discussed in the text, which were cloned by positional cloning are _____, and _____.

3. In humans, a very useful class of RFLP markers involving short tandemly repeated sequences that vary in number is called _____.

4. Clusters of cytosines and guanines that often precede human genes are called _____.

5. List three human diseases shown to be associated with an unstable trinucleotide repeats.

6. What is the most common inherited disease in humans of Northern European heritage?

7. An organism carrying a gene introduced by genetic engineering is called _____.

8. Which human genetic disease was the first to be treated by gene therapy?

9. Specific banding patterns on Southern blots of genomic DNA that has been cleaved with specific restriction enzymes and hybridized with appropriate DNA probes are called _____.

10. The first commercial success of recombinant DNA technology was the mass production of the _____ protein by genetically engineered bacteria.

11. The most common form of transformation in dicots is mediated by _____.

*12. In a search for the gene for a genetic disease using restriction fragment length polymorphisms, DNA was isolated from five people with the disease, and five people without the disease. The DNAs were treated with a restriction endonuclease and the digested fragments separated by agarose electrophoresis. Which band probably represents the fragment of DNA most closely linked to the gene for the disease? Why? (This problem is modified from problem 10.7 in N. L. Pruitt, *Problem Book and Study Guide for Cell and Molecular Biology* by G. Karp, 1996).

People without disease   People with disease

```
___   ___   ___   ___   ___      1      ___   ___   ___   ___   ___

___   ___   ___   ___   ___      2      ___   ___   ___   ___   ___

___   ___   ___   ___   ___      3      ___   ___   ___   ___   ___

___   ___   ___   ___   ___      4.     ___   ___   ___   ___   ___

___   ___   ___   ___   ___      5.     ___   ___   ___   ___   ___
```

*13. A man is about to go to trial for allegedly burgling a house. Although much circumstantial evidence suggests that the man committed the crime, the prosecuting attorney does not feel he can obtain a conviction unless it can be proven that the man was at the scene of the crime. Upon analyzing evidence provided by the police investigator he noticed a leaf was found lodged in the cuff of the pants worn by the defendant on the day of the crime. The prosecuting attorney remembered from his college botany class that the leaf was from a tree that was rarely found in that region of the country. Only four trees were known to exist in the city of the crime; one in the atrium of the burgled house and the others in city parks. After reviewing evidence, the defendant's lawyer insisted that the leaf must be from some source other than the burgled home. The prosecuting attorney, however, had a molecular genetics laboratory isolate and fingerprint the DNA of leaves of the four trees, and of the leaf found in the defendant's cuff. The resulting banding patterns (fingerprints) of the DNA from the leaves were as follows:

| Source of leaves | | | | |
defendant's pants	tree 1	tree 2	tree 3	burgled house

Based upon the DNA fingerprints of the leaves, should the prosecution of the case continue? Why or why not?

THOUGHT CHALLENGING EXERCISE

Are conventionally-bred organisms safer as food sources than those that have been genetically engineered?

SUMMARY OF KEY POINTS

The mutant genes responsible for Huntington's disease and cystic fibrosis were identified by positional cloning.

The nucleotide sequences of *huntingtin* and *CF* genes were used to predict the amino acid sequences of their polypeptide products and to obtain information about the functions of the gene products.

The characterization of the *huntingtin* and *CF* genes has led to the development of DNA tests that detect some of the mutations that cause Huntington's disease and cystic fibrosis.

Mutant genes that are responsible for inherited disorders can often be diagnosed by DNA tests.

The results of DNA tests for mutant genes that cause inherited diseases allow genetic counselors to inform families of the risk of affected children.

Gene therapy involves the addition of a normal (wild-type) copy of a gene to the genome of an individual who carries defective copies of the gene.

Although somatic-cell gene therapy effectively restored immunological function in boys with X-linked sever combined immunodeficiency disease, three of the boys subsequently developed leukemia or leukemia-like disorders.

Somatic-cell gene therapy holds promise for the treatment of many inherited human diseases; however, the results to date have been disappointing.

DNA fingerprints detect and record polymorphisms in the genomes of individuals.

DNA prints provide strong evidence of individual identity, evidence that may be extremely valuable in paternity and forensic cases.

Valuable proteins that could be isolated from eukaryotes only in small amounts and at great expense can now be produced in large quantities in genetically engineered bacteria.

Proteins such as human insulin and human growth hormone are valuable pharmaceuticals used to treat diabetes and pituitary dwarfism, respectively.

DNA sequences of interest can now be introduced into most plant and animal species.

The resulting transgenic organisms provide valuable resources for studies of gene function and biological processes.

The Ti plasmid of *Agrobacterium tumefaciens* is an important tool for transferring genes into plants.

Reverse genetic approaches use known nucleotide sequences to devise procedures for isolating null mutations of genes or inhibiting gene expression.

Antisense RNA, which is complementary to sense RNA (mRNA), can be used to shut off or reduce the expression of individual genes.

T-DNA or transposon insertions provide a source of null mutations of genes.

RNA interference—blocking gene expression with double-stranded RNA—can be used to dissect biological processes by inhibiting the functions of specific genes.

ANSWERS TO QUESTIONS AND PROBLEMS

1) positional cloning **2)** Huntington's disease and cystic fibrosis **3)** VNTRs (variable number of tandem repeats) **4)** CpG islands **5)** fragile-X syndrome, spino-bulbular muscular atrophy, and Huntington's disease **6)** cystic fibrosis **7)** transgenic **8)** severe combined immunodeficiency syndrome (SCID) **9)** DNA fingerprints **10)** insulin **11)** *Agrobacterium tumefaciens* **12 and 13)** see *Approaches to Problem Solving* section.

Some ideas concerning the thought challenging exercise:

There is no *a priori* reason to believe that natural or conventionally bred organisms produce safer food. Some plants, conventionally bred for food products, retain undesirable or suspected carcinogenic compounds. Also, genes coding for undesirable products, due to close linkage to the genes being selected, may piggyback undetected into a breeding line during introgression of genes from wild germplasm.

APPROACHES TO PROBLEM SOLVING

12. To solve this problem, look for a pair of bands that is associated with individuals with the disease, but is polymorphic among normal individuals. One form of band 4 is associated with the disease and may be closely linked to the gene for the disease. Since there is only a sample of five individuals, more sampling would be necessary to support this hypothesis.

13. Since the banding pattern of the DNA of the leaf found on the defendant's pants matches exactly that of the tree found in the atrium of the burgled house, but not that of the other three trees of the same kind growing in the town, this would be evidence that the defendant was at the scene of the crime. However, because a leaf could have been transported to another location, this would be circumstantial evidence that could not establish absolute proof that the defendant was at the scene of the crime. Whether the prosecution continued would depend upon how strong a case could be made on the cumulative evidence collected.

18

Transposable Genetic Elements

IMPORTANT CONCEPTS

  A. Transposable elements, or transposons, are present in the genomes of many different organisms. These move from one position to another in the genome and are responsible for inducing mutations and breaking chromosomes.
    1. Transposable elements are classified into three categories based upon how they transpose; cut-and-paste transposons, replicative transposons and retrotransposons.
        a. Cut-and-paste transposition is accomplished by excising an element from its position in a chromosome and inserting it into another position.
           (1) The excision and insertion of elements is catalyzed by transposase, which is encoded by the element.
        b. Replicative transposition is when an element's DNA is replicated and one copy of it is inserted at a new site and one copy remains at the original site.
           (1) A transposase encoded by the element mediates an interaction between the element and a potential insertion site.
        c. Replicative transposition is accomplished through a process that involves the insertion of copies of an element that were synthesized from the element's RNA.
           (1) Reverse transcriptase uses the element's RNA as a template to synthesize DNA molecules, which are then inserted into new chromosomal sites.
  B. In bacteria, there are three main types of transposons: the Insertion Sequences (*IS* elements), the composite transposons, which are formed when two *IS* elements capture unrelated DNA sequences between them, and the *Tn3* elements.
    1. All three types of transposons contain genes for proteins involved in transposition. In

addition, the composite transposons and the *Tn3* elements carry genes for antibiotic resistance.

C. In eukaryotes, there are many different families of transposons, including the *Ac/Ds* elements of maize and the *P* elements of *Drosophila*.
   1. Like bacterial transposons, these types of eukaryotic transposons have inverted DNA sequences at their termini and create target site duplications when they insert into DNA molecules.
   2. Some members of eukaryotic transposon families encode transposase, which catalyzes the movement of elements from one position to another.

D. Retroviruses have RNA as the genetic material and utilize an enzyme called reverse transcriptase, which allows these viruses to copy RNA into DNA. The DNA copy of their genome is integrated into the genome of the host.
   1. The retrovirus HIV is the human immunodeficiency virus that targets specialized cells of the immune system and causes AIDS, acquired immune deficiency syndrome.
      a. Because of its lethality and pandemic status, HIV/AIDS has been the focus of an enormous amount or research.
   2. The reproductive cycle of the HIV virus is well understood.
      a. HIV docks with target cell through an interaction between the viral protein gp120 and the cellular CD4 receptor protein.
      b. The viral and cellular membranes fuse, allowing the viral core to enter the cell.
      c. RNA and associated proteins are released from the viral core.
      d. Reverse transcriptase catalyzes the synthesis of double-stranded viral DNA from single-stranded viral RNA in the cytoplasm.
      e. Integrase catalyzes the insertion of viral DNA into the cellular DNA in the nucleus.
      f. Cellular RNA polymerase transcribes viral DNA into viral RNA.
      g. Some viral RNA serves as mRNA for the synthesis of viral proteins.
      h. Some viral RNA forms the genomes of progeny viruses.
      i. Progeny virus particles are assembled near the cellular membrane.
      j. Progeny virus particles are extruded from the cell by budding.
      k. The released progeny virus particles are free to infect other cells.

E. Eukaryotes possess different types of retrotransposons in their genomes.
   1. The movement of these elements depends on the reverse transcription of RNA into DNA by an enzyme encoded by the elements themselves.
   2. The retroviruslike elements resemble the integrated forms of retroviruses.
   3. The retroposons have a tract of A:T base pairs at one end.

F. At least 44 percent of the human DNA is derived from transposable elements.
   1. The most common transposable element is a retroposon called *L1* which belongs to a class of sequences known as Long Interspersed Nuclear elements, or LINEs.
   2. The Short Interspersed Nuclear Elements, or SINEs, are the second most abundant class of transposable elements in the human genome.
   3. Retroviruslike elements and DNA transposons also are present in the human DNA.
   4. The *L1* LINE and the *Alu* SINE are transpositionally active; the others are not.

G. In nature, transposons are responsible for a large fraction of all spontaneous mutations.

IMPORTANT TERMS

In the space allotted, concisely define each term.

cut-and-paste transposons:

replicative transposons:

retrotransposons:

insertion sequences (*IS* elements):

inverted terminal repeats:

transposase:

target site duplication:

composite transposons:

Tn3 elements:

cointegrate:

conjugative R plasmids:

resistance transfer factor (RTF):

R-determinant:

Ds (Dissociation):

Ac (Activator):

controlling elements:

hybrid dysgenesis:

P elements:

transposon tagging:

retrotransposon:

retrovirus:

reverse transcriptase:

human immunodeficiency virus (HIV):

acquired immune deficiency syndrome (AIDS):

long terminal repeates (LTRs):

endogenous retroviruses:

retroviruslike elements:

LTR retrotransposons:

Ty1 transposon:

retroposons:

HeT-A:

TART:

long interspersed nuclear elements (LINES):

short interspersed nuclear elements (SINEs):

ectotopic intrachromosomal exchanges:

ectotopic interchromosomal exchanges:

retroelements:

IMPORTANT NAMES

In the space allotted, concisely state the major contribution made by the individual or group.

Barbara McClintock:

Margaret and James Kidwell:

John Sved:

William Engels:

Michael Simmons and Johng Lim:

Donald Rio, Frank Laski, and Gerald Rubin:

David Baltimore, Howard Temin, and Satoshi Mizutani:

Hugh Robertson:

Mary Lou Pardue, Robert Levis, Harald Biessmann and James Mason:

Alan and Susan Kingsman:

Nancy Kleckner:

TESTING YOUR KNOWLEDGE

*In this section, answer the questions, fill in the blanks, and solve the problems in the space allotted. Problems noted with an * are solved in the Approaches to Problem Solving section at the end of the chapter.*

1. In *Drosophila,* a phenomenon called _____ occurs in progeny of M females mated to P males.

2. The simplest bacterial transposons, carrying inverted terminal repeats (9 to 40 nucleotide pairs), that can insert at many different sites and mutate genes are called _____.

3. The *IS* elements encode a protein called _____ that is necessary for excising the element from the chromosome or plasmid.

4. The staggered cleavage of the double-stranded DNA molecule during the insertion of an *IS* element results in a short direct repeat called _____.

5. A _____ results when two *IS* elements insert near each other and capture a DNA sequence between them.

6. The enzyme encoded by retrotransposons and retroviruses that catalyzes the formation of DNA from RNA is called _____.

7. The transposition of *Ty* elements occurs through an _____ intermediate.

8. The *Ds* element of maize cannot transpose unless a second element called _____ is present.

9. In maize, two structurally related transposons that can create mutation by inserting in or near a gene were named _____ and _____ by McClintock.

10. The procedure of cloning a transposon-mutated gene is called _____ .

11. The two major types of retrotransposons are _____ and _____.

12. Transposable elements, characterized by long terminal repeats which are oriented in the same direction, are called _____.

13. The only retroposons known to be active in the human genome are _____ and _____.

14. The best-studied retrovirus-like element in yeast is called _____.

15. A widely distributed class of retrotransposons that do not have inverted or direct repeats as integral parts of their termini, but instead have a homogeneous sequence of A:T base pairs at one end are called _____.

16. How do the *Ds* elements differ from the *Ac* elements of maize?

*17. A cross is made in maize between a female parent (lacking Ac elements) of the genotype *CC* with a male homozygous for a C^i allele (dominant inhibitor of color formation). The male parent contains a *Ds* element (but no *Ac* elements) proximal to the *C* locus on chromosome 9. What is the genotype and phenotype of the aleurone layer of the endosperm?

*18. A second cross is made between a female parent of the genotype *CC* (carrying *Ac* elements) with the same male as in the previous problem. What is the genotype and phenotype of the aleurone layer of the endosperm?

THOUGHT CHALLENGING EXERCISE

One way scientists are trying to control the HIV virus and treat AIDS is to target genes that regulate the HIV infection process and look for ways to disrupt the viral life cycle. Another way is to try to obtain a vaccine that would induce antibodies to destroy key proteins of the HIV infection process. Why has the production of a vaccine to protect against infection by the HIV virus been so difficult? You may want to do some additional reading on the replication properties of reverse transcriptase.

SUMMARY OF KEY POINTS

A cut-and paste transposon is excised from one genomic position and inserted into another by an enzyme, the transposase, which is usually encoded by the transposon itself.

A replicative transposon is copied during the process of transposition.

A retrotransposon produces RNA molecules that are reverse-transcribed into DNA molecules; these DNA molecules are subsequently inserted into new genomic positions.

Insertion sequences (IS elements) are cut-and-paste transposons that reside in bacterial chromosomes and plasmids.

Composite transposons consist of two IS elements flanking a region of that contains one or more genes for antibiotic resistance.

Tn3 is a replicative transposon that transposes by temporarily fusing DNA molecules into a cointegrate; when the cointegrate is resolved, each of the constituent DNA molecules emerges with a copy of *Tn3*.

Bacterial transposons are demarcated by inverted terminal repeats; when they insert into a DNA molecule, they create a duplication at the insertion site.

Conjugative plasmids can move transposons that contain genes for antibiotic resistance from one bacterial cell to another.

The maize transposable element *Ds*, discovered through its ability to break chromosomes, is activated by another transposable element, *Ac*, which encodes a transposase.

Transposable *P* elements are responsible for hybrid dysgenesis, a syndrome of germ-line abnormalities that occurs in the offspring of crosses between P and M strains of *Drosophila*.

The *mariner* transposons, which were originally discovered in *Drosophila*, are widespread in the animal kingdom, and appear to have spread horizontally among different species.

Retrovirus genomes are composed of single-stranded RNA comprising at least three genes: *gag* (coding for structural proteins of the viral particle), *pol* (coding for a reverse transcriptase/integrase protein), and *env* (coding for a protein imbedded in the virus' lipid envelope).

The human retrovirus HIV infects cells of the immune system and causes the life-threatening disease AIDS.

Retroviruslike elements possess genes homologous to *gag* and *pol*, but not to *env*.

Retroviruslike elements and the DNA forms of retroviruses inserted in cellular chromosomes are demarcated by long terminal repeat (LTR) sequences.

Retroposons lack LTRs; however, at one end they have a sequence of A:T base pairs derived from the reverse transcription of a poly(A) tail attached to the retroposon's RNA.

The retroposons *HeT-A* and *TART* are components of the ends of *Drosophila* chromosomes.

The human genome contains four basic types of transposable elements: LINEs, SINEs, retroviruslike elements, and cut-and-paste transposons.

The *L1* LINE and the *Alu* SINE are transpositionally active; other human transposons appear to be inactive.

Transposons are found in the genomes of many types of organisms.

Transposons are a major cause of mutations and chromosomal rearrangements.

Some transposons confer a selective advantage on the organisms that carry them; others are genetic parasites.

ANSWERS TO QUESTIONS AND PROBLEMS

1) hybrid dysgenesis **2)** *IS* elements **3)** transposase **4)** target site duplication **5)** composite transposon **6)** reverse transcriptase **7)** RNA **8)** *Ac* **9)** activator, dissociation **10)** transposon tagging **11)** retro-like elements and retroposons **12)** retro-like elements **13)** *L1* LINE and *Alu* SINE **14)** *Ty* transposon **15)** retroposons **16)** The *Ds* elements lack internal sequences that are present in *Ac*. Since *Ds* also lacks the ability to code for transposase, it requires *Ac* for transposition **17, 18)** see *Approaches to Problem Solving* section.

Some ideas concerning the thought challenging exercise:

A major difficulty in controlling the HIV virus is its very rapid rate of evolution and the occurrence of multiple strains. This is due, in part, to the fact that reverse transcriptase does not have proofreading capabilities. Therefore, mistakes made during the synthesis of DNA from RNA are not corrected and appear as mutations. The occurrence of multiple strains complicates the development of vaccines. Another problem in the development of vaccines is that in the intact HIV virus, the targeted key antigens (proteins) may not be exposed where they can be recognized by antibodies.

APPROACHES TO PROBLEM SOLVING

17. The cross *CC* (female) X *C^iC^i* male will produce triploid endosperm with the genotype *CC C^i*, because the endosperm is formed from the fusion of the two polar nuclei of the female gametophyte with a sperm nucleus. There is a *Ds* element proximal to *C^i* on chromosome 9, but no *Ac* element is present to activate it. Therefore the endosperm will be colorless because the *C^i* allele is dominant to the *C* allele.

18. In this cross, the genotype of the endosperm is the same as above, i.e., *CC C^i*. However, the *Ac* element enters with the female chromosomes. This *Ac* element can allow *Ds* to leave its site on

chromosome 9 and transpose to a new site. When this occurs, the chromosome may break at the Ds site, losing the chromosomal region distal to the break. In this case, the lost chromatin contains the C^i allele. When the C^i allele is lost the color allele, C, will be expressed. The endosperm will be primarily colorless, but will have colored spots or stripes.

19

The Genetics of Mitochondria and Chloroplasts

IMPORTANT CONCEPTS

A. The transmission of mitochondria and chloroplasts from cell to cell is through the cytoplasm.
    1. During sexual reproduction, these organelles are preferentially inherited through the cytoplasm of the gametes of one sex, which in higher eukaryotes is usually the female. This results in inheritance patterns of mitochondria and chloroplasts characterized by unequal contributions of the two parents and by an irregular segregation of alleles.
        a. Examples of these non-Mendelian patterns are the inheritance of leaf variegation in ornamental plants, cytoplasmic male-sterility in maize, antibiotic resistance in *Chlamydomonas* and metabolic defects in yeast.
            (1) All these traits appear to be controlled by genes located in either mitochondrial (mt) or chloroplast (cp) DNA.
B. The RNAs and proteins of mitochondria are encoded by genes found either in nuclear or mitochondrial DNA.
    1. Mitochondrial DNA molecules vary in size and structure.
        a. The circular mtDNA molecules encode some of the RNAs and polypeptides used within the mitochondrion, including proteins involved in aerobic metabolism.
        b. Most mitochondrial proteins are encoded by nuclear genes.
    2. In some organisms, the transcripts of mtDNA are spliced to remove introns and edited to alter nucleotide sequences. In others, mitochondrial mRNAs are formed by trans-splicing exons of different transcripts.
    3. The translational machinery within mitochondria employs a genetic code that is slightly different from that used in the cytosol.
    4. Mutations in mtDNA are responsible for at least two human diseases, Leber's hereditary optic neuropathy and Pearson marrow-pancreas syndrome.
C. Chloroplast DNA molecules also vary in size and structure, and most are circular.

1. Chloroplast DNA molecules encode some of the RNAs and proteins used within the chloroplast, including some proteins that are involved in photosynthesis.
D. Both mitochondria and chloroplasts apparently originated when prokaryotic organisms were internalized by primitive eukaryotic cells.
    1. In the course of evolution, these endosymbionts have exchanged genes with the nucleus.
    2. Mitochondria and chloroplasts cannot exist by themselves and are dependent on nuclear genes.

IMPORTANT TERMS

In the space allotted, concisely define each term.

chloroplasts:

mitochondria:

endosymbiosis:

heteroplasmy:

homoplasmy:

maternal inheritance:

non-Mendelian, biparental inheritance:

petite mutants:

mitochondrial DNA (mtDNA):

editing:

trans-splicing:

Leber's hereditary optic neuropathy (LHON):

Pearson marrow-pancreas syndrome:

plastids:

chloroplast DNA (cpDNA):

biogenesis:

phytochromes:

symbiosis:

IMPORTANT NAMES

In the space allotted, concisely state the major contribution made by the individual or group.

Carl Correns and Erwin Baur:

Ruth Sager:

Boris Ephrussi:

Lynn Margulis:

Allan Wilson:

Matthias Krings:

TESTING YOUR KNOWLEDGE

In this section, answer the questions, fill in the blanks, and solve the problems in the space allotted. Problems noted with an * are solved in the Approaches to Problem Solving section at the end of the chapter.

1. The presence of a single genotype of an organelle within a cell is called _____.

2. The exception given in the text to the typical inheritance of chloroplasts only through the female gametophyte in plants is _____.

3. The presence of two organelle genotypes within a cell is called _____.

4. Mitochondrial mutations in yeast that result in tiny colonies when grown on rich glucose-containing medium are called _____ mutants.

5. In humans, a condition characterized by the sudden onset of blindness in adults due to mitochondrial mutations that reduce the efficiency of oxidative phosphorylation is called _____.

6. A frequently lethal human disorder caused by a large deletion of the mtDNA and loss of bone marrow cells during childhood is called _____.

7. A CO_2 fixing enzyme that has one subunit coded by a nuclear gene and the other subunit coded by a chloroplast gene is called _____.

*8. Name, with examples, the types of genes encoded by mtDNA.

*9. Name, with examples, the types of genes encoded by cpDNA.

*10. What is the molecular basis of suppressive petite mutants and neutral petite mutants of yeast?

*11. Name three peculiarities of plant mitochondrial gene expression.

*12. Distinguish between a maternal effect and maternal inheritance.

*13. Discuss observations suggesting that genes may have moved from chloroplasts to the nucleus and from chloroplast to mitochondria in higher plants?

*14. Why is the genetic code not completely universal?

*15. How would you determine whether a variegated plant is due to mutants of chloroplasts or mutant nuclear genes?

*16. In the snail, *Limnaeae peregra,* shell coiling may be to the left (dextral) or to the right (sinistral). When snails from a sinistral line (female parent) are crossed with snails from a dextral line (male parent), the F_1 progeny all had sinistral shells. The F_2 progeny produced by self-fertilization of the F_1s were all sinistral. When self-fertilized, about 3/4 of the F_2 snails produced progeny with sinistral shells and 1/4 of the F_2 snails produced progeny with dextral shells. In the reciprocal cross, dextral female X sinistral male, all the F_1 progeny were dextral. All the F_2 progeny were sinistral. When self-fertilized, about 3/4 of the F_2 snails produced progeny with sinistral shells and 1/4 of the F_2 snails produced progeny with dextral shells. Is the direction of shell coiling in snails due to cytoplasmic inheritance? Why or why not?

THOUGHT CHALLENGING EXERCISE

A geneticist crossed two independently derived variegated *Pelargonium zonale* plants. In one of the variegated progeny, a dark green branch occurred from a pure white region of the plant. Seed obtained from flowers of this dark green branch produce dark green plants that in turn bred true for dark green. The geneticist then isolated chloroplast DNA from dark green regions and white regions of the plants and digested it with a variety of restriction endonucleases. It was established that the cpDNA of dark green regions of the parental variegated plants had similar restriction patterns, but the white regions of the two variegated plants each had a distinct restriction pattern. The DNA isolated from the white sectors of the hybrid plants possessed a restriction pattern that was additive of the white regions of the parental plants. The dark green plants derived from the white sector of the hybrid variegated plant had a DNA restriction pattern identical to the dark green regions of the parental plants. Explain these observations and the origin of the chloroplast DNA of derived dark green plants.

SUMMARY OF KEY POINTS

Organelle heredity is characterized by unequal parental contributions and by irregular phenotypic segregations.

The non-Mendelian features of organelle heredity are due, in part, to the preferential transmission of chloroplasts or mitochondria through the gametes of one sex, which in higher eukaryotes is usually the female.

Mitochondrial DNA (mtDNA) molecules range from 6 kb to 2500 kb in size, and most appear to be circular.

Mitochondrial DNA molecules contain genes for some of the ribosomal RNAs, transfer RNAs, and polypeptides used within the mitochondrion.

The structure, organization, and expression of mitochondrial genes vary among species.

In some organisms, the transcripts of mitochondrial genes are edited after they are synthesized.

Both mitochondrial and nuclear gene products are needed for normal mitochondrial function.

Leber's hereditary optic neuropathy and Pearson marrow-pancreas syndrome are due to mutations of mtDNA.

Chloroplast DNA (cpDNA) molecules are typically 120 to 292 kb in size, and they contain at least 100 genes.

The organization of cpDNA molecules varies among species of plants and algae.

Light induces chloroplasts to develop from unpigmented plastids through a process that involves the interplay of chloroplast and nuclear gene products.

Mitochondria and chloroplasts seem to have originated as bacteria that were incorporated into eukaryotic cells about a billion years ago.

As the endosymbiotic bacteria evolved into organelles, many of their genes were lost; other genes were transferred into the host cell's genome or into the genome of another endosymbiotic bacterium.

Organelle DNA, especially mtDNA, has been very useful in studying the evolution of different groups of organisms.

ANSWERS TO QUESTIONS AND PROBLEMS

1) homoplasmy 2) leaf variegation in *Pelargonium zonale* 3) heteroplasmy 4) petite 5) Leber's hereditary optic neuroplasmy 6) Pearson marrow-pancreas syndrome 7) ribulose 1, 5-bisphosphate carboxylase 8-16) see *Approaches to Problem Solving* section.

Ideas concerning thought challenging exercise:

The results suggest that the white region of the hybrid variegated plant likely had mutant chloroplasts derived from each parent. The dark green genotypes formed in the white region of the variegated plant may have resulted from genetic recombination between the DNA of two mutant chloroplasts.

APPROACHES TO PROBLEM SOLVING

8. Mitochondrial genes encode some of the RNAs and polypeptides that are used in the mitochondria. These include two to three types of ribosomal RNA, up to 25 different tRNAs and several

polypeptides, e.g., some of the ribosomal structural proteins and several enzymes involved in aerobic metabolism, e.g., subunits of cytochrome c oxidase.

9.   All cpDNA molecules carry the same basic set of genes including those for rRNA, some ribosomal proteins, various polypeptide components of the photosystems that are involved in capturing solar energy, the catalytically active subunit of 1, 5-bisphosphate carboxylase, and four subunits of a chloroplast-specific RNA polymerase.

10. Neutral petite mutants lack mitochondrial DNA. Suppressive petite mutants have a smaller DNA molecule that is relatively A:T-rich.

11. Mitochondrial DNA is organized into many separate transcriptional units, some containing information for more than one gene. RNA editing alters the composition of codons in some plant mitochondrial transcripts; some mitochondrial mRNAs are formed by the process of trans-splicing.

12. A maternal effect is when the genes of the mother are determining a phenotype of the progeny. Maternal inheritance is when genes (generally in organelle DNA) are inherited through the cytoplasm of the egg.

13. Several lines of evidence suggest that genes have transposed among the nucleus, chloroplasts, and mitochondria. This evidence includes (1) mtDNA in some higher plants contains a tRNA gene from cpDNA, (2) in one plant species, the mtDNA contains part of the chloroplast gene for the large subunit of 1, 5-bisphosphate carboxylase, and (3) in most plant species, the small subunit of 1, 5-bisphosphate carboxylase is encoded by a nuclear gene, but in some algae, it is encoded by a chloroplast gene.

14. In mammalian mitochondrial mRNA translation, AGA and AGG are termination codons, whereas in the cytosol they specify the incorporation of arginine into a polypeptide. The termination codon (UGA) in cytosol translation functions as a tryptophan codon in the mitochondria. The codon, AUA, codes for isoleucine in the cytosol, but is the methionine initiation codon in the mitochondria.

15. One would determine this by performing reciprocal crosses between variegated and green plants. If the variegation is due to mutants of the chloroplast, the trait will be maternally inherited. If the trait is determined by a nuclear gene, then some form of dominance will be displayed in the F_1 progeny. Segregation of the trait will be apparent in the F_2 generation, regardless of the direction in which the cross was made.

16. The data in this example on the direction of shell coiling do not adhere to a pattern of true cytoplasmic inheritance, because segregation of the traits is apparent. The results can be interpreted as conforming to what is called a maternal effect, i.e., the genotype of the mother determines a phenotype of the progeny. The crosses can be genetically diagrammed as follows:

P	*DD* (female) X *dd* (male)			P	*dd* (female) X *DD* (male)		
	sinistral dextral				dextral sinistral		
F_1	*Dd*			F_1	*Dd*		
	sinistral				dextral		
F_2	all sinistral			F_2	all sinistral		
	1/4 *DD*	1/2 *Dd*	1/4 *dd*		1/4 *DD*	1/2 *Dd*	1/4 *dd*
	↓	↓	↓		↓	↓	↓
F_3	*DD*	3/4 *D_* 1/4 *dd*	*dd*	F_3	*DD*	3/4 *D_* 1/4 *dd*	*dd*
	sinistral	sinistral	dextral		sinistral	sinistral	dextral

The sinistral (*D*) allele is dominant to the dextral (*d*) allele. The phenotypic ratio lags the genotypic ratio by one generation due to the maternal effect.

20

Regulation of Gene Expression in Prokaryotes and Their Viruses

---

IMPORTANT CONCEPTS

  A. Gene expression is frequently under the control of regulator genes.
    1. Regulator genes act primarily at the level of transcription, often affecting the ability of RNA polymerase to bind to promoter sequences.
    2. The effects of regulator gene products, in turn, are controlled by the presence or absence of specific effector molecules in the environment.
    3. In prokaryotes, genes with related functions are frequently present in coordinately regulated units called operons.
      a. Each operon is one unit of transcription, i.e., a single mRNA carries the coding sequences of all the genes in the operon.
    4. Certain regulatory genes encode proteins called repressors, which function by means of their sequence-specific binding to DNA.
      a. Some repressors (those for repressible operons) bind to DNA only in the presence of effector molecules called co-repressors; others (those for inducible operons) bind to DNA only in the absence of effector molecules called inducers.
      b. Repressors are allosteric proteins, i.e., proteins that undergo conformational shifts and correlated changes in activity in response to the binding of specific effector molecules.
        (1) Repressors bind to their DNA binding sites in one conformation, but not in other conformations.
      c. Repressors act by binding to DNA sequences called operators, which are located adjacent

to the structural genes whose transcription they control.

        (1) When a repressor, or the complex of repressor and co-repressor, is bound to the operator sequence of an operon, it prevents RNA polymerase from binding to the contiguous promoter sequence and initiating the transcription of the operon.

   5. Operons containing genes that encode enzymes involved in catabolic pathways often are under positive control by the catabolite activator protein (CAP) and cyclic AMP (cAMP).

      a. The binding of the CAP-cAMP complex to a site within the promoter region of an operon is required for the efficient binding of RNA polymerase to its binding site in the promoter region.

        (1) High concentrations of glucose result in low intracellular concentrations of cAMP.

        (2) Such operons cannot be induced in the presence of high concentrations of glucose, a phenomenon known as catabolite repression.

   6. Operons controlling enzymes involved in amino acid biosynthetic pathways are frequently controlled by attenuation.

      a. Attenuation occurs by premature termination of transcription at an attenuator site located within the mRNA leader sequence.

        (1) Attenuator sites contain transcription-termination sequences; however, the ability of the attenuator RNA sequences to form the secondary structures that lead to termination of transcription depends on the presence of the amino acid that is the end-product of the pathway controlled by the operon in question.

B. A temperate bacteriophage such as lambda can follow either the lytic cycle, during which it reproduces and lyses the host cell, or the lysogenic cycle, during which its chromosome exists as a dormant prophage inserted into the chromosome of the host bacterium.

   1. During lysogeny, the lytic genes of the prophage are kept turned off by a repressor-operator-promoter-circuit similar to those of bacterial operons.

   2. The decision between lysogeny and lytic development involves an elegant genetic switch that is controlled by two regulatory proteins, λ repressor and Cro protein, which bind to operator sites that control transcription of the λ genome.

      a. If repressor is bound at these sites, lysogeny ensues.

        (1) The lysogenic state is maintained by the autogenous control of the λ repressor.

      b. If Cro protein occupies these sites, lytic development occurs.

        (2) Lytic development is controlled by a regulatory cascade in which transcriptional antiterminators play a central role.

C. Preprogrammed sequences of viral gene expression occur in bacteriophage-infected cells.

   1. The products of genes expressed early after infection interact with RNA polymerase to change its promoter specificity and switch transcription to a second set of genes.

      a. In the case of some of the more complex bacteriophages, this switching process may be repeated several times during the life cycle of the virus.

D. Translational control of gene expression may occur by (1) unequal efficiencies of translational initiation, (2) altered efficiencies of ribosome movement , and (3) differential rates of mRNA degradation.

E. Posttranslational regulatory mechanisms include feedback inhibition or end-product inhibition.

IMPORTANT TERMS

In the space allotted, concisely define each term.

constitutive genes:

induction:

inducible genes:

inducible enzymes:

catabolic pathways:

repression:

derepression:

anabolic pathways:

regulator gene:

positive control mechanisms:

negative control mechanisms:

activators:

repressors:

effector molecules:

inducers:

co-repressors:

allosteric transitions:

operon model:

operator:

operon:

catabolite repression (glucose effect):

CAP (catabolite activator protein):

cyclic AMP (cAMP):

sequence-specific protein-nucleic acid interactions:

attenuation:

attenuator:

transcription-termination signals:

DNA binding domain:

dimerization domain:

connector region:

N protein:

Q protein:

Cro protein:

transcriptional antiterminator:

negative self-regulation:

negative autogenous regulation:

feedback (end-product) inhibition:

end-product or regulatory binding site:

IMPORTANT NAMES

In the space allotted, concisely state the major contribution made by the individual or group.

Felix d'Herelle:

Frederick W. Twort:

Francois Jacob and Jacques Monod:

Charles Yanofsky:

TESTING YOUR KNOWLEDGE

In this section, answer the questions, fill in the blanks, and solve the problems in the space allotted. Problems noted with an * are solved in the Approaches to Problem Solving section at the end of the chapter.

1. β-galactosidase cleaves lactose to form _____ and _____.

2. In inducible bacterial operons the inactivation of induction by elevated levels of glucose, even though the inducer is present, is called _____.

3. The two individuals who received a Nobel Prize for their research leading to the operon model of gene regulation in bacteria were _____ and _____.

4. Name the two protein complexes that bind to the promoter of the *lac* operon.

5. What would be the effect of a deletion of the promoter of the *lac* operon?

6. What would be the effect of a deletion of the promoter of the *i* gene of the *lac* operon?

7. Whether or not a repressor will bind to the operator and turn off the transcription of the structural genes in the operon is determined by the presence or absence of molecules called _____.

8. The enzymes produced by the coordinate regulation of the three genes in the *lac* operon are _____, _____, and _____.

9. The effector molecule of the *lac* operon is _____.

10. To bind to the promoter of the *lac* operon, the catabolite activator protein must be complexed with _____.

11. The best understood repressible operon of *E. coli* is the _____operon.

12. A nucleotide sequence in the 5′ region of a prokaryotic gene (or in its RNA) that causes premature termination of transcription is called a (an) _____.

13. In the *trp* operon, the co-repressor is _____.

14. Which pathway, lysogeny or lytic development will occur if the C_1 repressor is bound to both the O_L and O_R operator regions of the λ prophage?

15. For the λ bacteriophage, the determination of lysogeny or lytic growth depends upon a balance between two proteins called _____ and _____.

16. The inhibition of translation of an mRNA molecule by one of the products that it encodes is called _____.

17. The inhibition of the activity of an enzyme by an end product of the biosynthetic pathway is called

   _____.

18. The enhancement of enzyme activity by a substrate or other effector molecule is called _____.

*19. For each of the following merozygotes indicate the conditions under which β-galactosidase is produced.

   (a) $i^-p^+o^+z^+ / i^+p^+o^+z^-$
   (b) $i^+p^+o^+z^+ / i^+p^+o^cz^+$
   (c) $i^+p^+o^cz^- / i^+p^+o^+z^+$
   (d) $i^sp^+o^+z^+ / i^+p^+o^+z^+$
   (e) $i^sp^+o^+z^+ / i^+p^+o^cz^+$

THOUGHT CHALLENGING EXERCISE

You are given a *lac⁻* mutant of *E. coli*. How could you determine if the uninducible phenotype is due to a defective repressor (i^s), a defective promoter, or to a defective structural gene?

SUMMARY OF KEY POINTS

In prokaryotes, genes that specify housekeeping functions such as rRNAs, tRNAs, and ribosomal proteins are expressed constitutively. Other genes are expressed only when their products are needed.

Genes that encode enzymes in catabolic pathways often are expressed only in the presence of the substrates of the enzymes; their expression is inducible.

Genes that encode enzymes involved in anabolic pathways usually are turned off in the presence of the end product of the pathway; their expression is repressible.

Although gene expression can be regulated at many levels, transcriptional regulation is the most common.

Gene expression is controlled by both positive and negative regulatory mechanisms.

In positive control mechanisms, the product of a regulatory gene, an activator, is required to turn on the expression of the structural gene(s).

In negative control mechanisms, the product of a regulator gene, a repressor, is necessary to turn off the expression of the structural gene(s).

Activators and repressors regulate gene expression by binding to sites adjacent to the promoters of structural genes.

Whether or not the regulator proteins can bind to their binding sites depends on the presence or absence of small effector molecules that form complexes with the regulator proteins.

The effector molecules are called inducers in inducible systems and co-repressors in repressible systems.

In bacteria, genes with related functions frequently occur in coordinately regulated units called operons.

Each operon contains a set of contiguous structural genes, a promoter (the binding site for RNA polymerase) and an operator (the binding site for a regulatory protein called a repressor).

When a repressor is bound to the operator, RNA polymerase cannot transcribe the structural genes in the operon. When the operator is free of repressor, RNA polymerase can transcribe the operon.

The *E. coli lac* operon is a negative inducible and catabolite repressible system; the three structural genes in the *lac* operon are transcribed at high levels only in the presence of lactose and the absence of glucose.

In the absence of lactose, the *lac* repressor binds to the *lac* operator and prevents RNA polymerase from initiating transcription of the operon.

Catabolic repression keeps operons such as *lac* encoding enzymes involved in carbohydrate catabolism from being induced in the presence of glucose, the preferred energy source.

The *E. coli trp* operon is a negative repressible system; transcription of the five structural genes in the *trp* operon is repressed in the presence of significant concentrations of tryptophan.

Operons such as *trp* that encode enzymes involved in amino acid biosynthetic pathways often are controlled by a second regulatory mechanism called attenuation.

Attenuation occurs by the premature termination of transcription at a site in the mRNA leader sequence (the sequence 5' to the coding region) when tryptophan is prevalent in the environment in which the bacteria are growing.

Temperate bacteriophages can follow either of two pathways: (1) lytic growth, during which they reproduce and kill the host cells, or (2) lysogeny, during which their chromosomes exist as a dormant prophages covalently inserted in the chromosomes of the host bacteria.

During lysogeny, the lytic genes of the prophage are kept turned off by a repressor-operator-promoter circuit similar to those of bacterial operons.

Whether a lambda prophage will enter the lysogenic state or undergo lytic development is determined by which of two regulatory proteins, λ repressor or Cro protein, occupies key operator sites that control transcription of the λ genome.

The lysogenic state is maintained by the autogenous control of λ repressor, whereas lytic development is controlled by a regulatory cascade in which transcriptional antiterminators play a major role.

Preprogrammed temporal sequences of viral gene expression occur in bacteriophage-infected cells.

The first viral genes expressed in an infected cell are transcribed by unmodified bacterial RNA polymerase.

Subsequent sets of expressed viral genes are transcribed either by an RNA polymerase encoded by the phage genome or by bacterial RNA polymerase modified by the addition of viral protein(s).

Regulatory fine-tuning frequently occurs at the level of translation by modulation of the rate of polypeptide chain initiation or chain elongation.

Sometimes gene expression is regulated by the differential degradation of specific regions of polygenic mRNAs.

Feedback inhibition occurs when the product of a biosynthetic pathway inhibits the activity of the first enzyme in the pathway, rapidly arresting the biosynthesis of the product.

Enzyme activation occurs when a substrate or other effector molecule enhances the activity of an enzyme, increasing the rate of synthesis of the product of the biosynthetic pathway

ANSWERS TO QUESTIONS AND PROBLEMS

1) galactose and glucose **2)** catabolite repression **3)** Jacob and Monod **4)** RNA polymerase and cAMP-CAP **5)** uninducibility of the operon **6)** constitutive synthesis of *lac* operon proteins **7)** effectors **8)** β-galactosidase, β-galactoside permease, β-galactoside transacetylase **9)** allolactose **10)** cyclic AMP **11)** tryptophan **12)** attenuator **13)** tryptophan **14)** lysogeny **15)** λ repressor and Cro protein **16)** negative self-regulation or negative autogenous regulation **17)** feedback (end-product) inhibition **18)** substrate activation **19)** see *Approaches to Problem Solving* section below.

Some ideas concerning the thought challenging exercise:

First of all, it would be best to assay for the inducibility of the other two genes of the operon, i.e., permease and transacetylase. If the other two genes are inducible, but the i gene is not, then the z gene is defective. Then, if all three structural genes of the operon were uninducible, it could be due to either a mutation of the repressor gene or to a mutant promoter. It could be determined if the lac^- mutant was defective in the i gene by forming a merozygote between the mutant and an $i^- p^+ o^+ z^+$ cell. If the production of β-galactosidase is lacking in the merozygote and cannot be induced, then the mutant is most likely a mutant of the i gene, i.e., i^s. If all three genes of the operon are uninducible, and it is not due to a defective regulator gene, then the mutant is either a defective promoter or a mutation or deletion of the structural gene(s). The simplest approach to distinguish between these two possibilities would be to genetically map the position of the mutant gene.

APPROACHES TO PROBLEM SOLVING

19. The key to solving problems involving operons is first to fully learn how the different components interact. The operon model is presented in Figure 20.3 of the text.

(a) $i^- p^+ o^+ z^+ / i^+ p^+ o^+ z^-$

In this merozygote, the DNA strand indicated on the left has a mutant regulator gene i^-, and the other genes are wild-type. This would show constitutive synthesis of β-galactosidase, but i^+ is dominant to i^-; it produces a normal repressor that will fuse to o^+ unless lactose is added. The DNA strand on the right does not produce β-galactosidase because of the mutant gene z^-. This merozygote produces β-galactosidase only inducibly.

(b) $i^+ p^+ o^+ z^+ / i^+ p^+ o^c z^+$

In this merozygote, all genes are wild-type except for the operator gene o^c of the DNA strand indicated on the right. Due to the o^c gene, this merozygote will produce β-galactosidase constitutively. Since the DNA on the left is normal, it can be induced by lactose. Therefore, although the cell has constitutive production of β-galactosidase, the amount of β-galactosidase is even higher in induced cells.

(c) $i^+p^+o^cz^-$ / $i^+p^+o^+z^+$

The DNA strand on the left has an o^c gene and a z^- gene. No β-galactosidase will ever be produced following transcription of this strand. Remember, operator genes are binding sites, and therefore can only affect structural genes of the operon in the cis position. The DNA on the right is wild-type. It will produce β-galactosidase when induced.

(d) $i^s p^+o^+z^+$ / $i^+p^+o^+z^+$

This merozygote is normal except for the presence of a mutant gene i^s that produces a superrepressor. The superrepressor cannot be removed from the operator gene by the inducer and therefore this merozygote is permanently suppressed.

(e) $i^s p^+o^+z^+$ / $i^+p^+o^cz^+$

This merozygote has a superrepressor mutant of the i gene in the DNA on the left. The operon contained in this DNA molecule is permanently suppressed. However, the superrepressor cannot bind to the mutant operator o^c of the DNA strand on the right. The constitutive operator gene is epistatic to the i^s mutant. Therefore, this strand is transcribed all of the time and the production of β-galactosidase is constitutive.

21

Regulation of Gene Expression in Eukaryotes

IMPORTANT CONCEPTS

    A. Gene expression is spatially and temporally regulated in multicellular eukaryotes; it can occur in the nucleus or in the cytoplasm.

        1. In the nucleus, gene expression can be regulated at transcription or by RNA processing.

        2. In the cytoplasm, it can be regulated by mRNA stability.

        3. The transcription of some eukaryotic genes is induced by environmental stimuli such as heat shock or light.

        4. Other eukaryotic genes are transcribed in response to signaling molecules such as hormones and growth factors.

a. These factors are mediated by protein transcription factors that bind to enhancer or silencer sequences located in the vicinity of a gene.
   (1) Transcription factors have characteristic structural motifs such as the zinc finger, leucine zipper, helix-turn-helix, or helix-loop-helix.
5. Alternate splicing of eukaryotic gene transcripts provides a way for a single gene to encode several different polypeptides.
6. The stability of an mRNA depends on sequences in the 5′ untranslated region and on polyadenylation at the 3′ end.

B. Posttranscriptional mechanisms play important roles in regulating gene expression of eukaryotic cells.
1. Short interfering (siRNAs) and microRNAs (miRNAs) may regulate the expression of eukaryotic genes by interacting with messenger RNAs produced by these genes. This type of posttranscriptional gene regulation is called RNA interference (RNAi) and is mediated by the RNA-Induced Silencing Complex (RISC).
   a. Large double stranded RNA molecules are diced into small double-stranded interfering RNAs by dicer enzymes.
   b. Some of the small RNA molecules that induce RNAi are derived from transcripts of microRNA genes.
   c. RNAs that induce RNAi may also be derived from the transcription of other elements in the genome such as transposons, transgenes, and RNA viruses.
2. In RISCs, siRNAs and miRNAs become single-stranded and pair with complementary sequences in messenger RNA molecules.
   a. If the base pairing between the RNA within RISC and the target sequence in the mRNA is perfect or nearly so, the RISC cleaves the target mRNA and it is degraded.
      (1) RISC-associated RNAs that result in mRNA cleavage are called short interfering RNAs.
   b. If the RNA within the RISC pairs imperfectly with its target sequence, the mRNA is usually not cleaved; but translation of the mRNA is inhibited.
      (1) RNA within a RISC that inhibits translation is called microRNA.

C. Eukaryotic gene expression is influenced by chromosome organization and chromatin remodeling.
1. Transcription typically occurs in open chromatin configurations such as amphibian lampbrush chromosome loops or dipteran polytene chromosome puffs.
2. Transcriptionally active DNA is more sensitive to digestion with the enzyme DNase I.
3. The modification of nucleosomes by multi-protein complexes is called chromatin remodeling.
   a. Two general types of chromatin-remodeling complexes have been identified that prepare chromatin for transcription; those that transfer acetyl groups to the amino acid lysine at specific positions in the histones of the nucleosomes, and those that disrupt nucleosome structure in the vicinity of gene promoters.
   b. Chromatin can be remodeled into inactive chromatin by deacetylation and methylation of histones in nucleosomes, and by methylation of nucleotides in DNA.
4. Chromosomes may be organized into domains of gene expression by specialized chromatin structure.
   a. The transcription of eukaryotic genes that have been transposed into heterochromatin is repressed, causing the phenomenon called position-effect variegation.
   b. In a eukaryotic cell, most genes are transcriptionally silent.
      (1) In *Drosophila*, the *Polycomb Group* (*PcG*) genes are involved in gene silencing by encoding proteins that regulate the expression of transcription factors.
      (2) In yeast, the mating type locus (*MAT*) is involved with specific gene silencing.
      (3) In trypanosomes, a form of gene silencing controls the expression of the variant surface glycoprotein (*vsg*) genes.
   c. Chemical modification of DNA, i.e., methylation, is associated with gene silencing in mammals.
      (1) Methylation may lead to imprinting or the expression of a gene that is conditioned by its parental origin.
   d. Some genes such as those encoding the ribosomal RNAs in amphibian oocytes are amplified to increase their expression.

e. Large arrays of genes on an entire X chromosome are regulated to achieve equal X-linked gene expression in males and females.
   (1) These genes may be coordinately inactivated, hyperactivated or hypoactivated by RNAs or proteins that associate with the X chromosome.

IMPORTANT TERMS

In the space allotted, concisely define each term.

pseudogenes:

transcription factors:

alternate splicing:

inducer:

heat-shock proteins:

heat-shock transcription factor (HSTF):

heat-shock response elements (HSEs):

hormone:

steroid hormone:

hormone receptors:

peptide hormones:

membrane-bound receptor proteins:

signal transduction:

hormone response elements (HREs):

basal transcription factors:

special transcription factors:

enhancers:

zinc finger:

helix-turn-helix:

homeodomain:

leucine zipper:

helix-loop-helix:

RNA interference (RNAi):

dicer:

RNA-induced silencing complex (RISC):

short interfering RNAs:

microRNAs

chromomeres:

puffs:

ecdysone:

DNase I hypersensitivity sites:

locus control region (LCR):

chromatin remodeling:

heterochromatin:

euchromatin:

position-effect variegation:

silencers:

expression site (ES):

CpG-islands:

imprinted:

gene amplification:

nucleolar organizer:

dosage compensation:

X inactivation center (XIC):

IMPORTANT NAMES

In the space allotted, concisely state the major contribution made by the individual or group.

Pamela Geyer and Victor Corces:

Su Guo and Kenneth Kemphues:

M. Groudine and H. Weintraub:

A. Fire, S. Xu, M. K. Montgomery, S. A. Kostas, S. E. Driver, and C. C. Mello:

TESTING YOUR KNOWLEDGE

*In this section, answer the questions, fill in the blanks, and solve the problems in the space allotted. Problems noted with an * are solved in the Approaches to Problem Solving section at the end of the chapter.*

1. The tissue-specific expression of genes results from _____ regulation.

2. The _____ regulation of genes results in different genes being expressed at different times during development.

3. Positive and negative regulator proteins that bind to specific regions of the DNA and stimulate or inhibit transcription are called _____.

4. A mutant, nonfunctional duplicate of a normal gene is called a (an) _____.

5. The different rat troponin T proteins result from a phenomenon called _____.

6. Name the three levels at which eukaryotic gene expression can be regulated.

7. In animals the small lipid-soluble molecules derived from cholesterol, e.g., estrogen, progesterone, and testosterone, form a class of hormones called _____.

8. For estrogen to enter the nucleus and act as a transcription factor it first must bind to a cytoplasmic protein called a (an) _____.

9. A class of hormones consisting of polypeptides, e.g., insulin, is called _____.

10. The process of transmitting a hormonal signal through the cell and into the nucleus is called _____.

11. Proteins that bind to a sequence within a promoter and facilitate the proper alignment of the RNA polymerase are called _____.

12. What are specific DNA sequences that mediate hormone-induced gene expression called?

13. Peptide hormones are transmitted through the cell membrane by _____.

14. DNA sequences that can act over relatively large distances and influence gene regulation independent of orientation and position are called _____ and _____.

15. A structural motif found in some eukaryotic transcription factors that plays an important role in DNA binding, consisting of a protein loop formed when two cysteines in one part of the polypeptide and two histidines in another part nearby jointly bind to a zinc ion, is called a (an)_____.

16. The highly conserved region of approximately 60 amino acids in many transcription factors that coincides with a helix-turn-helix motif is called a _____.

17. A steroid hormone that has been demonstrated to induce puffing of polytene chromosome bands in *Drosophila* is called _____.

18. A mutant that results in a body part developing in an inappropriate position is called a (an) _____ mutant.

19. Highly compacted, dark staining chromatin with few genes and generally rich in repetitive DNA is called _____.

20. A structural motif found in some transcription factors that consists of a stretch of amino acids with leucine at every seventh position is called a _____.

21. The conditioning of the expression of a gene by its parental origin is called _____.

22. The region of eukaryotic chromosomes that contains numerous 18S-28S ribosomal RNA genes is called the _____.

23. The phenomenon of equalizing the activity of X-linked genes in the two sexes of organisms with XX/XY or XX/XO sex determination systems is called _____.

THOUGHT CHALLENGING EXERCISE

How could it be cytologically determined if the transcriptionally-active regions of chromosomes involved in puffs on polytene chromosomes and loops on lampbrush chromosomes are from regions containing protein-coding genes?

SUMMARY OF KEY POINTS

The α and β tubulin genes in *Arabidopsis* are expressed in a tissue-specific manner.

The α and β globins in vertebrates are expressed at different times during development.

Proteins called transcription factors interact with DNA to control the transcription of eukaryotic genes.

Eukaryotic gene transcripts may be alternatively spliced to produce messenger RNAs that encode distinct, but related, polypeptides.

The stability of eukaryotic messenger RNAs can influence the level of polypeptide synthesis.

Transcription of the *hsp70* genes in response to increased temperature is mediated by a heat-shock transcription factor.

Transcription of the gene for the photosynthetic enzyme ribulose 1,5-bisphosphate carboxylase (RBC) is induced by exposure to light.

Steroid hormones and their receptor proteins form complexes that act as transcription factors to regulate the expression of specific genes.

Peptide hormones interact with membrane-bound receptor proteins to activate a signaling system that regulates the expression of specific genes.

Enhancers act in an orientation-independent manner over considerable distances to regulate transcription from a gene's promoter.

Transcription factors recognize and bind to specific DNA sequences within enhancers.

Transcription factors possess characteristic structural motifs such as zinc finger, the helix-turn helix, the leucine zipper, and the helix-loop-helix.

Short interfering RNAs and microRNAs are produced from larger double-stranded precursors by the action of Dicer-type endonucleases.

In RNA-Induced Silencing Complexes (RISCs), siRNAs and miRNAs become single-stranded so they can target complementary sequences in messenger RNA molecules.

Messenger RNA that has been targeted by siRNA is cleaved, and mRNA that has been targeted by miRNA is prevented from serving as a template for polypeptide synthesis.

Hundreds of genes for miRNAs are present in eukaryotic genomes.

Transposons and transgenes may stimulate the synthesis of siRNAs.

RNA interference is used as a research tool to knock out or knock down the expression of genes in cells and whole organisms..

Transcription occurs preferentially in loosely organized chromosome regions exemplified by loops of lampbrush chromosomes and the puffs of polytene chromosomes.

Transcriptionally active DNA tends to be more sensitive to digestion with DNase I.

During transcriptional activation, chromatin is remodeled by multiprotein complexes.

Heterochromatin is associated with the repression of transcription.

Protein complexes are responsible for gene silencing in *Drosophila* and yeast.

In trypanosomes, only one telomeric variant surface glycoprotein (*vsg*) gene is expressed; all other *vsg* genes, including those at telomeres, are silent.

Methylation of DNA is associated with gene silencing in mammals.

The expression of a gene that is imprinted is conditioned by the gene's parental origin.

Increased gene expression may be achieved by amplifying the DNA, either within chromosomes or on extrachromosomal DNA molecules.

Inactivation of an X chromosome in XX female mammals is mediated by a noncoding RNA transcribed from the *XIST* gene on that chromosome.

Hyperactivation of the single X chromosome in male *Drosophila* is mediated by an RNA-protein complex that binds to many sites on that chromosome and stimulates the transcription of its genes.

Hypoactivation of the two X chromosomes in *C. elegans* hermaphrodites is mediated by proteins that bind to these chromosomes and reduce the transcription of their genes.

ANSWERS TO QUESTIONS AND PROBLEMS

1) spatial 2) temporal 3) transcription factors 4) pseudogene 5) alternate splicing
6) transcriptional, processing, translational 7) steroid hormones 8) hormone receptor 9) peptide hormones 10) signal transduction 11) basal transcription factors 12) hormone response elements
13) membrane-bound receptor proteins 14) enhancers and silencers 15) zinc finger 16) homeodomain 17) ecdysone 18) homeotic 19) heterochromatin 20) leucine zipper 21) imprinting
22) nucleolus organizer 23) dosage compensation

Ideas concerning thought challenging exercise:

In situ hybridization is a direct way to determine if the transcriptionally active regions of chromosomes involved in puffs on polytene chromosomes and loops on lampbrush contain protein-coding genes. A probe consisting of cDNAs could be hybridized to chromosomal preparations. If cDNAs hybridized to the DNA of a specific puff or loop, it could be concluded that the transcriptionally active region of the chromosome contained the polynucleotide sequences complementary to that of the cDNA.

22

The Genetic Control of Animal Development

IMPORTANT CONCEPTS

  A. Development of an animal results from cell divisions and differentiation that are directed by differential gene activity.
    1. The blastula of an animal embryo is formed by cleavage divisions in a zygote.
      a. This sphere of cells is reorganized by cell movements (gastrulation) to produce a body with primitive tissues (ectoderm, endoderm, and mesoderm).
      b. During morphogenesis, ectoderm, endoderm, and mesoderm differentiate into mature tissues and organs.
  B. The genetic analysis of invertebrate animal development has concentrated on two model organisms, the fruit fly *Drosophila melanogaster* and the nematode *Caenorhabditis elegans*.
    1. In these organisms, geneticists are able to dissect developmental pathways by identifying genes whose products are involved in differentiation.
      a. Some of the genes involved in sexual differentiation have been identified in both species.
    2. The course of development can be influenced by maternal-effect products that are transported into the egg during oogenesis.
      a. In *Drosophila,* maternal-effect products are responsible for establishing the dorsal-ventral and anterior-posterior axes of the embryo.
        (1) Subsequent embryonic development, including the formation of body segments and

the differentiation of specific cell types, requires the expression of zygotic genes.
- C. Genes that are involved in vertebrate development have been identified by virtue of their homology to invertebrate genes.
  - 1. Mutations that affect vertebrate development have been studied in the mouse.
  - 2. Transgenic mice, which are useful in the study of development, can be created by injecting DNA into developing embryos.
- D. The zebrafish is another model vertebrate organism for developmental studies.
  - 1. Since eggs of the zebrafish are fertilized externally, events of embryological development can be studied without having to extract eggs from females.
    - a. Disruption of gene expression can be achieved by injecting zebrafish eggs or embryos with specific synthetic molecules (morpholinos) that bind to the 5' untranslated region and the start codon in messenger RNA and prevent translation.
      - (1) Morpholinos are typically 20-25 subunits long. Each subunit, consisting of a morpholine moiety bonded to a nitrogenous base (A, T, U, G or C), is linked to each other by phosphorodiaminate bonds instead of phosphodiester bonds.
      - (2) The animals that develop from injected eggs are phenocopies of mutations in the targeted gene.
- E. Stem cells are unspecialized precursors of specialized cells.
  - 1. Since stem cells are pluripotent (able to differentiate in many ways), they have potential uses in curing human diseases that result from loss of specific cell types, such as *diabetes mellitus* (loss of pancreatic islet cells) and Parkinson's disease (loss of certain types of neurons in a particular region of the brain).

IMPORTANT TERMS

In the space allotted, concisely define each term.

blastula:

gastrulation:

ectoderm:

mesoderm:

endoderm:

differentiation:

cleavage divisions:

determination:

primary germ layers:

morphogenesis:

chorion:

micropyle:

syncytium:

cellular blastoderm:

pole cells:

imaginal disks:

larva:

pupa:

imago:

invariant lineages:

numerator elements:

denominator elements:

maternal-effect mutations:

maternal-effect genes:

zygotic gene expression:

segments:

homeotic mutations:

homeosis:

bithorax complex (BX-C):

Antennapedia Complex(ANT-C):

selector genes:

segmentation genes:

gap genes:

pair-rule genes:

parasegments:

segment-polarity genes:

ligand:

induction:

colinearity:

embryonic stem cells (ES cells):

chimeras:

"knockout" mutation:

targeting:

morpholino:

phencopies:

stem cells:

pluripotent:

embryoid bodies:

therapeutic cloning:

reproductive cloning

IMPORTANT NAMES

In the space allotted, concisely state the major contribution made by the individual or group.

T.H. Morgan:

Sidney Brenner:

John Sulston:

Christiane Nusslein-Volhard, Eric Weischaus, Trudi Schupbach, and Gerd Jurgens:

Calvin Bridges:

Edward Lewis:

Thomas Kaufman and Matthew Scott:

Gerald Rubin:

Walter Gehring:

George Streisinger:

TESTING YOUR KNOWLEDGE

*In this section, answer the questions, fill in the blanks, and solve the problems in the space allotted. Problems noted with an * are solved in the Approaches to Problem Solving section at the end of the chapter.*

1. The sphere of relatively nondescript cells that results from a rapid series of mitotic divisions of the zygote is called a _____.

2. The process in which undifferentiated cells are assigned developmental fates is called _____.

3. The combination of cell movements and differentiation through which the body of an animal takes shape is called _____.

4. The phenomenon of cells migrating to new positions in the embryo, therefore converting the nondescript blastula into an embryo with three distinct cell layers is called _____.

5. The single layer of cells formed at the surface of a *Drosophila* embryo after 13 cycles of mitosis that is equivalent to the blastula of vertebrates is called the _____.

6. X-linked genes which code for proteins in the *Drosophila* embryo that "count" the number of X chromosomes present are called _____.

7. Genes located on autosomes in *Drosophila* that affect the nominator of the X:A ratio are called _____.

8. The X-linked _____ gene is the master regulator of the sex-determination pathway in *Drosophila*.

9. The SLX protein allows the synthesis of a functional _____ protein in XX embryos but not in XY embryos.

10. The autosomal gene in *Drosophila* that is regulated by a protein encoded by the *transformer* 2 (*tra*2) gene and produces two proteins (through alternate splicing of its RNA) which determine whether male or female development will proceed is called the _____ gene.

11. What is the phenotype of embryos that are homozygous for loss-of-function mutants in *Sxl*?

12. What is the sex of an XY fly that is hemizygous for a loss-of-function mutation of the *Sxl* gene?

13. What is the sex of flies homozygous for loss-of-function mutants of *tra* or *tra* 2?

14. Mutations in genes that contribute to the formation of healthy eggs, but have no effect on the phenotype of the female making the eggs are called _____ mutations.

15. In *Drosophila*, the formation of the dorsal-ventral axis is determined by the selective induction and repression of genes by action of a maternally synthesized transcription factor encoded by the _____ gene.

16. In *Drosophila*, the asymmetric synthesis of a protein coded by the *hunchback* gene determines the _____ axis of the embryo.

17. The *hunchback* gene is activated by a transcription factor encoded by the _____ gene.

18. Genes that encode transcription factors and define segmental regions in the early embryo, e.g., *Kruppel, giant, hunchback* and *knirps*, are called _____ genes.

19. Genes that are regulated by gap genes, e.g., *tarazu* and *even-skipped*, and define a pattern of segments within the embryo are called _____ genes.

20. Genes that define the anterior and posterior compartments of individual segments along the anterior-posterior axis, e.g., *gooseberry, wingless*, and *engrailed*, are called _____ genes.

21. Masses of cells in the larvae of *Drosophila* that give rise to specific adult organs such as antennae, eyes, and wings are called _____.

THOUGHT CHALLENGING EXERCISE

What are some major features that differ between animal and plant development?

SUMMARY OF KEY POINTS

The development of a multicellular animal involves the assignment of fates to cells (determination), and the subsequent realization of those fates (differentiation).

Key materials for the early phases of animal development are transferred into the egg during oogenesis.

After fertilization, the egg proceeds through several cleavage divisions to form a sphere of cells (the blastula), which is subsequently reorganized by cell movements (gastrulation) to form the primary germ layers (ectoderm, endoderm, and mesoderm).

During morphogenesis, tissues and organs are formed from the primary germ layers.

In *Drosophila*, the developmental sequence is egg, embryo, larva, pupa, adult.

The lineages of all the cells in an adult *C. elegans* can be traced back to the single-celled zygote.

In *Drosophila* the pathway that controls sexual differentiation involves some genes that ascertain the X:A ratio, some that convert this ratio into a developmental signal, and others that respond to the signal by producing either male or female structures.

The *Sex-lethal (Sxl)* gene plays a key role in *Drosophila* sexual development by regulating the splicing of its own transcript and that of another gene (*tra*).

In *C. elegans*, the sexual-differentiation pathway involves genes that encode signaling proteins, their receptors, signal transducers, and transcription factors.

The proteins and RNAs encoded by maternal-effect genes such as *dorsal*, *hunchback*, *bicoid*, and *nanos* are transported into *Drosophila* eggs during oogenesis.

Maternal-effect gene products are involved in the determination of the dorsal-ventral and anterior-posterior axes in *Drosophila* embryos.

Recessive mutations in maternal-effect genes are expressed only in embryos produced by females homozygous for these mutations.

The zygotic genes are activated after fertilization in response to maternal gene products.

In *Drosophila*, the products of the segmentation genes regulate the subdivision of the embryo into a series of segments along the anterior-posterior axis.

The identity of each body segment is determined by the products of genes in the bithorax and Antennapedia homeotic gene complexes.

The formation of an organ may depend on the product of a master regulatory gene, such as the *eyeless* gene in *Drosophila*.

In *Drosophila* specific cell types differentiate after segmental identities have been established.

Many vertebrate genes—for example, the *Hox* genes—have been identified by homology with genes isolated from model organisms such as *Drosophila* and *C. elegans*.

Among vertebrates, the mouse and the zebrafish provide opportunities to study mutations that affect development.

Transgenic mice can be created by injecting DNA into eggs or embryos or by inserting transfected embryonic stem cells into developing embryos.

Transgenetic mice can be a source of insertional mutations, including knockout mutations, in genes that have a developmental significance.

Injecting antisense morpholinos into vertebrate eggs or embryos can disrupt gene expression and produce mutant phenotypes that reveal the developmental significance of the targeted gene.

Mammalian stem cells, especially those derived from embryos, can be cultured *in vitro* to study the mechanisms that underlie differentiation.

ANSWERS TO QUESTIONS AND PROBLEMS

1) blastula 2) gastrulation 3) morphogenesis 4) determination 5) cellular blastoderm 6) numerator elements 7) denominator elements 8) *Sex-lethal (Sxl)* 9) transformer 10) *double sex (dsx)* 11) males that die as embryos 12) male, because SXL protein is normally not made in males 13) both XX and XY embryos develop into males 14) maternal-effect 15) *dorsal* 16) anterior-posterior 17) *bicoid* 18) gap 19) pair-rule 20) segment-polarity 21) imaginal disks

Ideas concerning thought challenging exercise:

There are several major features that distinguish plant and animal development. One major feature is that animal embryogenesis is characterized by cell migration; plant embryogenesis lacks cell migration. Another major difference is that differentiation in plants occurs from meristems that may be active for the life of the plant. A plant meristem can differentiate shoots and leaves throughout much of the life of a plant and then switch to a floral meristem and differentiate flowers. Animal germ lines are set aside early in development. Although both plant and animal development are influenced by hormones, animal hormones are often synthesized in specialized cells or tissues. Plant hormones are not synthesized in specialized cells that function solely for the purpose of hormone production. Of course, the evolution of much more complex organs and tissues has occurred during animal, relative to plant evolution.

23

The Genetic Control of the Vertebrate Immune System

IMPORTANT CONCEPTS

A. The immune system of vertebrate animals is complex and protects them from invasion by pathogenic microorganisms and other foreign substances.
   1. Three different types of white blood cells are involved in the immune system.
      a. B lymphocytes differentiate into antibody-producing plasma cells.
      b. T lymphocytes develop into killer T cells that produce T cell receptors and use them to seek out and destroy cells carrying foreign antigens.
      c. Macrophages carry out phagocytosis of antibody-antigen complexes, viruses, bacteria, fungi, and protozoa.
   2. T cell recognition of cells that contain foreign antigens requires the presence of specific histocompatibility antigens encoded by the major histocompatibility complex (the HLA locus in humans).
B. A seemingly limitless variety of antibodies and T cell receptor proteins can be synthesized in response to antigens that the animal has not previously encountered.
   1. It is now understood how this phenomenal diversity of antibody specificity and T-cell receptor diversity is produced.
      a. The genetic information encoding antibody and T cell receptor chains is stored in several sets of gene segments, and the segments are put together in the appropriate sequences by

genome rearrangements that occur during the development of the antibody-producing plasma cells.

C. Antibodies are composed of two light chains (either kappa or lambda) and two heavy chains.
   1. Each antibody chain contains a variable region that forms the antigen-binding site and a constant region that anchors the antibody to the cell surface in the case of membrane-bound antibodies.
      a. The variable regions of antibodies are encoded by gene segments that are present in the germ-line DNA in multiple copies.
      b. The constant region of antibodies is encoded by gene segments that are present in the genome in only one or a few copies per cell.
   2. In humans, kappa light chain genes are assembled by genome rearrangements that occur during B cell differentiation from about 300 $L_K V_K$ gene segments, 5 J_K gene segments, and one C_K gene segment.
   3. The use of alternate sites of gene segment joining during the genomic rearrangement events and somatic hypermutation within the variable region gene segments contributes to the production of additional antibody diversity.

D. Antibodies are of five different immunoglobulin classes: IgM, IgD, IgG, IgE, and IgA.
   1. The class of antibody is determined by its heavy chain constant region, which in turn is determined by the C segment that was expressed during its synthesis.
   2. Class switching occurs when a B lymphocyte stops synthesizing one class of antibody and begins synthesizing another class of antibody with the same antigen specificity.
      a. Class switching involves the expression of the same variable region gene segments but a different heavy chain constant region gene segment.
      b. Class switching most often occurs by further genomic rearrangements similar to those that resulted in the synthesis of the original antibody chains.
      c. Class switching can also occur by alternate patterns of transcript splicing.

E. The productive rearrangement of antibody gene segments during B cell differentiation activates transcription of the assembled gene by bringing the promoter located upstream from the LV gene segment into the range of influence of a tissue-specific enhancer located in the intron between the J segment cluster and the C gene segment.

F. Clonal selection explains the production of large numbers of plasma cells all synthesizing the same antibody specific for a particular antigen present in the circulatory system.
   1. Only one productive light chain genomic rearrangement and one productive heavy chain genomic rearrangement can occur in a given B lymphocyte. This is known as allelic exclusion.
      a. The molecular basis of allelic exclusion remains unknown, but must be controlled by some type of feedback mechanism.

G. The class I antigens produced by the major histocompatibility genes (HLA genes in humans) usually are responsible for tissue rejections during tissue and organ transplant operations.
   1. The genes encoding class I histocompatibility antigens are highly polymorphic, thus different individuals (other than identical twins) are highly unlikely to carry histocompatibility antigens that are identical.

H. There are both inherited and acquired forms of immunodeficiencies.
   1. Inherited immunodeficiencies such as X-linked agammaglobulinemia and severe combined immunodeficiency syndrome (SCID) are usually fatal early in life if untreated.
      a. The adenosine deaminase deficient type of SCID was the first human disease to be treated by somatic cell therapy.
   2. The most severe of the acquired immunodeficiencies is acquired immunodeficiency syndrome (AIDS), which is caused by a retrovirus, the human immunodeficiency virus (HIV).
      a. HIV destroys the helper T cells of the infected individuals so that they are unable to combat infections by microorganisms.

IMPORTANT TERMS

In the space allotted, concisely define each term.

nonspecific immune response:

specific immune response:

antigen:

immunogen:

stem cells:

B lymphocytes (B cells):

T lymphocytes (T cells):

phagocytes:

macrophage:

plasma cells:

memory B cells:

cytotoxic (killer) T cells:

helper T cells:

suppressor T cells:

memory T cells:

severe combined immunodeficiency syndrome (SCID):

antibodies:

immunoglobulins:

light chains:

heavy chains:

variable region:

constant region:

domains:

antigen-binding sites:

effector function domain:

kappa chains:

lambda chains:

class switching:

T cell receptors:

histocompatibility antigens:

major histocompatibility complex (MHC):

HLA antigens:

HLA locus:

HLA class I genes:

transplantation antigens:

HLC class II genes:

HLC class III genes:

complement proteins:

autoimmune disease:

antibody-mediated (humoral) immune response:

epitope:

clonal selection:

X-linked agammaglobulinemia:

T cell-mediated (cellular) immune response:

primary immune response:

secondary immune response:

activated B or T cells:

variable region:

joining sequence:

constant region:

somatic hypermutation:

class switching:

allelic exclusion:

monoclonal antibody:

IMPORTANT NAMES

In the space allotted, concisely state the major contribution made by the individual or group.

Cesar Milstein and George Kohler:

William Dreyer and Claude Bennett:

Nobumichi Horzumi and Susuma Tonegawa:

TESTING YOUR KNOWLEDGE

*In this section, answer the questions, fill in the blanks, and solve the problems in the space allotted. Problems noted with an * are solved in the Approaches to Problem Solving section at the end of the chapter.*

1. The cells of the immune system that respond to the display of an antigen by a macrophage by stimulating B lymphocytes to produce antibodies and T lymphocytes to produce T cell receptors are called _____.

2. The undifferentiated bone marrow cells that give rise to the various cells of the immune system are called _____.

3. Lymphocytes that carry T cell receptors and kill cells displaying the recognized antigens are called _____.

4. Cells of the immune system that differentiate in the thymus are _____.

5. Antibody-producing white blood cells derived from B lymphocytes are called _____.

6. T cells that assist in down regulating the production of antibodies and T cell receptors by B cells and T cells, respectively, are called _____.

7. Large cells that capture, ingest, and destroy invading foreign agents such as viruses, bacteria, and fungi are called _____.

8. The substances that trigger an immune response are called _____.

9. Proteins produced by killer T cells that bind antigens with the help of major histocompatibility antigens are called _____.

10. Cells that differentiate in the bone marrow to form antibody-producing plasma cells and memory B cells are called _____.

11. T cells that facilitate the rapid production of a given T cell receptor in repeated encounters with an antigen are called _____.

12. Cell surface proteins that allow immune system cells to distinguish foreign substances from self and facilitate communication between cells of the immune system are called _____.

13. Phagocytic cells that ingest antigens and display them on the cell surface for interactions with other cells of the immune system are called _____.

14. The most predominant class of antibodies present in humans is _____.

15. The second most abundant class of human immunoglobulins that are secreted in milk, saliva and tears is _____.

16. An abnormal amount or functioning of which immunoglobin class is related to allergies in humans?

17. What is a group of approximately twenty soluble proteins called that circulate in the bloodstream and form large protein complexes, which in turn kill the cells involved in antibody-antigen binding?

18. The first class of antibody produced in a developing B cell is always _____.

19. The two types of light antibody chains are _____ and _____.

*20. A given mammal has 300 L_H-V_H gene segments, 10 D gene segments, 4 J_H gene segments, and 8 C_H gene segments on chromosome 12. How many different heavy antibody genes could be formed?

*21. Assume that the above mammal also has 290 L_K-V_K gene segments, 5 J_K gene segments, and 1 C_K region. How many different kappa light antibody genes could be assembled?

*22. Using the numbers calculated for problems 21 and 22, how many different antibodies could be formed during the differentiation of B lymphocytes?

THOUGHT CHALLENGING EXERCISE

Discuss the nature of the genetic information encoding antibody and T cell receptor chains in light of previously presented concepts on the nature of a gene.

SUMMARY OF KEY POINTS

The immune response in mammals involves three steps: (1) recognition of the foreign substance, (2) communication of this recognition to the responding cells, and (3) elimination of the invading agent.

Nonspecific responses include the recruitment of phagocytes to ingest and destroy viruses and microorganisms.

Specific responses include the synthesis of special proteins and cells that work together to remove the foreign entity from the body.

The immune response of mammals involves the coordinated activities of several specialized white blood cells.

After exposure to a foreign substance (antigen), B lymphocytes differentiate into plasma cells that produce antigen-binding proteins called antibodies.

Similarly, T lymphocytes develop into killer T cells that carry T cell antigen receptors on their surfaces and kill cells displaying the recognized antigens.

Highly polymorphic major histocompatibility antigens on cells allow killer T cells to distinguish foreign cells from "self" cells, the cells of the individual mounting the immune response.

Individuals with autoimmune diseases produce antibodies against "self" antigens.

When presented with antigens by macrophages, helper T cells stimulate B lymphocytes to differentiate into antibody-producing plasma cells.

The synthesis of large amounts of antibodies specific for a particular antigen results from clonal selection, the stimulation of a B cell producing an antibody that recognizes the antigen to multiply and produce a population of plasma cells all producing the same antibody.

After recognizing antigens displayed by macrophages, helper T-cells stimulate T lymphocytes to differentiate into cytotoxic or killer T cells that recognize and lyse cells displaying the antigens.

Long-lived memory cells facilitate a faster immune response during a second encounter with a foreign substance.

The vast number of different antibodies produced by mammals results largely from the combinatorial joining of antibody gene segments by somatic recombination during the differentiation of B lymphocytes into antibody-producing plasma cells.

Additional antibody diversity is produced by variability in the sites of gene-segment joining and by a high frequency of mutation—somatic hypermutation—in the DNA sequences that encode antigen-binding sites.

Developing B lymphocytes can switch from the production of IgM antibodies to other classes of antibodies by genome rearrangements or by alternate pathways of transcript splicing.

The genes that encode the receptor proteins on killer T cells are assembled from gene segments during T lymphocyte differentiation by genome rearrangements analogous to those involved in the assembly of antibody genes.

Antibody gene segments are not transcribed at significant levels in stem cells.

Their transcription is activated by the genome rearrangements that occur during the differentiation of B lymphocytes.

Only one productive antibody gene rearrangement occurs per cell.

Transcription of assembled heavy chain genes is activated by transfer of their promoters to positions near a strong tissue-specific enhancer.

ANSWERS TO QUESTIONS AND PROBLEMS

1) helper T cell 2) stem cells 3) cytotoxic or killer T cells 4) T lymphocytes 5) plasma cells 6) suppressor T cell 7) phagocytes 8) antigens 9) T cell receptors 10) B lymphocytes 11) memory T cells 12) MHC antigens 13) macrophages 14) IgG 15) IgA 16) IgE 17) complement 18) IgM 19) kappa and lambda 20-22) see *Approaches to Problem Solving* section.

Ideas concerning thought challenging exercise:

The amino acid sequence of a protein is determined by a colinear nucleotide sequence in the DNA or gene. Except for allelic differences, this sequence is the same in all cells of the organism and individuals of the population. Generally, there is a one gene-one polypeptide relationship. The processing and removing of introns, if present, occurs after transcription. The final gene product of a normal gene generally has one specific function or catalytic activity. A mutation in a protein-coding gene, which will probably be harmful, will be selected at the organism level.

The genetic organization of the immune system, e.g., that of B lymphocytes, is very different compared to genes coding for most proteins. There are many representatives of different sections of the final gene sequence. These sections are processed by somatic rearrangement of DNA to generate a random combination for each allele. Since there are many millions of possible rearrangements, the combination generated in each cell will in all probability be different. Further mutational events may superimpose even more variability potential. Selection among B cells occurs in the circulatory system by a process called clonal selection.

APPROACHES TO PROBLEM SOLVING

20. A heavy chain will have 1 of the 300 L_H-V_H regions, 1 of 10 D regions, 1 of 4 J_H regions, and 1 of 8 C_H regions. Assuming that the final gene results from random assemblage of these four regions, the different number of antibody heavy chains that could be produced is (300)(10)(4)(8) = 96,000.

21. A light kappa chain will have a C_K region, a J_K region, and an L_K-V_K region. Assuming that the final gene results from random assemblage of these three regions, the different number of genes for kappa light antibody chains is (290)(5)(1) = 1450.

22. An antibody contains two identical heavy chains and two identical light chains. The number of different antibodies capable of being formed by the somatic recombinational method alone is (96,000)(1450) = 13,920,000.

24

The Genetic Basis of Cancer

IMPORTANT CONCEPTS

A. Cancer is a collection of diseases in which cell growth and division are unregulated.
  1. Unregulated cells may divide ceaselessly and form tumors.
     a. A tumor is malignant when its cells detach and invade surrounding tissues.
     b. The spread of malignant tumors to other locations in the body is called metastasis.

B. Transitions between different phases of the cell cycle (G_1, S, G_2, and M) are regulated at "checkpoints."
  1. In the checkpoint model, progression through the cell cycle is halted at key points until critical processes such as DNA replication and repair of damaged DNA are completed.
  2. Cyclins and cyclin-dependent kinases (CDKs) are important checkpoint proteins.
     a. CDKs regulate the activity of other proteins by transferring phosphates to them.
        (1) The phosphorylation activity of CDKs requires cyclin-CDK complexes.
     b. Passage through checkpoints requires alternate formation and degradation of cyclin-CDK complexes.
  3. An important cell-cycle checkpoint, at mid-G_1, is called START.
     a. Cells in which the START checkpoint is dysfunctional are prone to become cancerous.
        (1) Normal cells are programmed to pause at the START checkpoint to repair DNA before entering S phase.
        (2) If the START checkpoint is dysfunctional the cells move into S phase before DNA repair is complete. Mutations that result from replication of unrepaired DNA may accumulate, thus causing further deregulation of the cell cycle that may lead to cancer.
  4. Programmed cell death (apoptosis) is an important check against abnormal cells that could otherwise proliferate uncontrollably.

C. Abnormal gene expression is the basis for many forms of cancer.
  1. Cancer-causing genes, or oncogenes, were originally discovered in the genomes of retroviruses.
     a. These *v-onc* genes have normal cellular counterparts, the *c-onc* genes which play

important roles in regulating cell growth and differentiation.

    (1) The first tumor-inducing virus that was discovered carried an oncogene, *v-src*, which encodes a protein kinase, the overproduction of which results in sarcoma of the connective tissue of chickens.

    (2) At least 20 different viral oncogenes have been discovered. Some of these are related to cellular genes that encode growth factors, others encode proteins that are similar to growth-factor and hormone receptors, and another group encodes proteins that apparently function as transcription factors.

2. Cancer can be caused by a mutant of cellular *c-onc* genes.

    a. The first evidence linking cancer to mutant *c-onc* gene, *c-H-ras*, came from a study of human bladder cancer.

        (1) The protein coded by the mutant *c-H-ras* gene has a valine-for-lysine substitution at position 12 which impairs the ability of the mutant protein to hydrolyze one of its substrates, GTP. The mutant protein is kept in its active signaling mode and transmits information that stimulates the cell to divide in an uncontrolled manner.

3. Cancer can also be caused by the inappropriate expression of normal cellular oncogenes that have been relocated by chromosomal rearrangements.

    a. Chronic myelogenous leukemia (CML) results from an aberration involving chromosome 22 (*Philadelphia chromosome*); the distal portion of the long arm of chromosome 22 has been reciprocally translocated with the tip of the long arm of chromosome 9.

        (1) The translocation breakpoints are in the *c-abl* oncogene and in a gene, *bcr*, which encodes an tyrosine kinase.

4. Cancer can result by inactivation of tumor-suppressor genes.

D. Some forms of cancer are inherited; other forms appear to be caused by the occurrence of a series of mutations in the somatic tissues.

    1. Knudson's two-hit hypothesis explains sporadic and inherited forms of retinoblastoma.

        a. Retinoblastoma occurs when two copies of the *RB* gene have been inactivated by mutation.

        b. In inherited cases of retinoblastoma, one of the mutations in the *RB* gene is inherited and the other occurs during development. In sporadic cases, two mutations in the *RB* gene occur during development.

    2. Over 20 different inherited cancers have been identified involving tumor suppressor genes rather than oncogenes.

        a. The proteins encoded by tumor-suppressor genes function in a diverse array of cellular processes, including division, differentiation, programmed cell death, and DNA repair.

            (1) Tumor suppressor proteins that have been studied include: the 105 kD nuclear protein which is encoded by the *RB* gene and is involved in cell-cycle regulation; the TP53 protein that plays a key role in cellular response to stress; the 310 kD pAPC protein that regulates the opposing forces of cell proliferation and cell death within the intestinal epithelium; the phMSH2 protein involved in human hereditary nonpolyposis colorectal cancer (HNPCC); and the BRCA1 and BRCA2 polypeptides which have been implicated in breast and ovarian cancer.

E. For most cancers, tumor formation, growth, and metastasis depend on the accumulation of mutations in several different genes, and involves genetic pathways that are diverse and complex.

    1. The development of intestinal carcinoma requires at least seven independent mutations (two inactivating hits in the *APC* gene, one activating mutation in the *K-ras* gene, two inactivating hits in a tumor-suppressor gene on chromosome 18, and two inactivating hits in the *TP53* gene), and still more mutations are likely required for metastasis.

    2. Mutations in the hereditary prostate cancer gene *HPC1*, located in the long arm of chromosome 1, lead to prostate tumors; mutations in tumor-suppressor genes in chromosomes 13, 16, 17, and 18 can transform prostate tumors into metastatic cancers; and overexpression of the *BCL-2* proto-oncogene can make these cancers immune to androgen deprivation therapy.

    3. Studies on the development and metastasis of different types of tumors have demonstrated that somatic mutations are almost always involved.

F. Six hallmarks of the pathways leading to malignant cancer have been proposed.

    1. Cancer cells acquire self-sufficiency in the signal processes that stimulate division and

growth.
2. Cancer cells are abnormally insensitive to signals that inhibit growth.
3. Cancer cells can evade programmed cell death.
4. Cancer cells acquire limitless replicative potential.
5. Cancer cells develop ways to nourish themselves.
6. Cancer cells acquire the ability to invade other tissues and colonize them.

IMPORTANT TERMS

In the space allotted, concisely define each term.

metastasis:

carcinogens:

checkpoint:

apoptosis:

retroviruses:

oncogenes:

proto-oncogenes (normal cellular oncogenes):

tumor-suppressor genes:

angiogenesis:

IMPORTANT NAMES

In the space allotted, concisely state the major contribution made by the individual or group.

Peyton Rous:

Robert Weinberg:

Alfred Knudson:

Douglas Hanahan and Robert Weinberg:

TESTING YOUR KNOWLEDGE

In this section, answer the questions, fill in the blanks, and solve the problems in the space allotted. Problems noted with an * are solved in the Approaches to Problem Solving section at the end of the chapter.

1. What is the process called by which a malignant tumor spreads to other locations in the body?

2. Agents such as radiation and mutagenic chemicals that induce cancer are called _____.

3. A _____ is a mechanism that halts progression through the cell cycle until a critical process such as DNA synthesis is completed, or until damaged DNA is repaired.

4. Why do mutagenenic chemicals and ionizing radiation induce tumors in experimental animals?

5. Why does a *c-onc* have introns wheras a *v-onc* does not?

6. How might chromosomal rearrangements cause human cancer?

7. A process of activation of genes whose products ultimately cause cells with damaged DNA to die is called _____.

8. Why is the risk of developing a cancer greater as we age?

9. Why could it be said that cancer in monoclonal?

10. A unique feature of tumorigenic Rous sarcoma virus is the presence of the oncogene *v-src* that encodes a polypeptide with _____ activity.

*11. Why do familial forms of retinoblastoma generally involve both eyes and occur early in life, whereas, non-familial forms generally involve only one eye and occur later in life?

THOUGHT CHALLENGING EXERCISE

Why is there often a long lag period (up to 20 years) between the exposure to a carcinogen and the actual development of cancer?

SUMMARY OF KEY POINTS

Cancer is a group of diseases in which the cellular cycle of growth and division is unregulated.

Cancers may develop if the mechanism for programmed cell death (apoptosis) is impaired.

Cancers are due to the occurence of mutations in genes whose protein products are involved in the control of the cell cycle.

Some viruses carry genes (oncogenes) that can induce the formation of tumors in animals.

Viral oncogenes are homologous to cellular genes (proto-oncogenes), which can induce tumors when they are overexpressed or when they are mutated to produce abnormally active protein products.

Mutations in proto-oncogenes actively promote cell proliferation.

Some cancers are associated with chromosome rearrangements that enhance the expression of proto-oncogenes or that alter the nature of their protein products.

Tumor-suppressor genes were discovered through their association with rare, inherited cancers such as retinoblastoma.

Mutational inactivation of various tumor-suppressor genes is characteristic of most forms of cancer.

Two mutational hits are required to eliminate both functional copies of a tumor-suppressor gene within a cell.

The proteins encoded by tumor-suppressor genes play key roles in regulating the cell cycle.

Different types of cancer are associated with mutations in different genes.

Cancer cells may stimulate their own growth and division.

Cancer cells do not respond to factors that inhibit cell growth.

Cancer cells can evade the natural mechanisms that kill abnormal cells.

Immortalized cancer cells can divide endlessly.

Tumors can expand when they induce the in-growth of blood vessels to nourish their cells.

Metastatic cancer cells can invade other tissues and colonize them.

ANSWERS TO QUESTIONS AND PROBLEMS

1) metastasis 2) carcinogens 3) checkpoint 4) Both damage DNA, induce mutation and cause chromosomal breaks; all of which can lead to cancer. 5) The *v-onc* probably derived from a *c-onc* by the insertion of a fully processed *c-onc* mRNA into the genome of a retrovirus. 6) The chromosomal breaks during the formation of a chromosomal aberration might be in an oncogene, thereby inactivating it; they may juxtapose the oncogene next to different chromatin and thereby modify its expression, or even fuse portions of an oncogene with a different gene. 7) apoptosis 8) The formation, growth and metastasis of a tumor depends upon the accumulation of mutations in several different genes. As we age, there is more time for this process to occur. 9) Because cancer may arise from just one mutated cell. 10) protein kinase 11) see *Approaches to Problem Solving* section.

Ideas concerning thought challenging exercise:

The long lag period that is observed between the initial exposure to a carcinogen and the development of cancer is associated with the diverse and complex pathways leading to most cancers. Tumor formation, growth, and metastasis usually depend on the accumulation of mutations in several different genes. An exposure to the carcinogen may mutate some, but not all, of the genes that are involved in cancer formation. Additional required gene mutations may accumulate over time, therefore creating a lag between the exposure to a carcinogen and the formation of cancer.

APPROACHES TO PROBLEM SOLVING

11. Retinoblastoma is a cancer that is determined by homozygosity for a recessive allele. Individuals who inherit one recessive allele from their parents require only one more mutation for retinoblastoma to develop. Since there are millions of cells in a retina, there is a high probability that the additional mutation will occur in at least one cell in each retina. Therefore, retinoblastoma of the familial type generally affects both eyes and at a young age. Individuals who do not inherit a mutant allele will still have a high probability of mutations occurring in retinal cells. In this case, for retinoblastoma to develop two mutations of the allele must occur in the same cell, or in descendants of a cell carrying the first mutant. Consequently, retinoblastoma of the nonfamilial type generally affects only one eye, and occurs later in life.

25

Inheritance of Complex Traits

IMPORTANT CONCEPTS

A. After the rediscovery of Mendel's principles in 1900, biologists struggled for nearly twenty years to reconcile the complexity of quantitative traits with the simplicity of Mendel's theory.
1. Some scientists such as the statistician Karl Pearson, asserted that Mendelian principles did not apply to complex inheritance patterns.
2. Through several genetic experiments including those of Johannsen, Nilssen-Ehle, and East, it was demonstrated that continuous variation could be explained by Mendelian segregating alleles at several loci (polygenes) whose expression is influenced by the environment.
3. A simple polygenic model for the inheritance of quantitatively varying traits can be summarized as follows:
    a. Alleles at two or more loci determine a quantitative trait.
    b. Each loci may have two kinds of alleles, those that contribute to the phenotype (contributing alleles) and those that are neutral (noncontributing alleles) and do not contribute to the phenotype.
    c. The effects of contributing alleles are equal and additive.
    d. The alleles of different loci are nonepistatic, and segregate independently.
    e. The expression of a polygenic trait is influenced by the environment.

4. Examples in the text of polygenic traits are corolla length of tobacco flowers, kernel color of wheat, and the weight of bean seeds.
5. Some traits, such as cleft lip in humans, are polygenic with a threshold effect.
B. Statistics are used in the analysis of quantitative traits.
1. The mean and mode are statistics that point to the center of a frequency distribution.
2. The variance and the standard deviation are statistics that indicate the extent to which data are scattered around the mean.
3. Heritability is the proportion of a population's phenotypic variation attributable to genetic factors.
   a. To estimate heritability, researchers must determine the total variation in a trait (Vr), and then partition this into the variance due to genetic differences (V_g) and the variance due to environmental differences (V_e).
   b. Broad-sense heritability (H^2) is that fraction of the total variance that is due to genetic differences among individuals in a population.
   c. Narrow-sense heritability is based on the variance due to additive effects of alleles.
C. A gene that influences a quantitative trait is called a quantitative trait locus (QTL).
1. The identification of QT loci has become feasible with the advent of techniques to detect and analyze genetic variation at the molecular level.
D. Inbreeding is the preferential mating between close relatives, whereas outbreeding is the preferential mating between unrelated individuals.
1. Inbreeding leads to an increase in the frequency of homozygotes and a decrease in the frequency of heterozygotes in a population.
2. The effects of inbreeding are proportional to the inbreeding coefficient, which is the probability that two gene copies in an inbred individual are identical by descent from a common ancestor.
3. Geneticists use a simple statistic, the inbreeding coefficient, to analyze the effects of mating between relatives.
   a. The coefficient of relationship is the fraction of genes that two individuals share by virtue of common ancestry.
4. Inbreeding often leads to inbreeding depression, whereas outbreeding can lead to hybrid vigor.
E. Quantitative analyses of the resemblance between relatives, e.g., monozygotic and dizygotic twins reared together and apart, can provide estimates of broad- and narrow-sense heritabilites.

IMPORTANT TERMS

In the space allotted, concisely define each term.

quantitative traits:

multifactorial:

threshold trait:

monozygotic (MZ):

dizygotic (DZ):

concordance rate:

sample:

frequency distribution:

modal class:

normal distribution:

variance:

multiple factor hypothesis:

broadsense heritability:

narrowsense heritability:

artificial selection:

quantitative trait locus (QTL):

consanguineous mating:

inbred line:

inbreeding depression:

heterosis (hybrid vigor):

identity by descent:

inbreeding coefficient:

coefficient of relationship:

IMPORTANT NAMES

In the space allotted, concisely state the major contribution made by the individual.

Charles Darwin:

Karl Pearson:

W. Johannsen:

Herman Nilsson-Ehle:

Edward M. East:

Ronald A. Fisher:

Sewell Wright:

TESTING YOUR KNOWLEDGE

*In this section, answer the questions, fill in the blanks, and solve the problems in the space allotted. Problems noted by an * are solved in the Approaches to Problem Solving section at the end of the chapter.*

1. A statistical association among variables is called _____.

2. Quantitative variation not represented by distinct classes that is generally binomially distributed is called _____.

3. Phenotypic variation involving distinct classes such as axial or terminally located flowers is called _____.

4. For quantitative traits, the proportion of the phenotypic variance that is attributed to genotypic variance is called _____.

5. For quantitative traits, the proportion of the phenotypic variance that is due to the additive effects of alleles is called _____.

6. A measure of variation in a population expressed as the square of the standard deviation is called the _____.

*7. Assume that three polygenes (A, B, C) control the hypocotyl length in sunflower seedlings. A plant with 4 cm hypocotyls (*aa bb cc*) is crossed with a plant that has 10 cm hypocotyls (*AA BB CC*). What are the phenotypes and their frequencies among the F_1 and F_2 progeny?

*8. A quantitative trait is controlled by four polygenes. If F_1 individuals, heterozygous for the four genes, are crossed, what proportion of the F_2 progeny will have the same phenotype as the F_1?

*9. A cross between two homozygous lines of wheat having mean seed weights of 40 mg and 60 mg, respectively, produced F_1 progeny with seeds that uniformly weighed 50 mg. The F_2 consisted of 4,080 plants with seed weights binomially distributed between 40 mg and 60 mg, and a mean of 50 mg. If one of these F_2 plants produced seeds that were 40 mg and another F_2 plant produced seed weighing 60 mg, how many pairs of polygenes segregated in the F_1?

*10. The frequencies of genotypes in a plant population involving lactic dehydrogenase alleles, A and B, are .36 AA, .48 A B, and .16 B B. What would be the frequency of the A B genotype after three generations of self-pollination?

*11. A plant breeder is selecting safflower plants for oil content in seeds. In the breeding population the mean oil content is 10%. Plants with 15% oil content were crossed. If the narrow-sense heritability is 0.1, what will the average oil content be in the plants of the next generation?

*12. The following table presents measurements of the corolla length of two inbred lines (A & B) of fox-gloves, their F_1, and F_2 progeny.

Inbred line A								Inbred line B			
8	50	49	10					6	47	53	7

F_1 Progeny

5   45 39 7

F_2 Progeny

1   7   29 55 71 57 26 8   1

21 22 23 24 25 26 27 28 29 30 31 32 33 34 35 36

COROLLA LENGTH (mm)

Assuming that each parental variety was homozygous and that the genes are additive and equal in effect, how many genes are apparently segregating in the F_1?

*13. The data shown below are from a series of crosses between two homozygous rodent lines to study hair length. All animals were grown under the same uniform environmental conditions.

Line A (hair length)	Line 2 (hair length)
20 mm	36 mm

F_1 (hair length)

28 mm

F_2

Hair length (mm)	# of animals
36	2
34	14
32	60
30	108
28	140
26	114
24	52
22	18
20	2

Assume that all the contributing alleles have equal and additive effects.

(a) Determine the number of polygene loci at which the parents carry different alleles.

(b) Provide a genotype for the parents and the F_1.

(c) What is the phenotypic value of each contributing allele?

(d) List the expected phenotypes and their frequencies among the progeny of a cross between the F_1 and a homozygous 24 mm hair length F_2.

*14. The Jumping Frenchman of Maine disorder (caused by homozygosity for a rare recessive allele) is characterized by an abnormal, exaggerated, startle reflex reaction. It occurs in about 25 out of a million people in Maine and southeastern Canada. Study the following pedigree.

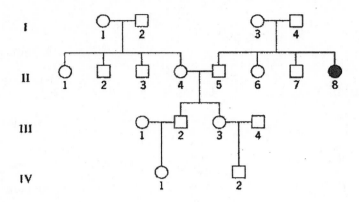

(a) If IV-1 marries at random in the population and does not marry a relative, what is the chance that her first child will be afflicted with the jumping disorder? The frequency of heterozygous individuals in the population is about 0.01.

(b) If IV-1 marries her cousin IV-2, what is the probability that her first child will be afflicted with the jumping disorder?

THOUGHT CHALLENGING EXERCISE

Intelligence is a complex polygenic trait in humans. Some psychologists and sociologists have been interested in differences in intelligence among various human ethnic and racial groups. Sizable differences (ca. 10%-15%) in mean IQ have been detected among racial and ethnic groups in the United States. Generally, mean IQ and estimates of heritability are lower in economically disadvantaged groups. Discuss what these differences in IQ and heritability may mean. Base your discussion on a sound genetic basis.

SUMMARY OF KEY POINTS

Resemblances between relatives and responses to selective breeding indicate that complex traits have a genetic basis.

Some complex traits can be quantified to permit genetic analysis based on Mendelian principles.

Many genetic and environmental factors influence the variation observed in quantitative traits.

Phenotypic segregations may provide a way to estimate the number of genes that influence a quantitative trait

Traits that are manifested when an underlying continuous variable (the "liability") reaches a threshold value may be influenced by genetic factors.

In humans, evidence that a threshold trait has a genetic basis comes from studies with twins.

The concordance rate is the fraction of twin pairs in which both twins show a trait among pairs in which at least one of them does.

The mean $\overline{X} = (\sum X_k)/n)$ and the modal class point to the center of a frequency distribution.

The variance ($s^2 = \sum(X_k - \overline{X})^2/(n-1)$ and standard deviation $\sqrt{s^2}$ are statistics that indicate the extent to which data are scattered around the mean in a frequency distribution.

The total phenotypic variance can be partitioned into genetic and environmental components: $V_T = V_g + V_e$.

The phenotypic variance in a population that is genetically uniform estimates V_e.

The broad-sense heritability is the proportion of the total phenotypic variance that is genetic variance: $H^2 = V_g / V_T$.

The genetic variance can be subdivided into additive genetic, dominance, and epistatic variances: $V_g = V_a + V_d + V_i$.

The narrow-sense heritability is the proportion of the total phenotypic variance that is due to the additive effects of alleles: $h^2 = V_a/V_T$.

The narrow-sense heritability is used to predict the phenotypes of offspring (T_o) given the average phenotype of the parents (T_p) and the mean phenotype in the population (μ) from which the parents came: $T_o = \mu + h^2(T_p - \mu)$.

The response to artificial selection can be predicted from the narrow-sense heritability and the selection differential: $R = h^2 S$.

By using molecular markers, geneticists are identifying quantitative trait loci.

Inbreeding increases the frequency of homozygotes and decreases the frequency of heterozygotes.

The effects of inbreeding are proportional to the inbreeding coefficient, which is the probability that two gene copies in an individual are identical by descent from a common ancestor.

The coefficient of relationship is the fraction of genes that two individuals share by virtue of common ancestry.

The correlation coefficient summarizes the degree of association between paired measurements, X_k and Y_k: $r = \sum[(X_k - \overline{X})(Y_k - Y)]/[(n-1)s_x s_y]$.

A correlation coefficient can be used to estimate the proportion of total variance in a quantitative trait that is due to genetic and environmental factors shared by relatives.

The correlation between monozygotic twins reared apart provides an estimate of broad-sense heritability.

The correlation between dizygotic twins reared apart provides an estimate of narrow-sense heritability.

Studying monozygotic and dizygotic twins, reared together or apart, has been useful in assessing the extent to which genes influence behavior in the general human population.

The broad-sense heritability for intelligence, as measured by IQ tests, is estimated to be 70 percent.

The broad-sense heritability for personality traits is estimated to be between 34 and 50 percent.

ANSWERS TO QUESTIONS AND PROBLEMS

1) correlation 2) continuous variation 3) discontinuous variation 4) broad-sense heritability
5) narrow-sense heritability 6) variance 7-14) see *Approaches to Problem Solving* section.

Ideas concerning the thought challenging exercise:

Unfortunately, the study of inheritance of human intelligence has not been free of racism and individual social and political agendas. Although it is generally agreed by geneticists and psychologists that intelligence is a quantitative and polygenic trait, there is no general agreement as to what the commonly used intelligence tests are measuring and the extent of educational and cultural bias in them. There is even greater disagreement concerning the role of the environment. The heritability component of "intelligence" is difficult to measure and the meaning of heritability is often not understood by the investigators. For example, some social scientists misuse the term heritability, thinking that high heritability means genetic superiority. If environment (education, economic, and social conditions) greatly influences IQ, then a lower mean IQ with a lower heritability would be expected in a group influenced by adverse environmental factors. Two groups could be genetically identical but have different mean IQs and different heritabilities simply due to differences in the environment.

APPROACHES TO PROBLEM SOLVING

7. In the cross *aa bb cc* (4 cm) x *AA BB CC* (10 cm), there are 6 cm differences in height between the parents. The 10 cm parent has six contributing alleles (capital letters) and the 4 cm plant has no contributing alleles (lower case letters). Therefore, each contributing allele adds 1 cm to a residual phenotype of 4 cm. The F_1 progeny are all of the genotype *Aa Bb Cc*. Since they have three contributing alleles, they have a residual phenotype of 4 cm plus the effects of three contributing allele worth 1 cm each. The F_1 progeny therefore have 7 cm hypocotyls.

To determine the phenotypes and their frequencies in the F_2s, one could simply determine all the genotypes in regard to contributing alleles and their frequencies by methods presented in chapter 3, such as the Punnett square or forked-line method. However, there is a much simpler way to obtain the F_2 phenotypic ratio. Consider the segregation of each gene pair independently. From the first gene pair *AA* x *aa*, the F_1 progeny are all *Aa*. The crossing of the F_1 progeny, *Aa* x *Aa*, leads to the following F_2 genotypic array: *AA* + 2*Aa* + *aa*. This is a binomial distribution represented by $(A + a)^2$. The F_2 genotypic array of the second gene pair is $(B + b)^2$, and that of the third gene pair is $(C + c)^2$. The F_2 genotypic array that represents the independent assortment of the three gene pairs is $(A + a)^2 (B + c)^2 (C + c)^2$.

Since all contributing alleles (*A*, *B*, *C*) have equal and additive effects, and the noncontributing alleles (*a*, *b*, *c*) do not contribute to the phenotype, all the contributing alleles may be designated by the same capital letter, e.g., H, and all noncontributing alleles may be designated by the corresponding lower case letter, h. By substituting these in place of $A\,a$, $B\,b$, and $C\,c$, the F_2 genotypic array may now be designated as follows:

$$(H + h)^2(H + h)^2(H + h)^2 = (H + h)^6.$$

The binomial $(H + h)^6$ is then expanded to give

$$1\,H^6 + 6\,H^5h^1 + 15\,H^4h^2 + 20\,H^3h^3 + 15\,H^2h^4 + 6\,H^1h^5 + 1\,h^6.$$

The coefficients indicate the proportions of the phenotypic classes. The exponents of H indicate the number of contributing alleles as indicated below.

$$1\,H^6 + 6\,H^5h^1 + 15\,H^4h^2 + 20\,H^3h^3 + 15\,H^2h^4 + 6\,H^1h^5 + 1\,h^6$$

number of contributing alleles 6 5 4 3 2 1 0

By adding 1 cm to a residual phenotype (no contributing alleles) of 4 cm, the following phenotypic ratio is obtained:

1/64 (10 cm): 6/64 (9 cm): 15/64 (8 cm): 20/64 (7 cm): 15/64 (6 cm): 6/64 (5 cm): 1/64 (4 cm).

The binomial distribution of the F_2 phenotypes is expected from the segregation of polygene loci.

The above approach can be utilized to depict phenotypic distributions from other crosses involving segregation of polygene loci.

8. To solve this problem, set it up like problem 7 and then look for the appropriate term in the expanded binomial representing the F_2s. In this case, there are four polygenes segregating in the F_1. The genotype of the F_1 can be initially presented as $Aa\,Bb\,Cc\,Dd$. The F_2 genotypic array is $(A + a)^2(B + b)^2(C + c)^2(D + d)^2$. Substituting common letters for contributing and noncontributing alleles, e.g., X and x, the F_2 genotypic array may be represented by $(X + x)^8$. From the expanded binomial we look for the term that has four contributing alleles (the same as the F_1). This term is $70\,X^4x^4$. Therefore, 70/256 of the F_2 progeny would be expected to have the same phenotype as the F_1.

9. The key to solving this problem is the binomially distributed F_2 population. Of 4,080 F_2 plants only one produced seeds as small as the 40 mg parent and only one produced seeds as large as the 60 mg parent plants. In the binomially distributed F_2 we expect $(1/4)^n$, where n is the number of segregating polygene loci, to have segregated to each parental extreme (homozygous for contributing or noncontributing alleles). Since $(1/4)^6 = 1/4096$, which is close to 1/4,080, an estimate of six segregating polygene pairs in the F_1 is obtained.

10. Inbreeding reduces the frequency of heterozygotes in a population. Self-pollination (self-fertilization) is the strictest form of inbreeding and reduces the frequency of heterozygotes by 1/2 each generation. This occurs because a heterozygous gene pair, e.g., Aa, upon selfing produces the following genotypes: $1/4\,AA + 1/2\,Aa + 1/4\,aa$. The initial frequency of individuals heterozygous for the A and B alleles of lactic dehydrogenase is .48. It follows that after three generations of self-pollination the frequency of heterozygotes should be $(.48)(0.5)(0.5)(0.5) = .06$.

11. A narrow-sense heritability of 0.1, determined by selection responses, indicates that the mean phenotype of the next generation should be moved 0.1 of the distance between the mean of the population and the mean of the plants selected. In this problem the selected population has 15% oil content, whereas the main population has 10% oil content. The mean oil content in the progeny of the selected plants should be $10\% + 0.1\,(5\%) = 10.5\%$.

12. The data presented in the table are typical for the inheritance of a quantitative trait. The two parents differ in corolla length and the F_1 is intermediate. Although the parents are inbred and highly homozygous, there is still phenotypic variability due in part to the influence of the environment. The

F_1 are heterozygous, but are all of the same genotype. The variability observed in the F_1 is again attributable to the environment and is of about the same magnitude as that present in each of the parental lines. The increase in variability and the binomial distribution observed in the F_2 is due to both genetic segregation of additive alleles and environmental components. Of 255 F_2 progeny, only one had a phenotype within the range of short corolla length parent and only one of the F_2 progeny had a corolla length within the range of the long corolla length parent. The number of genes segregating in the F_1 can be estimated by using the formula $(1/4)^n$. If n = 4, then we would expect about 1/256 to segregate to each parental extreme. Therefore, an estimate of four segregating gene pairs best fits the data.

13. (a) The inheritance in this cross indicates a quantitative trait. The F_1 is intermediate to the parents. The F_2 shows an increase in variability that is binomially distributed. To determine the number of polygene loci at which the parents carry different alleles, analyze the F_2 data to obtain an estimate of the number of polygene loci that segregated in the F_1. An estimate can be obtained from two parameters of the F_2 data. First, 2/510 or 1/255 F_2 progeny have the phenotype of each original parent. This is approximately $(1/4)^4$. Therefore, it is estimated that four gene pairs segregated in the F_1. The parents must have carried different polygene alleles at four polygene loci.

   (b) The genotype for the 20 mm parent can be designated as *aa bb cc dd*, and that of the 36 mm parent as *AA BB CC DD*. The genotype of the F_1 is *Aa Bb Cc Dd*.

   (c) There is a 16 mm (36 mm - 20 mm) difference between the hair length of the parents. This is determined by a total of eight contributing alleles. Therefore, each contributing allele is responsible for adding 2 mm to hair length.

   (d) The genotype of the F_1 is Aa Bb Cc Dd. The gametes produced by this F_1 can be represented as $(A + a)^1(B + b)^1(C + c)^1(D + d)^1$. An individual that is 24 mm has 2 contributing alleles (20 mm + 2 mm + 2 mm). If it is homozygous, both of these alleles must be of the same gene. The 24 mm rodent can be assigned the genotype *AA bb cc dd*. The gametes produced by this individual are of the genotype *A b c d*. Let's arbitrarily substitute H for contributing alleles and h for noncontributing alleles. This can be done because the contributing alleles are assumed to be equal and additive in effect. The gametes produced by the F_1 can now be represented by the binomial expression $(H + h)^1(H + h)^1(H + h)^1(H + h)^1 = (H + h)^4$. The phenotypes among progeny of this cross can be represented by the binomial $(H + h)^4 = 1H^4 + 4H^3h^1 + 6H^2h^2 + 4H^1h^3 + 1h^4$, plus one H allele obtained from the gametes of the homozygous 24 mm rodents.

$$1H^4 + 4H^3h^1 + 6H^2h^2 + 4H^1h^3 + 1h^4$$

	$1H^4$	$4H^3h^1$	$6H^2h^2$	$4H^1h^3$	$1h^4$
Proportion	1/16	4/16	6/16	4/16	1/16
Contributing alleles	5	4	3	2	1
Hair length (mm)	30	28	26	24	22

14. (a) To solve this problem, first determine the probability that IV-1 is heterozygous. To do this, we must analyze the pedigree. When analyzing rare genetic disorders, assume that all members marrying into a pedigree are homozygous for the normal allele (in this case *J J*) unless there is evidence to the contrary. Since II-8 is homozygous for the jumping allele (*j j*), her parents, I-3 and I-4, were both heterozygous *J j*. Individual II-5 would have a 2/3 chance of being heterozygous. If heterozygous, he would have a 1/2 chance of passing the allele to his son, III-2, who in turn would have a 1/2 chance of passing the allele to his daughter, IV-1. The probability that IV-1 is heterozygous is $(2/3)(1/2)(1/2) = 1/6$ or 0.167. To produce a child with the jumping disorder, she would have to marry another individual who was carrying the allele. Since the disorder is very rare, we consider only the probability that she will marry a heterozygote for the jumping allele. The frequency of heterozygotes for the jumping allele in the population is 0.01. If she is heterozygous (P = 0.167) and marries another heterozygote (P = 0.01), then she will have a .25 chance of having a child (j j) with the jumping disorder. The total probability of IV-1 having a

child afflicted with the jumping disorder if she marries at random is $(.167)(0.01)(0.25) = 0.0004$ or $1/2500$.

(b) To solve this part of the problem, determine the probability that IV-1 and her cousin IV-2 are both heterozygous for the jumping allele. If II-5 is heterozygous (we have already taken the 2/3 probability into account when determining the probability that IV-1 is heterozygous) he has a $(1/2)(1/2) = 1/4$ chance of passing the recessive allele to his grandson IV-2. Therefore, the probability that they are both heterozygous is $(1/6)(1/4) = 1/24$. If IV-1 marries her cousin IV-2, and if they are both heterozygous, they would have a 1/4 probability of having a child afflicted with the jumping disorder. Therefore, the total probability of having a child afflicted with the jumping disorder is $(1/24)(1/4) = 1/96$. This problem illustrates how inbreeding can increase the frequency of rare alleles becoming homozygous. Related individuals share a proportion of common alleles.

26

Population Genetics

IMPORTANT CONCEPTS

    A. Population genetics, the discipline that studies changes in gene frequency in populations and permits the formulation of mechanisms that account for the processes of evolution, had its origins early in the 1900s.

        1. In 1908, G. H. Hardy and W. Weinberg independently developed the basic principle of population genetics known today as the Hardy-Weinberg theorem.

            a. The Hardy-Weinberg theorem states that an equilibrium is established for allelic and genotypic frequencies in large, randomly mating, sexually reproducing populations, and that this constancy is maintained generation after generation as long as no disruptive forces are acting on the population.

            b. The Hardy-Weinberg principle can be used to estimate allelic and genotypic frequencies in populations.

            c. Departure from Hardy-Weinberg equilibrium can result from nonrandom mating, unequal survival of various genotypes, population subdivision, and migration.

    B. Knowledge of gene frequencies in specific populations is extremely important in genetic counseling.

    C. Natural selection is the primary force that alters allele frequencies (disrupts Hardy-Weinberg equilibrium) and brings about evolutionary change.

        1. The essence of natural selection is the increased production, by some genotypes relative to other genotypes, of reproductively successful progeny.

        2. Fitness, the capacity of a phenotype to donate genes to the next generation, is an important component to natural selection.

            a. The selection coefficient is a measure of fitness and varies from 0 to 1.

            b. The larger the selection coefficient, the more rapid is the rate of change in frequencies of alleles.

        3. Three modes of selection are directional selection, stabilizing selection, and disruptive selection.

a. Directional selection favors an extreme phenotype, thus shifting the population mean in one or the other direction.
b. Stabilizing selection favors an optimal phenotype and selects against both extremes, e.g., the heterozygote over the two homozygote classes.
c. Disruptive selection favors the more extreme variants and selects against the intermediates, e.g., the two homozygote classes over the heterozygote.
D. Forces, in addition to selection, can cause a departure from Hardy-Weinberg equilibrium.
  1. Inbreeding, the mating between genetically related individuals, results in an increase in the frequency of homozygous genotypes, i.e., a change in genotypic frequencies, but does not cause a change in allelic frequencies.
  2. Genetic drift is a random fluctuation in allelic frequency from generation to generation resulting generally from small population size.
  3. The allelic frequency of a population may change due to the introduction of alleles from another population by migration.
  4. Mutation introduces alleles into a population and provides variation on which natural selection acts.
E. Opposing evolutionary forces can create a dynamic equilibrium within a population.
  1. One type of dynamic equilibrium occurs through balancing selection, and at the balance point the frequency of A is $p = t/(s + t)$, and the frequency of a is $q = s/(s + t)$.
  2. If the mutant allele is deleterious, a mutation-selection equilibrium (for a recessive mutation, q = square root of u/s) is achieved when the rate of introduction of new alleles is balanced by the rate of elimination by selection.
  3. The opposing forces of mutation and drift come into balance and a dynamic equilibrium is established.
    a. The equilibrium heterozygosity at the point of mutation-drift balance is designated as
       $\hat{H} = 4Nu/(4Nu + 1)$.
      (1) Drift dominates over mutation in small populations.
      (2) In large populations mutation dominates over drift.

IMPORTANT TERMS

In the space allotted, concisely define each term.

polymorphism:

Hardy-Weinberg principle:

Hardy-Weinberg equilibrium:

panmictic:

natural selection:

relative fitness:

selection coefficient:

directional selection:

disruptive selection:

stabilizing selection:

random genetic drift:

balanced polymorphism:

IMPORTANT NAMES

In the space allotted, concisely state the major contribution of the individual or group.

G. H. Hardy and W. Weinberg:

H. B. D. Kettlewell:

R. A. Fisher:

Sewall Wright, R. A. Fisher, and Mottoo Kimura:

TESTING YOUR KNOWLEDGE

*In this section, answer the questions, fill in the blanks, and solve the problems in the space allotted. Problems noted with an * are solved in the Approaches to Problem Solving section at the end of the chapter.*

1. The frequency of individuals with PKU is 1/12,000 in a southwestern United States population. What is the frequency of the PKU allele?

*2. In a herd of 1000 cattle, the following phenotypes were observed: Red (RR), 100; roan (RR'), 260; white ($R'R'$), 640. The frequencies of the R and R' alleles are _____ and _____, respectively.

*3. The frequency of children homozygous for a certain recessive allele is 1/25000. Assuming Hardy-Weinberg equilibrium and that just two alleles of the gene occur in the population, what is the frequency of heterozygotes in the population?

4. From blood samples of 100 humans, 35 have blood antigen M, 50 have blood antigens M and N, and, 15 have blood antigen N. What are the frequencies of the L^M and L^N allele in the sample?

5. A population in Hardy-Weinberg equilibrium has two alleles of a gene, A and a. If the frequency of the a allele is 0.3, the frequency of the AA individuals is expected to be _____.

*6. For a human population in Hardy-Weinberg equilibrium, a sex-linked dominant trait is found in 91% of the females. The frequency at which this trait is expected to occur in males is _____.

*7. Assuming Hardy-Weinberg equilibrium in a population with $p = 0.8$, what is the expected frequency of the heterozygous individuals after four generations of self-fertilization?

*8. If, in a population obeying Hardy-Weinberg equilibrium, there are ten times more homozygous dominant individuals (AA) as heterozygous individuals (Aa), what are the expected frequencies of the three genotypes, AA, Aa, aa?

*9. A phenotype due to a sex-linked dominant allele occurs in males in the frequency of 0.4. Assuming that the population is in Hardy-Weinberg equilibrium, what is the expected frequency of the dominant phenotype among females of the population?

*10. In a Hardy-Weinberg population with $p = 0.7$, what is the expected frequency of the heterozygote four generations later?

*11. If the heterozygous individuals of a population are selected against, the frequency of the least frequent allele of a gene would be expected:
(a) to increase in frequency.
(b) to decrease in frequency.
(c) to change at the same rate as the more common allele.
(d) to remain at the same frequency.
(e) none of the above

*12. What is the equilibrium frequency of a detrimental recessive allele if the mutation rate is 5×10^{-5} and the selection coefficient is 0.9?

13. In a human population, a certain sex-linked recessive trait is found in females with a frequency of 0.81. Assuming Hardy-Weinberg equilibrium, what is the expected frequency of this trait among males?

14. The number of reproducing offspring donated to the next generation by a particular genotype, relative to that of another genotype, is called _____.

15. Selection that favors an extreme phenotype, thus shifting the population mean in one or the other direction is called _____ selection.

16. What term defines changes in allele frequencies, in small populations, due to chance fluctuations?

17. Selection that favors an optimal phenotype and selects against both extremes is called _____ selection.

THOUGHT CHALLENGING EXERCISE

1. Is it possible to eliminate or greatly reduce the frequency of individuals possessing relatively rare recessive traits by forbidding individuals possessing the recessive trait (*aa*) from reproducing, i.e., carrying out complete selection against the homozygous recessive genotypes?

SUMMARY OF KEY POINTS

Allele frequencies can be estimated by enumerating the genotypes in a sample from a population.

Under the assumption of random mating, the Hardy-Weinberg Principle allows genotype frequencies for autosomal and X-linked genes to be predicted from allele frequencies.

The Hardy-Weinberg Principle does not apply to populations with consanguineous or assortative mating, unequal survival among genotypes, geographic subdivision, or migration.

The Hardy-Weinberg Principle is useful in genetic counseling.

Natural selection occurs when genotypes differ in the ability to survive and reproduce—that is, when they differ in fitness.

The intensity of natural selection is quantified by the selection coefficient.

At the level of the gene, natural selection changes the frequencies of alleles in populations.

At the level of the phenotype, natural selection influences the distributions of quantitative traits.

Natural selection may be directional, disruptive, or stabilizing.

Genetic drift, the random change in allele frequencies in populations, is due to uncertainties in Mendelian segregation and to unpredictable variation in the number of offspring.

In diploid organisms, the rate at which genetic variability is lost by random drift is $1/2N$, where N is the population size.

Small populations are more susceptible to drift than large ones.

Drift ultimately leads to the fixation of one allele at a locus and the loss of all other alleles; the probability that an allele will ultimately be fixed is equal to its current frequency in the population.

Selection involving heterozygote superiority (balancing selection) creates a dynamic equilibrium in which different alleles are retained in a population despite their being harmful in homozygotes.

In humans sickle-cell anemia is associated with balancing selection at the locus for ß-globin.

Selection against a deleterious recessive allele that is replenished in the population by mutation leads to a dynamic equilibrium in which the frequency of the recessive allele is a simple function of the mutation rate and the selection coefficient: $q = \sqrt{\mu/s}$.

A population's acquisition of selectively neutral alleles through mutation is balanced by the loss of these alleles through genetic drift. At equilibrium, the frequency of heterozygotes involving these alleles is a function of the population's size and the mutation rate: $H = 4N\mu(4N\mu + 1)$.

ANSWERS TO QUESTIONS AND PROBLEMS

1) 0.009 **2) 4)** 0.6, 0.4 **5)** 0.49 **13)** 0.9 **14)** fitness **15)** directional **16)** genetic drift **17)** stabilizing **2, 3, 6-12,)** see *Approaches to Problem Solving* section below.

Ideas on thought challenging exercise:

Most single-gene detrimental genetic traits are caused by rare alleles; the efficacy of selection is very low because most of the recessive alleles occur in the heterozygous condition. Only those found in the homozygous state can be selected against. Therefore, selection against a recessive allele would be ineffective for all practical purposes. Furthermore, many recessive alleles may already be in an equilibrium between selection and mutation, and further selection could not eliminate the allele from the population.

APPROACHES TO PROBLEM SOLVING

2. To solve this problem the number of alleles of each type can be summed and divided by 2000. Since each individual is diploid, the total number of alleles of 1000 individuals is 2000.

    $$\text{Frequency of } R \text{ is } p = \frac{2(100)RR + 260\ R'R}{2000} = 0.23. \qquad \text{Since } p + q = 1, q\ (R') = 0.77.$$

3. The frequency of homozygous recessive individuals is $q^2 = 1/25000 = 0.00004$. Therefore q is the square root of $0.00004 \approx 0.006$. $p = 1 - q = 0.994$. The frequency of heterozygotes in the population is $2pq \approx 0.012$.

6. Sex-linked genes in equilibrium are distributed p^2 (AA) + $2pq$ (Aa) + q^2 (aa) among females. Since males are hemizygous, all genotypes are expressed and are distributed p (A) + q (a). Therefore, allelic frequencies can be estimated directly by looking at the frequencies of the male phenotypes, A and a. In this problem, the frequency of the dominant phenotype in females is 0.91 $= p^2 + 2pq$. It follows that $q^2 = 1 - (p^2 + 2pq) = 0.09$. Then $q = 0.3$ and $p = 0.7$. The frequency of this sex-linked dominant trait should occur in males at a frequency of $p = 0.7$.

7. Self-fertilization reduces the frequency of the heterozygote by 0.5 each generation. The initial frequency of heterozygotes in the population is $2pq = 2 \times 0.8 \times 0.2 = 0.32$. The frequency of the

heterozygote after four generations of self-fertilization is expected to be (0.32)(0.5)(0.5)(0.5)(0.5) = 0.02.

8. There are ten times as many homozygous dominant individuals (AA) as heterozygous individuals (Aa) in the population. This can be mathematically expressed as $p^2 = 10\ (2pq)$. Substituting $(1 - p)$ for q leads to the equation $p^2 = 10 \times 2(p)(1 - p)$. This is solved for $p = 0.952$. The value for $q = 1 - p = 0.048$. The expected frequencies of the genotypes are p^2 (AA) ≈ 0.906, $2pq$ (Aa) ≈ 0.091, and q^2 (aa) ≈ 0.002.

9. The sex-linked dominant allele occurs in males at a frequency of $p = 0.4$. It occurs in females at a frequency of $p^2 + 2pq = 1 - q^2$. Therefore, calculate the frequency of the recessive phenotype ($q^2 = 0.36$). The frequency of the dominant phenotype in females is 0.64.

10. The population is in equilibrium with the frequency of heterozygotes being $2pq = 0.42$. If no forces act to change the genotypic frequencies, the frequency of heterozygotes should still be 0.42 after four generations.

11. Each heterozygote removed from the population takes away one allele of each type. This will have a proportionally greater effect on the least frequent allele. Therefore, the frequency of the least frequent allele will decrease with selection against the heterozygotes of the population.

12. This problem can be solved by using the formula $q = \sqrt{\mu/s}$), where μ (5×10^{-5}) is the mutation rate and s (0.9) is the selection coefficient. $q = \sqrt{(0.00005/\ 0.9)} = 0.007$.

27

Evolutionary Genetics

IMPORTANT CONCEPTS

A. Modern evolutionary theory has its origin in the work of Charles Darwin who, in 1859, proposed a theory of evolution by natural selection.
   1. Darwin's theory of evolution was based upon several premises.
      a. Species change as a result of competition between variants within a species.
      b. Variants that favor survival in the struggle for existence tend to be preserved in the species, whereas, variants that impair survival tend to be lost.
      c. Over time, this process, which Darwin called natural selection, results in favorable variants replacing less favorable variants in the population.
   2. Darwin's theory offered no explanation for the origin of variation among individuals
      a. The rediscovery of Mendel's principles in 1900 provided the explanation.
B. Genetic variation in natural populations exists at the phenotypic, chromosomal, and molecular levels.
   1. Most of the early studies of genetic variability involved genes controlling phenotypic traits such as color of fur in mammals, eye color of *Drosophila*, flower color and morphology of plants, and wing patterns in butterfies.
   2. Members of a population of some species may exhibit considerable variation in chromosome structure.
      a. The pattern of bands on polytene chromosomes provided a powerful tool for studying chromosome polymorphisms and evolution within and among *Drosophila* species.

3. Genetic variation can be studied at the molecular level by analyzing proteins and nucleic acids.
    a. A new era in the study of genetic variation in natural populations emerged in 1966 with the application of gel electrophoresis to detect genetic polymorphism at the protein level.
    b. Over the past generation, many techniques have been developed that allow genetic variability to be analyzed at the DNA level. These include restriction fragment length polymorphisms (RLFPs), microsatellite polymorphisms, direct comparison of DNA sequences, and single-nucleotide polymorphisms detected by microarray technology.
C. DNA and protein sequences provide information on the phylogenetic relationships among different organisms, and on their evolutionary history.
    1. The descendants of an ancestral DNA or protein sequence are called homologous.
    2. Two sequences that resemble each other but are derived from different ancestral sequences called analogous.
    3. Evolutionary relationships among organisms can be obtained by representing comparison of amino acid and/or DNA sequences in the form of diagrams called phylogenetic trees.
        a. In some cases the branches in rooted phylogenetic trees can be superimposed in time by comparison to dated fossil records. This permits estimates to be made on the rate of species and molecular evolution.
D. The rate of amino acid substitutions for some proteins has been calculated to be fairly constant and follows what has been called a molecular clock.
    1. The molecular clock may vary among different lineages.
    2. When a molecular clock is operating, the estimated rate of evolution can be used to calculate the time since two lineages diverged from a common ancestor.
    3. The observed rate of amino acid substitutions among proteins ranges over three orders of magnitude.
        a. Variation in the rate of evolution among proteins may be the result of the amount of functional constraint. Slowly evolving proteins are more constrained than rapidly evolving proteins because amino acid changes are rigorously selected against.
E. Variation in evolutionary rate is also seen when DNA sequences are examined, which likely reflects variation in the functional restraints on them.
    1. Duplicated genes (pseudogenes) that do not encode functional proteins have the highest evolutionary rates.
        a. The nucleotides in a pseudogene are not constrained by selection because the pseudogene has lost its function.
    2. Evolutionary substitution of nucleotides in the first and second positions of a codon in a functional gene is constrained because change will generally result in a different specified amino acid.
    3. Nucleotides in the third position of codons within functional genes evolve faster than nucleotides in the first or second position because degeneracy of the genetic code. A third position nucleotide change would more likely not result in change in the encoded amino acid.
    4. Nucleotides in introns evolve more rapidly than nucleotides in the 3' untranslated regions, which in turn evolve more rapidly than nucleotides in the 5' untranslated regions.
F. Mutations may be beneficial, detrimental or neutral.
    1. Mutations that have little or no effect on fitness are called selectively neutral.
        a. Neutral mutations occur in regions that are not functionally constrained.
        b. The fate of selectively neutral mutations depends completely on random genetic drift. Most are rapidly lost from the population. A very small fraction survives and becomes fixed in the population. The rate of fixation of selectively neutral mutants is $2N\mu(1/2N) = \mu$.
        c. The Neutral Theory explains the variation in evolutionary rates that is observed among proteins and DNA regions by invoking differences in functional contraints.
        d. Although the Neutral Theory does not apply to traits that are adaptive, it has none-the-less had an enormous impact on the study of evolution at the molecular level.
G. Several types of molecular changes have accompanied evolution of adaptations and the diversification of organisms.
    1. Accumulations of nucleotide and amino acid substitutions do not assure that phenotypic evolution will occur.

2. Duplication of genes followed by subsequent divergence and the acquisition of different functions has apparently been an important evolutionary phenomenon.
3. Exon shuffling, the rearranging of exons from different genes in modular fashions, is apparently an important evolutionary process.
   a. Mixing and matching exons provides a nearly limitless possibility to form mosaic proteins.
4. Mutations in genes whose products regulate the expression of other genes has apparently been important to the evolutionary process.

H. In evolutionary biology a species may be defined as a group of interbreeding, or potentially interbreeding, populations that is reproductively isolated from other such groups.
1. Reproductive isolation between subpopulations is the key to the formation of species.
2. Numerous prezygotic and postzygotic mechanisms keep species reproductively isolated.
3. There are several mechanisms that lead to the formation of new species.
4. The allopatric model of speciation requires the physical separation of two populations to achieve reproductive isolation.
5. In the sympatric mode of speciation, geographic separation of two populations is not necessary to achieve reproductive isolation.

I. The evolutionary history of humans can be studied by investigating the fossil record and by analyzing genetic relationships among extant human populations.
1. Fossil evidence indicates that remote ancestors of human beings evolved in Africa, beginning about 4-5 million years ago.
2. The analyses of variation in genes, gene products, and DNA sequences allow relatedness of different racial and ethnic groups to be deduced.
3. Genetic analysis also permits researchers to decipher key events in human evolutionary history.
   a. Genetic evidence indicates that modern human populations emerged from Africa about 100-200 thousand years ago and subsequently spread to other continents.

IMPORTANT TERMS

In the space allotted, concisely define each term.

allozymes:

monomorphic:

microsatellite:

silent polymorphisms:

single-nucleotide polymorphisms (SNP):

phylogenetic trees:

phylogenies:

unrooted tree:

rooted tree:

homologous:

analogous:

principle of parsimony:

molecular clock:

functional constraint:

selectively neutral:

gene duplication:

exon shuffling:

reproductively isolated:

prezygotic isolating mechanisms:

postzygotic isolating mechanisms:

allopatric speciation:

sympatric speciation:

IMPORTANT NAMES

In the space allotted, concisely state the major contribution of the individual or group.

Charles Darwin:

Sewall Wright:

R. A. Fisher:

J. B. S. Haldane:

Theodosius Dobzhansky:

R. C. Lewontin, J. H. Hubby, and H. Harris:

Martin Kreitman:

Walter Gilbert:

Hermann Muller:

Moto Kimura:

Jack Lester King and Thomas H. Jukes:

TESTING YOUR KNOWLEDGE

In this section, answer the questions, fill in the blanks, and solve the problems in the space allotted. Problems noted with an * are solved in the Approaches to Problem Solving section at the end of the chapter.

1. Proteins that exhibit electrophoretic variation are designated as _____.

2. Nucleotide differences in the coding sequence of a gene that have no effect on the amino acid sequence of the polypeptide are referred to as _____.

3. The descendants of an ancestral DNA or protein sequence are said to be _____.

4. Two sequences that resemble each other, but evolved from an entirely different ancestral sequence, are called _____.

5. The phenomenon of maintaining the amino acid sequence of a protein to preserve function is called _____.

6. Duplicate genes that have lost their capability to encode functional products are called _____.

7. The _____ explains the variation in evolutionary rates that is observed among proteins and DNA regions by invoking differences in functional constraints.

8. The creation of novel proteins by combining exons from different genes in modular fashion is called _____.

9. The key event in the formation of species is _____.

10. A process by which reproductive isolation occurs among groups of individuals within a population is called _____.

11. The mode of speciation that may occur when two or more races of a species become geographically isolated is called _____ speciation.

*12. A systemist is studying a plant species (species A) widely dispersed on well-developed soils in the Sacramento River valley. Upon exploring the surrounding hills, two closely related species were found, one (species B) growing on serpentine outcrops and the second (species C) on very dry gravelly sites. Although there was no indication that hybridization occurred naturally, both species B and C could be crossed with species A (grown under greenhouse conditions) to produce partially fertile hybrids. Chromosome counts were obtained; root tips of species A had 18 chromosomes, species B had 18 chromosomes, and species C had 36 chromosomes. The hybrid between A and B had five bivalents and two quadrivalents at metaphase I of meiosis. The hybrid of species A and C had 9 bivalents and 9 univalents at metaphase I of meiosis. What do these data tell us about the evolution of species B and C?

*13. The cotton genus, *Gossypium,* contains approximately fifty species. The diploid species are classified into eight cytogenetic groups (A, B, C, D, E, F, G, and K) based on chromosome size and pairing in interspecific hybrids. Comparison of DNA sequences from several nuclear and chloroplast genes identified a set of natural lineages (Cronn and Wendel, New Phytologist 161:133-142, 2003) that mirror the genome designations as indicated in the following phylogenetic tree.

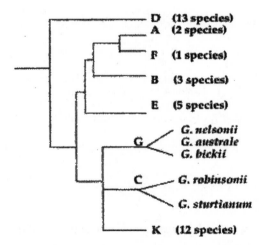

Based on nuclear genomic sequences, G. *bickii* was shown to share a more recent common ancestor with the morphologically similar G genome species, *G. australe* and *G. nelsonii,* than with the morphologically distant C genome species, *G. sturtianum* and *G. robinsonii.* However, the chloroplast genome of *G. bickii* was shown to be distinctly different than the other G genome species and nearly identical to the plastid genome of *G. sturtianum* and *G. robinsonii.* Propose evolutionary events that could explain the evolution of *G. bickii* in light of the incongruent chloroplast/nuclear DNA data.

*14. Four species of *Drosophila* reside on neighboring Pacific islands. Species A lives on the geologically oldest island. Morphological and genetic data indicate that the Species A is ancestral to Species B, C, and D. The following diagram presents the banding order of polytene chromosome 3 for the four species.

Species A	<u>a b c d e f</u>
Species B	<u>a f c d e f</u>
Species C	<u>a d c b e f</u>
Species D	<u>a d c f e b</u>

What is the evolutionary relationship of these species based on the chromosome comparison and assuming Species A to be ancestral?

THOUGHT CHALLENGING EXERCISE

The statement is often made that modern medicine, by allowing individuals of almost all genotypes to live and reproduce, is leading to an accumulation of detrimental alleles and the degeneration of the human gene pool. Based upon your understanding of population and evolutionary genetics, is there any scientific basis for such a viewpoint?

SUMMARY OF KEY POINTS

Charles Darwin formulated a theory in which species evolve through natural selection.

After the rediscovery of Mendel's work, Darwin's ideas became grounded on Mendelian principles of inheritance.

Genetic variation in natural populations can be detected at the phenotypic, chromosomal, and molecular levels.

Classical studies established the existence of genetic polymorphisms for conspicuous phenotypic traits, such as flower color, and for blood types.

Polymorphisms in chromosome structure have been documented in various species of *Drosophila* by analyzing banding patterns in the polytene chromosomes.

Polymorphisms in polypeptide structure have been detected by using the technique of protein gel electrophoresis.

Polymorphisms in DNA structure have been detected by analyzing restriction fragments and microsatellite loci, by sequencing cloned or PCR-amplified DNA, and by hybridizing samples to probes in diagnostic microarrays.

Phylogenetic trees based on the comparison of DNA and protein sequences show the evolutionary relationships among organisms.

The rate of molecular evolution can be determined by calculating the average number of amino acid or nucleotide changes that have occurred per site in a molecule since two or more evolving lineages diverged from a common ancestor.

The near uniformity of the rate of molecular evolution in different lineages is metaphorically described as a "molecular clock".

The rate of evolution varies among different protein and DNA sequences and appears to depend on the extent to which these sequences are constrained by natural selection to preserve their function.

Selectively neutral mutations are fixed in a population at a rate equal to the neutral mutation rate.

Gene duplication and exon shuffling have played important roles in evolution.

Changes in the spatial and temporal aspects of gene regulation may have contributed to the rapid evolution of some types of organisms.

In evolutionary genetics, a species is a group of populations that share a common gene pool.

The development of reproductive isolation between populations is the key event in the speciation process.

Speciation may occur when the populations are geographically separated (allopatric) or when they coexist in the same territory (sympatric).

The inviability or sterility of interspecific hybrids may be due to incompatibilities among genes that changed while the species were evolving.

Fossil evidence indicates that the remote ancestors of human beings evolved in Africa, beginning about 4-5 million years ago.

Genetic evidence indicates that modern human populations may have emerged from Africa about 100,000 to 200,000 years ago and subsequently spread to other continents.

ANSWERS TO QUESTIONS AND PROBLEMS

1) polymorphic **2)** silent polymorphism **3)** homologous **4)** analogous **5)** functional constraint **6)** pseudogenes **7)** neutral theory **8)** exon shuffling **9)** reproductive isolation **10)** sympatric speciation **11)** allopatric **12, 13, 14)** see *Approaches to Problem Solving* section below.

Ideas on thought challenging exercise:

If selection is relaxed within a large randomly mating population, the frequencies of alleles will remain the same generation after generation if no other factors are affecting gene frequency. The first generation after relaxed selection conditions will be an increase in the number of formerly lethal or sublethal individuals that survive in the population, but this won't increase in future generations. Over many generations, mutation may cause a slight increase in the frequency of detrimental alleles. However, the rate of addition of new alleles by mutation is very slow and will reach an equilibrium with the rate of back mutation. Based upon these population genetic principles, it can be clearly stated that modern medicine is not causing a degeneration of the human gene pool.

APPROACHES TO PROBLEM SOLVING

12. Species B grows on a very restricted and specialized habitat so we may suspect that it evolved from species A. The two quadrivalents in the hybrid of A × B indicates that the genome of species B differs from that of species A by two translocations. Therefore, the postulated evolution of species B from species A involved restructuring of the chromosomes in the form of two reciprocal translocations. Species B is tetraploid as is indicative by its chromosome number. The homology of nine chromosomes of species C with the nine chromosomes of species A (9 bivalents) in the hybrid indicates that the chromosomes of species A are homologous to the nine chromosomes of species C. It appears that A hybridized with another species to form the allotetraploid species C.

13. *Gossypium bickii* possesses a chloroplast genome that is uncharacteristic of its taxonomy. The discrepancy between the nuclear/cpDNA data suggests a biphyletic ancestry, whereby a *G. sturtianum*-like species was the female parent in an ancient hybridization with a male species possessing a G-genome. Continuous introgression of nuclear genes from the G-genome species resulted in plants with a nuclear G-genome and a *G. sturtianum*-like cytoplasm. The result of this introgression is represented today by *G. bickii*.

14. The chromosome arrangements that differentiate the four species are inversions. The three species evolved from Species A in the order A → C → D → B. The inverted chromosome regions are indicated in bold face.

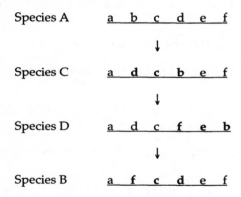

Species A	a b c d e f
	↓
Species C	a **d c b** e f
	↓
Species D	a d c **f e b**
	↓
Species B	a **f c d e** f

PART II

Complete Answers and Solutions to Text Questions and Problems

CHAPTER 1

1.1 Mendel postulated transmissible factors—genes—to explain the inheritance of traits. He discovered that genes exist in different forms, which we now call alleles. Each organism carries two copies of each gene. During reproduction, one of the gene copies is randomly incorporated into each gamete. When the male and female gametes unite at fertilization, the gene copy number is restored to two. Different alleles may coexist in an organism. During the production of gametes, they separate from each other without having been altered by coexistence.

1.2 Each nucleotide consists of a sugar, a nitrogen-containing base, and a phosphate.

1.3 The bases present in DNA are adenine, thymine, guanine, and cytosine; the bases present in RNA are adenine, uracil, guanine, and cytosine. The sugar in DNA is deoxyribose; the sugar in RNA is ribose.

1.4 A genome is the set of all the DNA molecules that are characteristic of an organism. Each DNA molecule forms one kind of chromosome in a cell of the organism.

1.5 TAACGGCAG.

1.6 There are 3 × 141 = 423 nucleotides in the gene's coding sequence. Its polypeptide product will contain 141 amino acids.

1.7 GAACGGUCT.

1.8 Transcription is the production of an RNA chain using a DNA chain as a template. Translation is the production of a chain of amino acids—that is, a polypeptide—using an RNA chain as a template.

1.9 Sometimes DNA is synthesized from RNA in a process called reverse transcription. This process plays an important role in the life cycles of some viruses.

1.10 The two mutant forms of the β-globin gene are properly described as alleles. Because neither of the mutant alleles can specify a "normal" polypeptide, an individual who carries each of them would probably suffer from anemia.

1.11 The human and cow α-globins are least different; therefore, on the assumption that differences in α-globin reflect the degree of phylogenetic relationship, the human and the cow are the most closely related organisms among those mentioned. The next closest "relative" of humans is the carp, and the most distant relative is the shark.

1.12 The gene for the human clotting factor could be isolated from the human genome and transferred into bacteria, which could then be grown in vats to produce large amounts of the gene's protein product. This product could be isolated from the bacteria, purified, and then injected into patients to treat hemophilia. Another approach would be to transfer a normal copy of the clotting factor gene into the cells of people who have hemophilia. If expressed properly, the transferred normal gene might be able to compensate for the mutant allele these people naturally carry. For this approach to succeed, the normal clotting factor gene would have to be transferred into the cells that produce clotting factor, or into their precursors.

CHAPTER 2

2.1 Sugars combine to form carbohydrates; amino acids combine to form proteins.

2.2 Cell membranes are made of lipids and proteins; they have a fluid structure. Cell walls are made of more rigid materials such as cellulose.

2.3 In a eukaryotic cell the many chromosomes are contained within a membrane-bounded structure called the nucleus; the chromosomes of prokaryotic cells are not contained within a special subcellular compartment.

Eukaryotic cells usually possess a well-developed internal system of membranes and they also have membrane-bounded subcellular organelles such as mitochondria and chloroplasts; prokaryotic cells do not typically have a system of internal membranes (although some do), nor do they possess membrane-bounded organelles.

2.4   In the haploid state, each chromosome is represented once; in the diploid state, each chromosome is represented twice. Among multicellular eukaryotes, gametes are haploid and somatic cells are diploid.

2.5   Prokaryotic chromosomes are typically (but not always) smaller than eukaryotic chromosomes; in addition, prokaryotic chromosomes are circular, whereas eukaryotic chromosomes are linear. For example, the circular chromosome of *E. coli*, a prokaryote, is about 1.4 mm in circumference. By contrast, a linear human chromosome may be 10 to 30 cm long. Prokaryotic chromosomes also have a comparatively simple composition: DNA, some RNA, and some protein. Eukaryotic chromosomes are more complex: DNA, some RNA, and lots of protein.

2.6   During interphase, the chromosomes duplicate. During M phase (mitosis), the duplicated chromosomes, each consisting of two identical sister chromatids, condense (a feature of prophase), migrate to the equatorial plane of the cell (a feature of metaphase), and then split so that their constituent sister chromatids are separated into different daughter cells (a feature of anaphase); this last process is called sister chromatid disjunction.

2.7   Interphase typically lasts longer than M phase. During interphase, DNA must be synthesized to replicate all the chromosomes. Other materials must also be synthesized to prepare for the upcoming cell division.

2.8   The microtubule organizing centers of animal cells have distinct centrosomes, whereas the microtubule organizing centers of plant cells do not.

2.9   (1) Anaphase: (f), (h); (2) metaphase: (e), (i); (3) prophase: (b), (c), (d); (4) telophase: (a), (g).

2.10   (c), (f), (a), (e), (b), (d).

2.11   There are 8 chromosomes in a *Drosophila* spermatogonial cell about to enter meiosis. There are 16 chromatids in a *Drosophila* spermatogonial cell at metaphase I of meiosis. There are 8 chromatids in a *Drosophila* cell at metaphase II of meiosis.

2.12   Chromosomes 11 and 16 would not be expected to pair with each other during meiosis; these chromosomes are heterologues, not homologues.

2.13   Crossing over occurs *after* chromosomes have duplicated in cells going through meiosis.

2.14   The chiasmata, which are visible late in prophase I of meiosis, indicate that chromosomes have crossed over.

2.15   Chromosome disjunction occurs during anaphase I. Chromatid disjunction occurs during anaphase II.

2.16   In flowering plants sporophytic tissue is diploid. Eight nuclei are present in the female gametophyte; three are present in the male gametophyte. All the gametophytic nuclei are haploid, although in the female gametophyte, two of these nuclei fuse to form a diploid endosperm nucleus.

2.17   Among eukaryotes, there doesn't seem to be clear relationship between genome size and gene number. For example, humans, with 3.2 billion base pairs of genomic DNA, have between 20,000 and 25,000 genes, and *Arabidopsis* plants, with about 150 million base pairs of genomic DNA, have roughly the same number of genes as humans. However, among prokaryotes, gene number is rather tightly correlated with genome size, probably because there is so little nongenic DNA.

2.18   The single surviving colony may represent a mutant strain that is resistant to infection by the bacteriophage. The mutation probably occurred in a single bacterial cell, which then divided to produce the colony. Cells from this colony could be grown on another plate in the presence of the bacteriophage to confirm that they are resistant to infection.

2.19   Yeast chromosomes are, on average, smaller than *E. coli* chromosomes. Thus, some eukaryotic chromosomes are smaller than some prokaryotic chromosomes.

2.20 No, there isn't any advantage associated with an even number of chromosomes. As long as the chromosomes come in pairs, they will be able to synapse during prophase I and then disjoin during anaphase I to distribute the genetic material properly to the two daughter cells.

2.21 One of the pollen nuclei fuses with the egg nucleus in the female gametophyte to form the zygote, which then develops into an embryo and ultimately into a sporophyte. The other genetically functional pollen nucleus fuses with two nuclei in the female gametophyte to form a triploid nucleus, which then develops into a triploid tissue, the endosperm; this tissue nourishes the developing plant embryo.

CHAPTER 3

3.1 (a) All tall; (b) 3/4 tall, 1/4 dwarf; (c) all tall; (d) 1/2 tall, 1/2 dwarf.

3.2 Round (WW) × wrinkled (ww) ---> F_1 round (Ww); F_1 self-fertilized ---> F_2 3/4 round (2 WW: 1 Ww), 1/4 wrinkled (ww). The expected results in the F_2 are 5493 round, 1831 wrinkled. To compare the observed and expected results, compute χ^2 with one degree of freedom: $(5474 - 5493)^2/5493 + (1850 - 1831)^2/1831 = 0.263$, which is not significant at the 5% level. Thus, the results are consistent with the Principle of Segregation.

3.3 The data suggest that coat color is controlled by a single gene with two alleles, C (gray) and c (albino), and that C is dominant over c. On this hypothesis, the crosses are: gray (CC) × albino (cc) → F_1 gray (Cc); $F_1 \times F_1 \to$ 3/4 gray (2 CC: 1 Cc), 1/4 albino (cc). The expected results in the F_2 are 203 gray, 67 albino. To compare the observed and expected results, compute χ^2 with one degree of freedom: $(198 - 203)^2/203 + (67 - 72)^2/72 = 0.470$, which is not significant at the 5% level. Thus, the results are consistent with the hypothesis.

3.4 (a) Woman's genotype Pp, father's genotype Pp, mother's genotype pp; (b) 1/2.

3.5 (a) Checkered, red ($CC\ BB$) × plain, brown ($cc\ bb$) → F_1 all checkered, red ($Cc\ Bb$); (b) F_2 progeny: 9/16 checkered, red (C- B-), 3/16 plain, red ($cc\ B$-), 3/16 checkered, brown (C- bb), 1/16 plain, brown ($cc\ bb$).

3.6 (a) colored, normal ($CC\ Vv$) × white, normal ($cc\ Vv$); (b) colored, normal ($Cc\ Vv$) × colored, normal ($Cc\ Vv$); (c) colored, normal ($Cc\ Vv$) × white, waltzing ($cc\ vv$).

3.7 Among the F_2 progeny with long, black fur, the genotypic ratio is 1 $BB\ RR$: 2 $BB\ Rr$: 2 $Bb\ RR$: 4 $Bb\ Rr$; thus, 1/9 of the rabbits with long, black fur are homozygous for both genes.

3.8 (a) all red; (b) 1/2 red, 1/2 roan; (c) all roan; (d) 1/4 red, 1/2 roan, 1/4 white.

3.9

	F_1 gametes	F_2 genotypes	F_2 phenotypes
(a)	2	3	2
(b)	$2 \times 2 = 4$	$3 \times 3 = 9$	$2 \times 2 = 4$
(c)	$2 \times 2 = 4$	$3 \times 3 \times 3 = 27$	$2 \times 2 \times 2 = 8$
(d)	2^n	3^n	2^n, where n is the number of genes

3.10 (a) $2 \times 2 \times 1 \times 2 \times 2 \times 2 = 32$; (b) $3 \times 2 \times 2 \times 2 \times 3 \times 2 = 144$; (c) $2 \times 2 \times 2 \times 2 \times 2 \times 2 = 64$; (d) $(1/2) \times (1/2) \times (1/2) \times (1/2) \times (1/4) \times (1/2) = 1/128$; (e) $(3/4) \times (1/2) \times (1/2) \times (1/2) \times (1/4) \times (1/2) = 3/256$.

3.11 On the hypothesis, the expected number in each class is 27.5; χ^2 with three degrees of freedom is calculated as $(31 - 27.5)^2/27.5 + (26 - 27.5)^2/27.5 + (27 - 27.5)^2/27.5 + (26 - 27.5)^2/27.5 = 0.618$, which is not significant at the 5% level. Thus, the results are consistent with the hypothesis of two independently assorting genes, each segregating two alleles.

3.12 (a) 1, reject; (b) 2, reject; (c) 3, accept; (d) 3, accept.

3.13 $\chi^2 = (30 - 25)^2/25 + (20 - 25)^2/25 = 2$, which is less than 3.84, the 5 percent critical value for a chi-square statistic with one degree of freedom; consequently, the observed segregation ratio is consistent with the expected ratio of 1:1.

3.14 If capsule shape is determined by a single gene with two alleles, the F_2 plants should segregate in a 3:1 ratio. To test for agreement between the observed segregation data and the expected ratio, compute the expected number of plants with either triangular or ovoid seed capsules: $(3/4) \times 80 = 60$ triangular and $(1/4) \times 80 = 20$ ovoid; then compute a χ^2 statistic with one degree of freedom: $\chi^2 = (72 - 60)^2/60 + (8 - 20)^2/20 = 9.6$, which exceeds the critical value of 3.84. Consequently, the data are inconsistent with the hypothesis that capsule shape is determined by a single gene with two alleles.

3.15 Half the children from $Aa \times aa$ matings would be have albinism. In a family of three children, the chance that one will be unaffected and two affected is $3 \times (1/2)^1 \times (1/2)^2 = 3/8$.

3.16 (a) $(3/4)^4 = 81/256$; (b) $4 \times (3/4)^3 \times (1/4)1 = 108/256$; (c) $6 \times (3/4)^2 \times (1/4)^2 = 54/256$; (d) $4 \times (3/4)^1 \times (1/4)^3 = 12/256$.

3.17 Man ($Cc\,ff$) $\times$ woman ($cc\,Ff$). (a) $cc\,ff$, $(1/2) \times (1/2) = 1/4$; (b) $Cc\,ff$, $(1/2) \times (1/2) = 1/4$; (c) $cc\,Ff$, $(1/2) \times (1/2) = 1/4$; (d) $Cc\,Ff$, $(1/2) \times (1/2) = 1/4$.

3.18 $9!/(7!\ 2!) \times (1/2)^7 \times (1/2)^2 = 0.07$.

3.19 $(1/2)^3 = 1/8$.

3.20 (a) $4 \times (1/2)^2 \times (1/2)^2 = 4/16$; (b) $(1/2)^4 = 1/16$; (c) 2 boys, 2 girls; (d) 1 – probability that all four are boys = $1 - (1/2)^4 = 15/16$.

3.21 $(20/64) + (15/64) + (6/64) + (1/64) = 42/64$.

3.22 (a) zero; (b) 1/2.

3.23 (a) $(1/2) \times (1/4) = 1/8$; (b) $(1/2) \times (1/2) \times (1/4) = 1/16$; (c) $(2/3) \times (1/4) = 1/6$; (d) $(2/3) \times (1/2) \times (1/2) \times (1/4) = 1/24$.

3.24 (a) Recessive; (b) dominant.

3.25 For III-1 $\times$ III-2, the chance of an affected child is 1/2. For IV-2 $\times$ IV-3, the chance is zero.

3.26 1/2.

3.27 (a) $(1/4)^3 = 1/64$; (b) $(1/2)^3 = 1/8$; (c) $3 \times (1/2)^1 \times (1/2)^2 = 3/8$; (d) 1 – probability that the offspring is not homozygous for the recessive allele of any gene = $1 - (3/4)^3 = 37/64$.

3.28 The researcher has observed what appears to be a non-Mendelian ratio because he has been studying only families in which at least one child shows albinism. In these families, both parents are heterozygous for the mutant allele that causes albinism. However, other couples in the population might also be heterozygous for this allele but, simply by chance, have failed to produce a child with albinism. If a man and a woman are both heterozygous carriers of the mutant allele, the chance that a child they produce will not have albinism is 3/4. The chance that four children they produce will not have albinism is therefore $(3/4)^4 = 0.316$. In the entire population of families in which two heterozygous parents have produced a total of four children, the average number of affected children is 1. Among families in which two heterozygous parents have produced at least one affected child among a total of four children, the average must be greater than 1. To calculate this *conditional* average, let's denote the number of children with albinism by x, and the probability that exactly x of the four children have albinism by $P(x)$. The average number of affected children among families in which at least one of the four children is affected—that is, the conditional average—is therefore.

$\Sigma\ x\ P(x)/(1 - P(0))$, where the sum starts at $x = 1$ and ends at $x = 4$

We start the sum at $x = 1$ because we must exclude those cases in which none of the four children is affected. The divisor $(1 - P(0))$ is the probability that the couple has had at least one affected child among their four children. Now $P(0) = 0.316$ and $\Sigma\ x\ P(x) = 1$. Therefore, the average we seek is simply $1/(1 - 0.316) = 1.46$. If, in the subset of families with at least one affected child, the average number of affected children is 1.46, then the average number of unaffected children is $4 - 1.46 = 2.54$. Thus, the expected ratio of unaffected to affected children in these families is 2.54:1.56, or 1.74:1, which is what the researcher has observed.

3.29 (a) The observed numbers, expected numbers, and chi-square calculation are laid out in the table below:

	Observed	Expected	$(Obs - Exp)^2/Exp$
Dominant homozygotes (SS)	29	$62 \times 1/3 = 20.7$	3.33
Heterozygotes (Ss)	33	$62 \times 2/3 = 41.3$	1.67
Total	62	62	5.00

The total chi-square value is greater than the critical value for a chi-square statistic with one degree of freedom (3.84). Therefore, we reject the hypothesis that the expected proportions are 1/3 and 2/3.

(b) The problem with the geneticist's classification procedure is that it allows for a heterozygote to be misclassified as a homozygote if none of its three progeny shows the recessive (short) phenotype. The probability of this event is 1/2 for any one offspring—therefore $(1/2)^3$ = 1/8 for all three offspring.

(c) The predicted frequencies must take into account the probability of misclassifying a heterozygote as a homozygote. The frequency of heterozygotes expected *a priori* (2/3) must be decreased by the probability of misclassification (1/8); thus, the predicted frequency of heterozygotes is $62 \times (2/3) \times (1 - 1/8) = 62 \times (7/12) = 36.2$. The predicted frequency of homozygotes is obtained by subtraction: $62 - 36.2 = 25.8$.

(d) The chi-square calculation is $(29 - 25.8)^2/25.8 + (33 - 36.2)^2/36.2 = 0.68$, which is much less than the critical value for a chi-square statistic with one degree of freedom. Therefore, we tentatively accept the idea that adjusting for the probability of misclassification explains the observed data.

CHAPTER 4

4.1 M and MN.

4.2 (a) all wild-type; (b) 3/4 wild-type, 1/4 albino; (c) 3/4 wild-type, 1/4 chinchilla; (d) 1/2 chinchilla, 1/2 albino; (e) 3/4 wild-type, 1/4 himalayan; (f) 1/2 himalayan, 1/2 albino.

4.3

	Parents	Offspring
(a)	yellow × yellow	2 yellow: 1 light belly
(b)	yellow × light belly	2 yellow: 1 light belly: 1 black and tan
(c)	black and tan × yellow	2 yellow: 1 black and tan: 1 black
(d)	light belly × light belly	all light belly
(e)	light belly × yellow	1 yellow: 1 light belly
(f)	agouti × black and tan	1 agouti: 1 black and tan
(g)	black and tan × black	1 black and tan: 1 black
(h)	yellow × agouti	1 yellow: 1 light belly
(i)	yellow × yellow	2 yellow: 1 light belly

4.4 (a) $S^1 S^2$, $S^1 S^3$, $S^2 S^3$; (b) $S^1 S^3$, $S^1 S^4$, $S^2 S^3$, $S^2 S^4$; (c) none; (d) $S^3 S^5$, $S^3 S^6$, $S^4 S^5$, $S^4 S^6$.

4.5 (a) all AB; (b) 1 A: 1 B; (c) 1 A: 1 B: 1 AB: 1 O; (d) 1 A: 1 O.

4.6 No. The woman must be *ii*; if her mate is $I^A I^B$; they could not have an *ii* child.

4.7 No. The woman is $I^A I^B$. One man could be either $I^A I^A$ or $I^A i$; the other could be either $I^B I^B$ or $I^B i$. Given the uncertainty in the genotype of each man, either could be the father of the child.

4.8 One gene with four alleles. (a) purple: $c^p c^p$, $c^p c^b$; $c^p c^t$, $c^p c^w$; (b) blue: $c^b c^b$, $c^b c^t$, $c^b c^w$; (c) turquoise: $c^t c^t$, $c^t c^w$; (d) light-blue: $c^t c^w$; (e) white: $c^w c^w$.

4.9 The woman is *ii* $L^M L^M$; the man is $I^A I^B L^M L^N$; the blood types of the children will be A and M, A and MN, B and M, and B and MN, all equally likely.

4.10 Cross homozygous *waltzing* with homozygous *tango*. If the mutations are alleles, all the offspring will have an uncoordinated gait; if they are not alleles, all the offspring will be wild-type. If the two mutations are alleles, they could be denoted with the symbols *v* (*waltzing*) and *vt* (*tango*).

4.11 The individuals III-4 and III-5 must be homozygous for recessive mutations in different genes; that is, one is *aa BB* and the other is *AA bb*; none of their children is deaf because all of them are heterozygous for both genes (*Aa Bb*).

4.12 9/16 dark red, 7/16 brownish-purple.

4.13 No. The test for allelism cannot be performed with dominant mutations.

4.14 The allele for yellow fur is homozygous lethal.

4.15 The mother is *Bb* and the father is *bb*. The chance that a daughter is *Bb* is 1/2. (a) The chance that the daughter will have a bald son is $(1/2) \times (1/2) = 1/4$. (b) The chance that the daughter will have a bald daughter is zero.

4.16 Dominant. The condition appears in every generation and nearly every affected individual has an affected parent. The exception, IV-2, had a father who carried the ataxia allele but did not manifest the trait–an example of incomplete penetrance.

4.17 (a) 3/4 walnut, 1/4 rose; (b) 1/2 walnut, 1/2 pea; (c) 3/8 walnut, 3/8 rose, 1/8 pea, 1/8 single; (d) 1/2 rose, 1/2 single.

4.18 *Rr pp × Rr Pp*.

4.19 12/16 white, 3/16 yellow, 1/16 green.

4.20 13/16 white, 3/16 colored.

4.21 9/16 dark red (wild-type), 3/16 brownish-purple, 3/16 bright red, 1/16 white.

4.22 9 black: 3 gray: 52 white.

4.23 9 black: 39 gray: 16 white.

4.24 (a) The simplest explanation for the inheritance of the trait is recessive epistasis combined with incomplete dominance, summarized in the following table:

Genotype	Phenotype	Frequency in F_2
AA B-	round	3/16
Aa B-	shield	6/16
aa B-	triangular	3/16
A- bb	no color	3/16
aa bb	no color	1/16

The cross was triangular *aa BB* male × no color *AA bb* female ---> shield *Aa Bb* F_1 ---> F_2 shown in the table. (b) The male without a colored patch must be either *A- bb* or *aa bb*; the female with the shield must be *Aa BB* or *Aa Bb*. Because none of the offspring lack a colored patch, the female's genotype is *Aa BB*, and the male's is *Aa bb*. Thus, shield male (*Aa BB*) × no color female (*Aa bb*) --> 1/4 round (*AA Bb*), 1/2 shield (*Aa Bb*), and 1/4 triangular (*aa Bb*).

4.25 (a) purple × red; (b) proportion white (*aa*) = 1/4; (c) proportion red (*A- B- C- dd*) = $(3/4)(3/4)(3/4)(1/2) = 27/128$, proportion white (*aa*) = 1/4 = 32/128, proportion blue (*A- B- cc Dd*) = $(3/4)(3/4)(1/4)(1/2) = 9/128$.

4.26 (a) Because the F_2 segregation is approximately 9 black: 3 chocholate: 4 yellow, coat color is determined by epistasis between two independently assorting genes: black = *B- E-*; chocholate = *bb E-*; yellow = *B- ee* or *bb ee*. (b) yellow pigment ---*E*--> brown pigment ---*B*--> black pigment.

4.27 (a) Because the F_2 segregation is approximately 9 red: 7 white, flower color is due to epistasis between two independently assorting genes: red = *A-B-* and white = *aa B-*, *A-bb*, or *aa bb*. (b) colorless precursor —*A*→ colorless product —*B*→ red pigment.

4.28 (a) proportion red = $(3/4)^3 \times (1/4)$ = 27/256; (b) proportion pink = $(3/4)^4 + [(3/4)^2 \times (1/4)]$ = 117/256; (c) proportion white = 1 – 144/256 = 112/256.

CHAPTER 5

5.1 The male-determining sperm carries a Y chromosome; the female-determining sperm carries an X chromosome.

5.2 Cross the singed male to wild-type females, and then intercross the offspring. If the singed bristle phenotype is due to an X-linked mutation, approximately half the F_2 males, but none of the F_2 females, will show it.

5.3 All the daughters will be green and all the sons will be rosy.

5.4 The cross is *go/go +/+* female × *+/Y bw/bw* male ---> F_1: *go/+ bw/+* females (wild-type eyes and body) and *go/Y bw/+* males (golden body, wild-type eyes). An intercross of the F_1 offspring yields the following F_2 phenotypes in both sexes:

Body	Eyes	Genotype	Proportion
golden	brown	*go/go* or Y *bw/bw*	$(1/2) \times (1/4)$ = 1/8
golden	wild-type	*go/go* or Y *+/bw* or +	$(1/2) \times (3/4)$ = 3/8
wild-type	brown	*+/go* or Y *bw/bw*	$(1/2) \times (1/4)$ = 1/8
wild-type	wild-type	*+/go* or Y *+/bw* or +	$(1/2) \times (3/4)$ = 3/8

5.5 XX is female, XY is male, XXY is female, XXX is female (but barely viable), XO is male (but sterile).

5.6

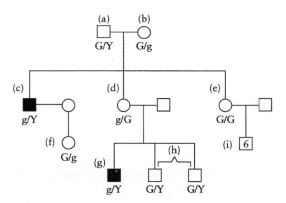

5.7 No. Defective color vision is caused by an X-linked mutation. The son's X chromosome came from his mother, not his father.

5.8 The risk for the child is P(woman transmits mutant allele) × P(child is male) = $(1/2) \times (1/2)$ = 1/4.

5.9 The risk for the child is P(mother is *C/c*) × P(mother transmits *c*) × P(child is male) = $(1/2) \times (1/2) \times (1/2)$ = 1/8; if the couple has already had a child with color blindness, P(mother is *C/c*) = 1, and the risk for each subsequent child is 1/4.

5.10 (a) The man is $X^c Y\ ii$; the woman is $X^+ X^c\ I^A\ I^B$. (b) Probability color blind = 1/2; probability type B blood = 1/2; combined probability = $(1/2) \times (1/2)$ = 1/4. (c) Probability color blind = 1/2; probability type A blood = 1/2; combined probability $(1/2) \times (1/2)$ = 1/4. (d) 0.

5.11   Each of the rare vermilion daughters must have resulted from the union of an X(v) X(v) egg with a Y-bearing sperm. The diplo-X eggs must have originated through nondisjunction of the X chromosomes during oogenesis in the mother. However, we cannot determine if the nondisjunction occurred in the first or the second meiotic division.

5.12   Each of the rare white-eyed daughters must have resulted from the union of an X(w) X(w) egg with a Y-bearing sperm. The rare diplo-X eggs must have originated through nondisjunction of the X chromosomes during the second meiotic division in the mother.

5.13   1/8.

5.14   3/8 wild-type (red), 5/8 brown for both male and female F_2 progeny.

5.15   Female.

5.16   Three-fourths will be phenotypically female (genotypically *tfm/Tfm, Tfm/Tfm,* or *tfm*/Y). Among the females, 2/3 (*tfm/Tfm* and *Tfm/Tfm*) will be fertile; the *tfm*/Y females will be sterile.

5.17   Male.

5.18   (a) Female; (b) intersex; (c) intersex; (d) male; (e) female; (f) male.

5.19   (a) 1/2 X^{bb} X^{bb} females, 1/2 X^{bb} Y^+ wild-type males; (b) 1/2 X^+ X^{bb} wild-type females, 1/2 X^{bb} Y^{bb} bobbed males; (c) 1/4 X^+ X^+ wild-type females, 1/4 X^+ X^{bb} wild-type females, 1/4 X^+ Y^{bb} wild-type males, 1/4 X^{bb} Y^{bb} bobbed males; (d) 1/4 X^+ X^{bb} wild-type females, 1/4 X^{bb} X^{bb} bobbed females, 1/4 X^+ Y^+ wild-type males, 1/4 X^{bb} Y^+ wild-type males.

5.20   Yes. The gene for feather patterning is on the Z chromosome. If we denote the allele for barred feathers as B and the allele for nonbarred feathers as b, the crosses are: *B/B* (barred) male × *b/W* (nonbarred) female ---> F_1: *B/b* (barred) males and *B/W* (barred) females. Intercrossing the F_1 produces *B/B* (barred) males, *B/b* (barred) males, *B/W* (barred) females, and *b/W* (nonbarred) females, all in equal proportions.

5.21   *Drosophila* does not achieve dosage compensation by inactivating one of the X chromosomes in females.

5.22   (a) Zero; (b) one; (c) one; (d) two; (e) three; (f) zero.

5.23

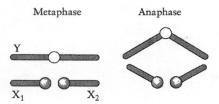

5.24   Color is determined by an autosomal gene (alleles A and a) and a sex-linked gene (alleles B and b) on the Z chromosome (females are ZW and males are ZZ) and the recessive alleles are mutually epistatic--that is, *aa, bb,* or *bW* birds have red eyes, and *A- B-* or *A- BW* birds have brown eyes.

Cross 1

P	Strain A red male	×	Strain B red female
	aa BB		*AA bW*
F_1	*Aa Bb* brown males	×	*Aa BW* brown females
F_2	*A- Bb* brown males (6/16)		*A- BW* brown females (3/16)
	aa Bb red males (2/16)		*A- bW* red females (4/16)
			aa BW red females (1/16)

Cross 2

P	Strain A red female	×	Strain B red male
	aa BW		*AA bb*
F_1	*Aa Bb* brown males	×	*Aa bW* red females
F_2	*A- Bb* brown males (3/16)		*A- bW* brown females (3/16)
	A- bb red males (3/16)		*A- bW* red females (3/16)
	aa b- red males (2/16)		*aa -W* red females (2/16)

5.25 Eye color in canaries is due to a gene on the Z chromosome, which is present in two copies in males and one copy in females. The allele for pink color at hatching (p) is recessive to the allele for black color at hatching (P). There is no eye color gene on the other sex chromosome (W), which is present in one copy in females and absent in males. The parental birds were genotypically p/W (cinnamon females) and P/P (green males). Their F_1 sons were genotypically p/P (with black eyes at hatching). When these sons were crossed to green females (genotype P/W), they produced F_2 progeny that sorted into three categories: males with black eyes at hatching ($P/-$, half the total progeny), females with black eyes at hatching (P/W, a fourth of the total progeny), and females with pink eyes at hatching (p/W, a fourth of the total progeny). When these sons were crossed to cinnamon females (genotype p/W), they produced F_2 progeny that sorted into four equally frequent categories: males with black eyes at hatching (genotype P/p), males with pink eyes at hatching (genotype p/p), females with black eyes at hatching (genotype P/W), and females with pink eyes at hatching (genotype p/W).

CHAPTER 6

6.1 Use one of the banding techniques.

6.2 46, XX, 22q- or 46, XY, 22q-, depending on the sex chromosome constitution.

6.3 In allotetraploids, each member of the different sets of chromosomes can pair with a homologous partner during prophase I and then disjoin during anaphase I. In triploids, disjunction is irregular because homologous chromosomes associate during prophase I by forming either bivalents and univalents or by forming trivalents.

6.4 (a) Species A is an allotetraploid with a genome from each of species C and species D; (b) 0 bivalents and 15 univalents; (c) 0 bivalents and 15 univalents.

6.5 The fertile plant is an allotetraploid with 7 pairs of chromosomes from species A and 9 pairs of chromosomes from species B; the total number of chromosomes is $(2 \times 7) + (2 \times 9) = 32$.

6.6 The F_1 hybrid had 5 chromosomes from species X and 7 from species Y, for a total of 12. When this hybrid was backcrossed to species Y, the few progeny that were produced had $5 + 7 = 12$ chromosomes from the hybrid and 7 from species Y, for a total of 19. This hybrid was therefore a triploid. Upon self-fertilization, a few F_2 plants were formed, each with 24 chromosomes. Presumably the chromosomes in these plants consisted of $2 \times 5 = 10$ from species X and $2 \times 7 = 14$ from species Y. These vigorous and fertile F_2 plants were therefore allotetraploids.

6.7 XX is female, XY is male, XO is female (but sterile), XXX is female, XXY is male (but sterile), XYY is male.

6.8 1/2.

6.9 The fly is a gynandromorph, that is, a sexual mosaic. The yellow tissue is X(y)/O and the gray tissue is X(y)/X(+). This mosaicism must have arisen through loss of the X chromosome that carried the wild-type allele, presumably during one of the early embryonic cleavage divisions.

6.10 Approximately half the progeny should be disomic $ey/+$ and half should be monosomic ey/O. The disomic progeny will be wild-type, and the monosomic progeny will be eyeless.

6.11 The daughter with Turner and Hunter syndrome in family A must have received her single X chromosome from her mother, who is heterozygous for the mutation causing Hunter syndrome. The daughter did not receive a sex chromosome from her father because sex chromosome nondisjunction must have occurred during meiosis in his germline. The son with Klinefelter syndrome in family B is karyotypically XXY, and both of his X chromosomes carry the mutant allele for Hunter syndrome. This individual must have received two mutant X chromosomes from his heterozygous mother due to X chromosome nondisjunction during the second meiotic division in her germline. The daughter with Hunter syndrome in family C is karyotypically XX, and both of her X chromosomes carry the mutant allele for Hunter syndrome. This individual received the two mutant X chromosomes from her heterozygous mother through nondisjunction during the second meiotic division in the mother's germline. Furthermore, because the daughter did not receive a sex chromosome from her father, sex chromosome nondisjunction must have occurred during meiosis in his germline too.

6.12   Nondisjunction must have occurred in the mother. The color blind woman with Turner syndrome was produced by the union of an X-bearing sperm, which carried the mutant allele for color blindness, and a nullo-X egg.

6.13   XYY men would produce more children with sex chromosome abnormalities because their three sex chromosomes will disjoin irregularly during meiosis. This irregular disjunction will produce a variety of aneuploid gametes, including the XY, YY, XYY, and nullo sex chromosome constitutions.

6.14   The animal is heterozygous for an inversion:

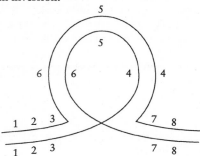

6.15   (a) Deletion:

(b) Duplication:

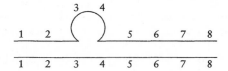

(c) A terminal inversion:

6.16   In translocation heterozygotes, only alternate segregation leads to euploid gametes, and the frequency of alternate segregation is typically around 50 percent.

6.17

6.18 The exceptional male, whose genotype is *bw*/+ *st*/+, is heterozygous for a translocation between chromosomes 2 and 3. It is not possible to determine whether the translocation is between the two mutant chromosomes or between the two wild-type chromosomes, that is, whether it is T(*bw*; *st*) or T(+; +); however it clearly is not between a mutant chromosome and a wild-type chromosome, that is, T(*bw*; +) or T(+; *st*). If it were, the progeny would be either brown or scarlet, not either wild-type or white.

6.19 The boy carries a translocation between chromosome 21 and another chromosome, say chromosome 14. He also carries a normal chromosome 21 and a normal chromosome 14. The boy's sister carries the translocation, one normal chromosome 14, and two normal copies of chromosome 21.

6.20 A compound chromosome is composed of segments from the same pair of chromosomes, as when two X chromosomes become attached to each other. A Robertsonian translocation involves a fusion of segments from two different pairs of chromosomes. These segments fuse at or near the centromeres, usually with the loss of the short arms of each of the participating chromosomes.

6.21 All the daughters will be yellow-bodied and all the sons will be white-eyed.

6.22 Zygotes produced by this couple will be either trisomic or monosomic for chromosome 21. Thus, 100% of their viable children will develop Down syndrome.

6.23 The three populations are related by a series of inversions:

```
P1    1 2 3 4 5 6 7 8 9 10
              └──────┘
P2    1 2 3 9 8 7 6 5 4 10
             └──────┘
P3    1 2 3 9 8 5 6 7 4 10
```

6.24 Arrangement (a) produced (c) by inversion of segment 2345; (c) produced (f) by a duplication of band 2; (f) produced (d) by a deletion of band 5; (d) produced (e) by inversion of segment 43226; (e) produced (b) by inversion of segment 622.

6.25 The mother is heterozygous for a reciprocal translocation between the long arms of the large and small chromosomes; a piece from the long arm of the large chromosome has been broken off and attached to the long arm of the short chromosome. The child has inherited the rearranged large chromosome and the normal small chromosome from the mother. Thus, because the rearranged large chromosome is deficient for some of its genes, the child is hypoploid.

6.26 The phenotype in the female offspring is mosaic because one of the X chromosomes is inactivated in each of their cells. If the translocated X is inactivated, the autosome attached to it could also be partially inactivated by a spreading of the inactivation process across the translocation breakpoint. This spreading could therefore inactivate the color-determining gene on the translocated autosome and cause patches of tissue to be phenotypically mutant.

6.27 The sons will have bright red eyes because they will inherit the Y chromosome with the *bw*⁺ allele from their father. The daughters will have white eyes because they will inherit an X chromosome from their father.

6.28 A reciprocal translocation between chromosomes 2 and 3. One translocated chromosome carries the wild-type alleles of *vg* and *h* on chromosomes 2 and 3, respectively, and the other carries the recessive mutant alleles of these genes. The chromosome that carries the *ey* gene (chromosome 4) is not involved in the rearrangement.

6.29 XX zygotes will develop into males because one of their X chromosomes carries the *SRY* gene that was translocated from the Y chromosome. XY zygotes will develop into females because their Y chromosome has lost the *SRY* gene.

CHAPTER 7

7.1 If Mendel had known of the existence of chromosomes, he would have realized that the number of factors determining traits exceeds the number of chromosomes, and he would have concluded that some factors must be linked on the same chromosome. Thus, Mendel would have revised the Principle of Independent Assortment to say that factors on different chromosomes (or far apart on the same chromosome) are inherited independently.

7.2 The class represented by 351 offspring indicates that at least two of the three genes are linked.

7.3 No. The genes *a* and *d* could be very far apart on the same chromosome—so far apart that they recombine freely, that is, 50 percent of the time.

7.4 A two-strand double crossover must have occurred.

7.5 1/19.

7.6 20%.

7.7 (a) Cross: $a^+ b^+ / a^+ b^+ \times ab/ab$
      Gametes: $a^+ b^+$ from one parent, ab from the other
      F_1: $a^+ b^+/ab$
  (b) 40% $a^+ b^+$, 40% ab, 10% $a^+ b$, 10% ab^+
  (c) F_2 from testcross: 40% $a^+ b^+/ab$, 40% ab/ab, 10% $a^+ b/ab$, 10% ab^+/ab
  (d) Coupling linkage phase
  (e) F_2 from intercross:

		Sperm		
	40% $a^+ b^+$	40% $a b$	10% $a^+ b$	10% $a b^+$
40% $a^+ b^+$	16% $a^+ b^+/a^+ b^+$	16% $a^+ b^+/a b$	4% $a^+ b^+/a^+ b$	4% $a^+ b^+/a b^+$
40% ab	16% $ab/a^+ b^+$	16% $a b/a b$	4% $a b/a^+ b$	4% $a b/a b^+$
10% $a^+ b$	4% $a^+ b/a^+ b^+$	4% $a^+ b/a b$	1% $a^+ b/a^+ b$	1% $a^+ b/a b^+$
10% ab^+	4% $a b^+/a^+ b^+$	4% $a b^+/a b$	1% $a b^+/a^+ b$	1% $a b^+/a b^+$

Eggs (row label, left margin)

Summary of phenotypes:
a^+ and b^+	66%
a^+ and b	9%
a and b^+	9%
a and b	16%

7.8 (a) Cross: $a^+ b/a^+ b \times ab^+/ab^+$
      Gametes: $a^+ b$ from one parent, ab^+ from the other
      F_1: $a^+ b/ab^+$
  (b) 40% $a^+ b$, 40% ab^+, 10% $a^+ b^+$, 10% ab
  (c) F_2 from testcross: 40% $a^+ b/ab$, 40% ab^+/ab, 10% $a^+ b^+/ab$, 10% ab/ab
  (d) Repulsion linkage phase
  (e) F_2 from intercross:

		Sperm		
	40% $a^+ b$	40% $a b^+$	10% $a^+ b^+$	10% $a b$
40% $a^+ b$	16% $a^+ b/a^+ b$	16% $a^+ b/a b^+$	4% $a^+ b/a^+ b^+$	4% $a^+ b/a b$
40% ab^+	16% $a b^+/a^+ b$	16% $a b^+/a b^+$	4% $a b^+/a^+ b^+$	4% $a b^+/a b$
10% $a^+ b^+$	4% $a^+ b^+/a^+ b$	4% $a^+ b^+/a b^+$	1% $a^+ b^+/a^+ b^+$	1% $a^+ b^+/a b$
10% ab	4% $a b/a^+ b$	4% $a b/a b^+$	1% $a b/a^+ b^+$	1% $a b/a b$

Eggs (row label, left margin)

Summary of phenotypes:
a^+ and b^+ 51%
a^+ and b 24%
a and b^+ 24%
a and b 1%

7.9   Coupling heterozygotes $a^+ b^+/ab$ would produce the following gametes: 30% $a^+ b^+$, 30% ab, 20% $a^+ b$, 20% ab^+; repulsion heterozygotes $a^+ b/ab^+$ would produce the following gametes: 30% $a^+ b$, 30% ab^+, 20% $a^+ b^+$, 20% ab. In each case, the frequencies of the testcross progeny would correspond to the frequencies of the gametes.

7.10   No. The leaf color and tassel seed traits assorf independently.

7.11   Yes. Recombination frequency = (24 + 26)/(126 + 24 + 26 + 124) = 0.167. Cross:

$\dfrac{b \ vg}{b^+ vg^+}$ female $\times$ $\dfrac{b \ vg}{b \ vg}$ male

$\dfrac{b \ vg}{b^+ vg^+}$	$\dfrac{b \ vg}{b \ vg}$	$\dfrac{b \ vg}{b^+ vg}$	$\dfrac{b \ vg}{b \ vg^+}$
126	124	24	26

7.12   Yes. Recombination frequency = (23 + 26)/(23 + 127 + 124 + 26) = 0.163. Cross:

$\dfrac{b^+ vg}{b \ vg^+}$ female $\times$ $\dfrac{b \ vg}{b \ vg}$ male

$\dfrac{b \ vg}{b^+ vg^+}$	$\dfrac{b \ vg}{b \ vg}$	$\dfrac{b \ vg}{b^+ vg}$	$\dfrac{b \ vg}{b \ vg^+}$
23	26	127	124

7.13   Yes. Recombination frequency is estimated by the frequency of black offspring among the colored offspring: 34/(66 + 34) = 0.34. Cross:

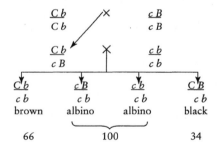

7.14   Plant I has the genotype $D \ P/d \ p$, and when crossed to a $d \ p/d \ p$ plant, produces four classes of progeny:

$\underline{D \ p}$	$\underline{d \ p}$	$\underline{D \ p}$	$\underline{d \ P}$
d p	d p	d p	d p
81	79	22	17

Plant II has the genotype $D \ p/d \ P$, and when crossed to a $d \ p/d \ p$ plant, produces four classes of progeny:

$\underline{D \ p}$	$\underline{d \ P}$	$\underline{D \ p}$	$\underline{d \ p}$
d p	d p	d p	d p
21	18	5	4

If the two plants are crossed ($D \ p/d \ p \times D \ p/d \ P$), the phenotypes of the offspring can be predicted from the following table.

		Gametes from plant I			
		D p 0.40	*d p* 0.40	*D p* 0.10	*d P* 0.10
	D p 0.40	*D p/D P* 0.16	*D p/d p* 0.16	*D p/D p* 0.04	*D p/d P* 0.04
Gametes from plant II	*d P* 0.40	*d P/D P* 0.16	*d P/d p* 0.16	*d P/D p* 0.04	*d P/d P* 0.04
	D P 0.10	*D P/D P* 0.04	*D P/d p* 0.04	*D P/D p* 0.01	*D P/d P* 0.01
	d p 0.10	*d p/D P* 0.04	*d p/d p* 0.04	*d p/D p* 0.01	*d p/d P* 0.01

Summary of phenotypes
tall, spherical   0.54
tall, pear   0.21
dwarf, spherical   0.21
dwarf, pear   0.04

7.15 Because the two chromosomes assort independently, the genetic makeup of the gametes (and, therefore, of the backcross progeny) can be obtained from the following table.

		Chromosome 3 in gametes			
		c d 0.425	*c⁺ d⁺* 0.425	*c d⁺* 0.075	*c⁺ d* 0.075
	a b 0.40	*a b c d* 0.17	*a b c⁺ d⁺* 0.17	*a b c d⁺* 0.03	*a b c⁺ d* 0.03
Chromosome 2 in gametes	*a⁺ b⁺* 0.40	*a⁺ b⁺ c d* 0.17	*a⁺ b⁺ c⁺ d⁺* 0.17	*a⁺ b⁺ c d⁺* 0.03	*a⁺ b⁺ c⁺ d* 0.03
	a⁺ b 0.10	*a⁺ b c d* 0.0425	*a⁺ b c⁺ d⁺* 0.0425	*a⁺ bcd⁺* 0.0075	*a⁺ b c⁺ d* 0.0075
	a b⁺ 0.10	*a b⁺ c d* 0.0425	*a b⁺ c⁺ d⁺* 0.0425	*a b⁺ c d⁺* 0.0075	*a b⁺ c⁺ d* 0.0075

7.16 (a) The F_1 females, which are *sr e⁺/sr⁺ e*, produce four types of gametes: 46% *sr e⁺*, 46% *sr⁺ e*, 4% *sr e*, 4% *sr⁺ e⁺*. (b) The F_1 males, which have the same genotype as the F_1 females, produce two types of gametes: 50% *sr e⁺*, 50% *sr⁺ e*; remember, there is no recombination in *Drosophila* males. (c) 46% striped, gray; 46% unstriped, ebony; 4% striped, ebony; 4% unstriped, gray. (d) The offspring from the intercross can be obtained from the following table.

		Sperm	
		sr e⁺ 0.50	*sr⁺ e* 0.50
	sr e⁺ 0.46	*sr e⁺/sr e⁺* 0.23	*sr e⁺/sr⁺ e* 0.23
Eggs	*sr⁺ e* 0.46	*sr⁺ e/sr e⁺* 0.23	*sr⁺ e/sr⁺ e* 0.23
	sr e 0.04	*sr e/sr e⁺* 0.002	*sr e/sr⁺ e* 0.002
	sr⁺ e⁺ 0.04	*sr⁺ e⁺/sr e⁺* 0.002	*sr⁺ e⁺/sr⁺ e* 0.002

Summary of phenotypes
striped, gray   0.25
unstriped, gray   0.50
striped, ebony   0
unstriped, ebony   0.25

7.17 (a) The F_1 females, which are *cn vg⁺/cn⁺ vg*, produce four types of gametes: 45% *cn vg⁺*, 45% *cn⁺ vg*, 5% *cn⁺ vg⁺*, 5% *cn vg*. (b) 45% cinnabar eyes, normal wings; 45% reddish-brown eyes, vestigial wings; 5% reddish-brown eyes, normal wings; 5% cinnabar eyes, vestigial wings.

7.18 In the enumeration below, classes 1 and 2 are parental types, classes 3 and 4 result from a single crossover between *st* and *ss*, classes 5 and 6 result from a single crossover between *ss* and *e*, and classes 7 and 8 result from a double crossover, with one of the exchanges between *st* and *ss* and the other between *ss* and *e*.

Class	Phenotypes	(a) Frequency with no interference	(b) Frequency with complete interference
1	scarlet, spineless	0.3784	0.37
2	ebony 0.3784	0.37	
3	scarlet, ebony	0.0616	0.07
4	spineless 0.0616	0.07	
5	scarlet, spineless, ebony	0.0516	0.06
6	wild-type 0.0516	0.06	
7	scarlet 0.0084	0	
8	spineless, ebony	0.0084	0

7.19 In the enumeration below, classes 1 and 2 are parental types, classes 3 and 4 result from a single crossover between *Pl* and *Sm*, classes 5 and 6 result from a single crossover between *Sm* and *Py*, and classes 7 and 8 result from a double crossover, with one of the exchanges between *Pl* and *Sm* and the other between *Sm* and *Py*.

Class	Phenotypes	(a) Frequency with no interference	(b) Frequency with complete interference
1	purple, salmon, pigmy	0.405	0.40
2	green, yellow, normal	0.405	0.40
3	purple, yellow, normal	0.045	0.05
4	green, salmon, pigmy	0.045	0.05
5	purple, salmon, normal	0.045	0.05
6	green, yellow, pigmy	0.045	0.05
7	purple, yellow, pigmy	0.005	0
8	green, salmon, normal	0.005	0

7.20 In the enumeration below, classes 1 and 2 are parental types, classes 3 and 4 result from crossing over between *Tu* and *J2*, and classes 5 and 6 result from crossing over between *J2* and *Gl3*; only the chromosome from the triply heterozygous F_1 plant is shown. Because interference is complete, there are no double crossover progeny.

Class	Genotype	Frequency
1	*tu j2 gl3*	0.445
2	*Tu J2 Gl3*	0.445
3	*tu J2 Gl3*	0.025
4	*Tu j2 gl3*	0.025
5	*tu j2 Gl3*	0.030
6	*Tu J2 gl3*	0.030

7.21 The double crossover classes, which are the two that were not observed, establish that the gene order is *y—w—ec*. Thus, the F_1 females had the genotype *y w ec/+ + +*. The distance between *y* and *w* is estimated by the frequency of recombination between these two genes: (8 + 7)/1000 = 0.015; similarly, the distance between *w* and *ec* is (18 + 23)/1000 = 0.041. Thus, the genetic map for this segment of the X chromosome is *y*—1.5 cM—*w*—4.1 cM—*ec*.

7.22 The last two classes, consisting of yellow, bar flies and vermilion flies, with a total of 50 progeny, result from double crossovers. Thus, the order of the genes is *y--v--B*, and the F_1 females had the genotype *y v B/+ + +*. The distance between *y* and *v* is the average number of crossovers between them: (244 + 50)/1000 = 29.4 cM; likewise, the distance between *v* and *B* is (160 + 50)/1000 = 21.0 cM. Thus, the genetic map is *y--29.4 cM--v--21.0 cM--B*.

7.23 (a) Two of the classes (the parental types) vastly outnumber the other six classes (recombinant types); (b) *st + +/+ ss e*; (c) *st—ss—e*; (d) [(145 + 122) × 1 + (18) × 2]/1000 = 30.3 cM; (e) (122 + 18)/1000 = 14.0 cM; (f) (0.018)/(0.163 × 0.140) = 0.789. (g) *st + +/+ ss e* females × *st ss e/st ss e* males → 2 parental classes and 6 recombinant classes.

7.24 The female will produce four kinds of gametes: 30% *w +*, 30% *+ dor*, 20% *w dor*, and 20% *+ +*; thus, 80% of the progeny will be mutant (either white or deep orange), and 50% will be pigmented (either red or deep orange).

7.25 Ignore the female progeny and base the map on the male progeny. The parental types are $+ + z$ and $x\ y\ +$. The two missing classes ($+ y +$ and $x + z$) must represent double crossovers; thus, the gene order is y—x—z. The distance between y and x is $(32 + 27)/1000 = 5.9$ cM and that between x and z is $(31 + 39)/1000 = 7.0$ cM. Thus, the map is y—5.9 cM—x—7.0 cM—z. The coefficient of coincidence is zero.

7.26 The F, females are genotypically $pn +/+ g$. Among their sons, 40 percent will be recombinant for the two x-linked genes, and half of the recombinants will have the wild-type alleles of these genes. Thus the frequency of sons with dark red eyes will be $1/2 \times 40\% = 20$ percent.

7.27 $(P/2)^2$.

7.28 5.

7.29 From the parental classes, $+ + c$ and $a\ b\ +$, the heterozygous females must have had the genotype $+ + c/a\ b\ +$. The missing classes, $+ b +$ and $a + c$, which would represent double crossovers, establish that the gene order is b—a—c. The distance between b and a is $(96 + 110)/1000 = 20.6$ cM and that between a and c is $(65 + 75)/1000 = 14.0$ cM. Thus, the genetic map is b—20.6 cM—a—14.0 cM—c.

7.30 2.4 chiasmata.

7.31 5.4 chiasmata.

7.32 100 cM.

7.33 The lethal mutation resides in band 7.

7.34 II-1 has the genotype $C\ h/c\ H$, that is, she is a repulsion heterozygote for the alleles for color blindness (c) and hemophilia (h). None of her children are recombinant for these alleles.

7.35 II-1 is either (a) $C\ h/c\ H$ or (b) $c\ h/C\ H$. Her four sons have the genotypes $c\ h$ (1), $C\ h$ (2), $c\ H$ (3), and $C\ H$ (4). If II-1 has the genotype $C\ h/c\ H$, sons 1 and 4 are recombinant and sons 2 and 3 are nonrecombinant. If II-1 has the genotype $c\ h/C\ H$, sons 2 and 3 are recombinant and sons 1 and 4 are nonrecombinant. Either way, the frequency of recombination is 0.5.

7.36 The woman is a repulsion heterozygote for the alleles for color blindness and hemophilia—that is, she is $C\ h/c\ H$. If the woman has a boy, the chance that he will have hemophilia is 0.5 and the chance that he will have color blindness is 0.5. If we specify that the boy have only one of these two conditions, then the chance that he will have hemophilia is 0.45 and the chance that he will have color blindness is 0.45. The reason is that the boy will inherit a nonrecombinant X chromosome with a probability of 0.9, and half the nonrecombinant X chromosomes will carry the mutant allele for color blindness and the other half will carry the mutant allele for hemophilia. The chance that the boy will have both conditions is 0.05 and the chance that he will have neither condition is 0.05. The reason is that the boy will inherit a recombinant X chromosome with a probability of 0.1, and half the recombinant X chromosomes will carry both mutant alleles and the other half will carry neither mutant allele.

7.37 The gene is on chromosome 6, which is present in lines B and D, but not in any of the other lines.

7.38 The gene for the protein is located in 15p.

7.39 M2 carries an inversion that suppresses recombination in the chromosome.

7.40 A two-strand double crossover within the inversion; the exchange points of the double crossover must lie between the genetic markers and the inversion breakpoints.

CHAPTER 8

8.1 Viruses reproduce and transmit their genes to progeny viruses. They utilize energy provided by host cells and respond to environmental and cellular signals like other living organisms. However, viruses are obligate parasites; they can reproduce only in appropriate host cells.

8.2 Bacteriophages reproduce in bacteria; other viruses reproduce in protists, plants, and animals.

8.3 Bacteriophage T4 is a virulent phage. When it infects a host cell, it reproduces and kills the host cell in the process. Bacteriophage lambda can reproduce and kill the host bacterium—the lytic response—just like phage T4, or it can insert its chromosome into the chromosome of the host and remain there in a dormant state—the lysogenic response.

8.4 The mature (packaged) lambda chromosome and the lambda prophage are circular permutations of one another (see Fig. 8.6).

8.5 The insertion of the phage λ chromosome into the host chromosome is a site-specific recombination process catalyzed by an enzyme that recognizes specific sequences in the λ and *E. coli* chromosomes. Crossing over between homologous chromosomes is not sequence-specific. It can occur at many sites along the two chromosomes.

8.6 Genetic crosses between viruses are performed by simultaneously infecting host cells with viruses of different genotypes and examining the progeny viruses for recombinant genotypes. Recombination occurs between viral chromosomes in infected cells; this recombination occurs within a population of replicating viral chromosomes rather than between the paired homologous chromosomes during meiosis in eukaryotes.

8.7 Linkage maps are based on population averages. With circularly permuted chromosomes, genes that are close together on some chromosomes are at opposite ends of other chromosomes. At the population level, these circularly permuted chromosomes yield a circular genetic map.

8.8 The linear chromosome replicates to produce two or more linear molecules, which undergo recombination within the terminally redundant regions to produce long DNA molecules—concatamers—many chromosomes in length (see Figure 8.12). When the concatameric DNA is packaged into phage heads, each head holds one complete chromosome plus a little bit extra (the terminally redundant region). Therefore, as each head is filled and the DNA cleaved, a population of circularly permutated and terminally redundant chromosomes is produced (see Figure 8.13).

8.9 Recombination has occurred, producing wild-type or prototrophic bacteria.

8.10 Perform two experiments: (1) determine whether the process is sensitive to DNase, and (2) determine whether cell contact is required for the process to take place. The cell contact requirement can be tested by a U-tube experiment (see Fig. 8.17). If the process is sensitive to DNase, it is similar to transformation. If cell contact is required, it is similar to conjugation. If it is neither sensitive to DNase nor requires cell contact, it is similar to transduction.

8.11

RECOMBINATION PROCESS	AGENT MEDIATING DNA TRANSFER	SIZE OF DNA UNITS TRANSFERRED	STATE OF DONOR DNA IN RECOMBINANT CELL
Transformation	Active uptake of free DNA	Small (about 1/200 to 1/100 of a chromosome)	Single strand integrated (see Fig. 8.20)
Transduction	Bacteriophage	Small (about 1/100 to 1/50 of a chromosome)	Integrated into host chromosome (except autonomous in abortive transduction)
Sexduction	F factor	Variable (see Fig. 8.34)	Initially added to the host cell as separate plasmid; may undergo recombination with host chromosome to yield stable sexductant

8.12 F⁻ cells, no F factor present; F⁺ cells, autonomous F factor; Hfr cells, integrated F factor (see Fig. 8.22). (b) F⁺ and Hfr cells have F pili; F⁻ cells do not. (c) F⁻ cells are converted to F⁺ cells by the conjugative transfer of F factors from F⁺ cells. Hfr cells are formed when F factors in F⁺ cells become integrated into the chromosomes of these cells. Hfr cells become F⁺ cells when the integrated F factors exit the chromosome and become autonomous (self-replicating) genetic elements.

8.13 (a) F′ factors are useful for genetic analyses where two copies of a gene must be present in the same cell, for example, in determining dominance relationships. (b) F′ factors are formed by abnormal excision of F factors from Hfr chromosomes (see Fig. 8.33). (c) By the conjugative transfer of an F′ factor from a donor cell to a recipient (F⁻) cell.

8.14 Generalized transduction: (1) transducing particles often contain only host DNA; (2) transducing particles may carry any segment of the host chromosome. Thus, all host genes are transduced. Specialized transduction: (1) transducing particles carry a recombinant chromosome, which contains both phage DNA and host DNA; (2) only host genes that are adjacent to the prophage integration site are transduced.

8.15 IS elements (or insertion sequences) are short (800–1400 nucleotide pairs) DNA sequences that are transposable—that is, capable of moving from one position in a chromosome to another position or from one chromosome to another chromosome. IS elements mediate recombination between nonhomologous DNA molecules—for example, between F factors and bacterial chromosomes.

8.16 By interrupting conjugation at various times after the donor and recipient cells are mixed (using a blender or other form of agitation), one can determine the length of time required to transfer a given genetic marker from an Hfr cell to an F⁻ cell. Since the chromosome is transferred in a linear sequence, the positions of genetic markers can be ordered relative to each other.

8.17 Cotransduction refers to the simultaneous transduction of two different genetic markers to a single recipient cell. Since bacteriophage particles can package only 1/100 to 1/50 of the total bacterial chromosome, only markers that are relatively closely linked can be cotransduced. The frequency of cotransduction of any two markers will be an inverse function of the distance between them on the chromosome. As such, this frequency can be used as an estimate of the linkage distance. Specific cotransduction-linkage functions must be prepared for each phage-host system studied.

8.18

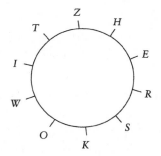

8.19 *lacY — lacZ — proC.*

8.20 *anth — A34 — A223 — A46.*

8.21 *anth — A487 — A223 — A58.*

8.22 *am* A453 — *nrd*11 — *nd*28 — *am*M69 — *am*N54.

8.23 PD >> NPD, so the genes are linked; the distance between the genes is estimated as $[(1/2) \times (23 + 3)]/48 = 30$ cM.

8.24 The distance is half the frequency of second division segregation asci: $(1/2) \times (84/200) = 21$ cM.

8.25 The distance between *a* and *b* is $[(1/2) \times (220 + 14)]/2000 = 5.85$ cM; the distance between *a* and the centromere is $(1/2) \times (220/2000) = 5.5$ cM; the distance between *b* and the centromere is $(1/2) \times (14/2000) = 0.35$ cM. Thus, the genetic map is *a*—5.5 cM—centromere—0.35 cM—*b*.

8.26 Genes *b* and *c* are linked and are 17 cM apart. Gene *a* is located on a different chromosome.

8.27 The *arg* and *thi* loci are unlinked; however, the *arg* and *leu* loci are linked. The distance between *arg* and *leu* is [(1/2) × 44]/200 = 11 cm; the distance between *arg* and its centromere is (1/2) × (1/300) = 0.17 cM; thus, the genetic map for the chromosome that carries *arg* and *leu* is centromere—0.17 cM—*arg*—11 cM—*leu*. The *thi* gene is very tightly linked to its centromere.

8.28 (a) 2 crossovers; (b) one exchange between *x* and *y*, the other between *y* and *z*; (c) 4-strand double crossover.

8.29 (a) 7 cM; (b) these genes are not linked.

8.30 (a) 8 cM; (b) 21.5 cM; (c) the *arg* and *w* loci are located in different arms of a chromosome.

CHAPTER 9

9.1 (a) Griffith's *in vivo* experiments demonstrated the occurrence of transformation in pneumococcus. They provided no indication as to the molecular basis of the transformation phenomenon. Avery and colleagues carried out *in vitro* experiments, employing biochemical analyses to demonstrate that transformation was mediated by DNA.

   (b) Griffith showed that a transforming substance existed; Avery et al. defined it as DNA.

   (c) Griffith's experiments did not include any attempt to characterize the substance responsible for transformation. Avery *et al.* isolated DNA in "pure" form and demonstrated that it could mediate transformation.

9.2 (a) no effect.

   (b) no effect.

   (c) DNase will destroy the capacity of the extract to transform type IIR cells to Type IIIS by degrading the DNA in the extract. Protease and RNase will degrade the proteins and RNA, respectively, in the extract. They will have no effect, since the proteins and RNA are not involved in transformation.

9.3 Purified DNA from Type III cells was shown to be sufficient to transform Type II cells. This occurred in the absence of any dead Type III cells.

9.4 About $\frac{1}{2}$ protein, $\frac{1}{2}$ DNA. A single long molecule of DNA is enclosed within a complex "coat" composed of many proteins.

9.5 (a) The objective was to determine whether the genetic material was DNA or protein.

   (b) By labeling phosphorus, a constituent of DNA, and sulfur, a constituent of protein, in a virus, it was possible to demonstrate that only the labeled phosphorus was introduced into the host cell during the viral reproductive cycle. The DNA was enough to produce new phages.

   (c) Therefore DNA, not protein, is the genetic material.

9.6 When tobacco leaves were infected with reconstituted virus particles containing RNA from type A viruses and protein from type B viruses, the progeny viruses were type A, showing that RNA, not protein, carries the genetic information in TMV.

9.7 (a) The RNA genome of a virus is packaged inside a protein coat, whereas a viroid is composed of naked RNA.

   (b) A prion is composed of protein, whereas a viroid is RNA.

   (c) Viroids and prions are both infectious; they can maintain the phenotype they produce from generation to generation.

9.8 (a) The ladderlike pattern was known from X-ray diffraction studies. Chemical analyses had shown that a 1:1 relationship existed between the organic bases adenine and thymine and between cytosine and guanine. Physical data concerning the length of each spiral and the stacking of bases were also available.

(b) Watson and Crick developed the model of a double helix, with the rigid strands of sugar and phosphorus forming spirals around an axis, and hydrogen bonds connecting the complementary bases in nucleotide pairs.

9.9 (a) A multistranded, spiral structure was suggested by the X-ray diffraction patterns. A double-stranded helix with specific base-pairing nicely fits the 1:1 stoichiometry observed for A:T and G:C in DNA.

(b) Use of the known hydrogen-bonding potential of the bases provided a means of holding the two complementary strands in a stable configuration in such a double helix.

9.10 (a) 400,000.

(b) 20,000.

(c) 400,000.

(d) 68,000 nm.

9.11 3'-C A G T A C T G-5'.

9.12 (a) DNA has one atom less of oxygen than RNA in the sugar part of the molecule. In DNA, thymine replaces the uracil that is present in RNA. (In certain bacteriophages, DNA-containing uracil is present.) DNA is most frequently double-stranded, but bacteriophages such as ϕX174 contain single-stranded DNA. RNA is most frequently single-stranded. Some viruses, such as the Reoviruses, however, contain double-stranded RNA chromosomes.

9.13 13%.

9.14 No. TMV RNA is single-stranded. Thus the base-pair stoichiometry of DNA does not apply.

9.15 (a) false.

(b) false.

(c) true.

(d) true.

(e) true.

(f) true.

(g) false.

(h) true.

(I) true.

(j) false.

(k) true.

(l) false.

(m) true.

9.16 (a) double-stranded DNA.

(b) single-stranded DNA.

(c) single-stranded RNA.

9.17 The B form of DNA helix is that proposed by Watson and Crick and is the conformation that DNA takes under physiological conditions. It is a right-handed double helical coil with 10 bases per turn of the helix and a diameter of 1.9 nm. It has a major and a minor groove. Z-DNA is left-handed, has 12 bases per turn, a single deep groove, and is 1.8 nm in diameter. Its sugar-phosphate backbone takes a zigzagged path, and it is G:C rich. A-DNA is a right-handed helix with 11 base pairs per turn. It is a shorter, thicker double helix with a diameter of 0.23 nm and has a narrow, deep major groove and a broad, shallow minor groove. A-DNA forms *in vitro* under high salt concentrations or in a partially dehydrated state.

9.18   The value of T_m increases with the GC content because GC base pairs, connected by three hydrogen bonds, are stronger than AT base pairs connected by two hydrogen bonds.

9.19   (a) 89.5° C.

   (b) about 39%.

9.20   (1) The nucleosome level; the core containing an octamer of histones plus 146 nucleotide pairs of DNA arranged as $1\frac{3}{4}$ turns of a supercoil (see Figure 9.24), yielding an approximately 10 nm diameter spherical body; or juxtaposed, a roughly 10 nm diameter fiber.

   (2) The 30 nm fiber observed in condensed mitotic and meiotic chromosomes; it appears to be formed by coiling or folding the 10-nm nucleosome fiber.

   (3) The highly condensed mitotic and meiotic chromosomes (for example, metaphase chromosomes); the tight folding or coiling maintained by a "scaffold" composed of nonhistone chromosomal proteins (see Figure 9.27).

9.21   (a) 3.2×10^{10} bp.

   (b) 3.2×10^{10} bp.

   (c) 1.6×10^{10} bp.

   (d) 0.8×10^{10} bp.

9.22   In the diploid nucleus of *D. melanogaster*, 1.65×10^6 nucleosomes would be present; these would contain 3.3×10^6 molecules of each histone, H2a, H2b, H3, and H4.

9.23   The satellite DNA fragments would renature much more rapidly than the main-band DNA fragments. In *D. virilus* satellite DNAs, all three have repeating heptanucleotide-pair sequences. Thus essentially every 40 nucleotide-long (average) single-stranded fragment from one strand will have a sequence complementary (in part) with every single-stranded fragment from the complementary strand. Many of the nucleotide-pair sequences in main-band DNA will be unique sequences (present only once in the genome).

9.24   It indicates that highly repetitive DNA sequences do not contain structural genes specifying RNA and polypeptide gene products.

9.25   (a) (1) euchromatin.

   (2) euchromatin.

   (3) heterochromatin.

   (b) yes; no.

   (c) (1) Most of the single-copy DNA sequences are believed to be structural genes encoding proteins—structural proteins and the vast repertoire of enzymes employed by living organisms.

   (2) Essentially nothing is known regarding the functions of the highly repetitive DNA sequences—your hypotheses are probably as valid as anyone else's.

   (d) Some moderately repetitive DNA sequences specify products such as ribosomal RNA molecules that are required by cells in large quantities. Others are believed to be binding sites for proteins that regulate gene expression and replication of the multiple replicons in the giant DNA molecules of eukaryotic chromosomes. Some moderately repetitive sequences probably play structural roles in chromosomes, especially during the condensations of mitosis and meiosis. Others undoubtedly carry out functions that are still unknown.

9.26   Interphase. Chromosomes are for the most part metabolically inactive (exhibiting little transcription) during the various stages of condensation in mitosis and meiosis.

9.27   (a) (1) Centromeres function as spindle-fiber attachment sites on chromosomes; they are required for the separation of homologous chromosomes to opposite poles of the spindle during anaphase I of meiosis and for the separation of sister chromatids during anaphase of mitosis and anaphase II of meiosis.

   (2) Telomeres provide at least three important functions: (i) prevention of exonucleolytic degradation of the ends of the linear DNA molecules in eukaryotic chromosomes, (ii) prevention of the fusion of ends of DNA

molecules of different chromosomes, and (iii) provision of a mechanism for replication of the distal tips of linear DNA molecules in eukaryotic chromosomes.

(b) Yes. Most telomeres studied to date contain DNA sequence repeat units (for example, TTAGGG in human chromosomes), and, at least in some species, telomeres terminate with single-stranded 3′ overhangs that form "hairpin" structures. The bases in these hairpins exhibit unique patterns of methylation that presumably contribute to the structure and stability of telomeres.

(c) Telomerase adds the terminal DNA sequences or telomeres to the linear chromosomes in eukaryotes.

(d) The broken ends resulting from irradiation will not contain telomeres; as a result, the free ends of the DNA molecules are apparently subject to the activities of enzymes such as exonucleases, ligases, and the like, which modify the ends. They can regain stability by fusing to broken ends of other DNA molecules that contain terminal telomere sequences.

9.28   Viscoelastometry is a procedure used to measure the viscosity of molecules in solution. In addition, viscoelastometric methods can be used to estimate the sizes of the largest DNA molecules present in aqueous solutions. By using viscoelastometry, scientists have obtained evidence which indicates that all of the DNA present in chromosomes of eukaryotes exists as giant, "chromosome-size" DNA molecules (one huge DNA molecule per chromosome). These data eliminated early models of chromosome structure with multiple DNA molecules joined end-to-end by protein or RNA "linkers."

9.29   (a) Two. The axial region of a "lampbrush" chromosome contains the two chromatids of one homologous chromosome (postreplication).

(b) One. Each lateral loop of a "lampbrush" chromosome is a segment of a single chromatid.

9.30   Nonhistone chromosomal proteins. The "scaffold" structures of metaphase chromosomes can be observed by light microscopy after removal of the histones by differential extraction procedures.

9.31   (a) Histones have been highly conserved throughout the evolution of eukaryotes. A major function of histones is to package DNA into nucleosomes and chromatin fibers. Since DNA is composed of the same four nucleotides and has the same basic structure in all eukaryotes, one might expect that the proteins that play a structural role in packaging this DNA would be similarly conserved.

(b) The nonhistone chromosomal proteins exhibit the greater heterogeneity in chromatin from different tissues and cell types of an organism. The histone composition is largely the same in all cell types within a given species—consistent with the role of histones in packaging DNA into nucleosomes. The nonhistone chromosomal proteins include proteins that regulate gene expression. Because different sets of genes are transcribed in different cell types, one would expect heterogeneity in some of the nonhistone chromosomal proteins of different tissues.

9.32   (a) There are four major classes of transposable genetic elements in the human genome: LINEs (long interspersed nuclear elements), SINEs (short interspersed nuclear elements), elements with LTRs (long terminal repeats), and DNA transposons (for details, see Chapter 18). (b) No, many of these elements are inactive derivatives that have lost components required for transposition. (c) Alu elements are named after the restriction enzyme AluI that cleaves at a specific site within the element. Restriction enzymes are named using the first letter of the genus and the first two letters of the species of the bacterium that produces them. AluI is synthesized in _Arthrobacter luteus_. (d) There is no clear answer to this question. One proposal is that transposable elements are just "selfish DNAs" that replicate and spread through genomes. Another possibility is that they facilitate the process of evolution by allowing more rapid changes in genomes, including the evolution of genes with new functions.

9.33   If a transposable genetic element inserts itself into the coding region of a gene, the gene will no longer be able to produce its normal product. If it inserts into a regulatory region controlling the time and place of expression, the gene may either not be expressed at all or may be over-expressed, under-expressed, or expressed in the wrong tissues of cells.

9.34   (a) One μg of human DNA will contain, on average, 3.04×10^5 copies of the genome. Using an average molecular weight per nucleotide pair of 660, the "molecular" weight of the entire human genome is 1.98×10^{12} ($3 \times 10^9 \times 660$). Thus, 1.98×10^{12} grams (1 "mole" = number of grams equivalent to the "molecular" weight) of human DNA will contain, on average, 6.02×10^{23} molecules [Avogadro's number = number of molecules (here, copies of the genome) present in one "mole" of a substance]. One gram will contain, on average,

$(3.04 \times 10^{11}) \left(\dfrac{6.02 \times 10^{23}}{1.98 \times 10^{12}} \right)$ copies of the genome; thus, 1 µg will contain, on average, 3.04×10^5 copies of the human genome.

(b) One copy of the human genome weighs approximately

$(3.3 \times 10^{-12} \text{ g}) \left(\dfrac{1.98 \times 10^{12} \text{ g per "mole"}}{6.20 \times 10^{23} \text{ molecules per "mole"}} \right)$ or 3.3×10^{-6} µg.

(c) By analogous calculations, 1 µg of *Arabidopsis thaliana* DNA contains, on average, 1.18×10^7 copies of the genome.

(d) Similarly, one copy of the *A. thaliana* genome weighs approximately 8.4×10^{-8} µg.

(e) In carrying out molecular analyses of the structures of genomes, geneticists frequently need to know how many copies of a genome are present, on average, in a given quantity of DNA.

9.35 The locations of the satellite DNA sequence in the chromosomes can be determined by *in situ* hybridization (see Focus on *In Situ* Hybridization).

CHAPTER 10

10.1 (a) (i) One-half of the DNA molecules with ^{15}N in both strands and $\frac{1}{2}$ with ^{14}N in both strands; (ii) all DNA molecules with one strand containing ^{15}N and the complementary strand containing ^{14}N; (iii) all DNA molecules with both strands containing roughly equal amounts of ^{15}N and ^{14}N. (b) (i) $\frac{1}{4}$ of the DNA molecules with ^{15}N in both strands and $\frac{3}{4}$ with ^{14}N in both strands; (ii) $\frac{1}{2}$ of the DNA molecules with one strand containing ^{15}N and the complementary strand containing ^{14}N and the other $\frac{1}{2}$ with ^{14}N in both strands; (iii) all DNA molecules with both strands containing about $\frac{1}{4}$ ^{15}N and $\frac{3}{4}$ ^{14}N. See Milestone Figure 1.

10.2 One-half of the DNA molecules fully heavy (^{15}N in both strands); the other half of the molecules "hybrid" (^{15}N in one strand, ^{14}N in the complementary strand).

10.3 (a) Both $3' \rightarrow 5'$ and $5' \rightarrow 3'$ exonuclease activities. (b) The $3' \rightarrow 5'$ exonuclease "proofreads" the nascent DNA strand during its synthesis. If a mismatched base pair occurs at the 3'-OH end of the primer, the $3' \rightarrow 5'$ exonuclease removes the incorrect terminal nucleotide before polymerization proceeds again. The $5' \rightarrow 3'$ exonuclease is responsible for the removal of RNA primers during DNA replication and functions in pathways involved in the repair of damaged DNA (see Chapter 13). (c) Yes, both exonuclease activities appear to be very important. Without the $3' \rightarrow 5'$ proofreading activity during replication, an intolerable mutation frequency would occur. The $5' \rightarrow 3'$ exonuclease activity is essential to the survival of the cell. Conditional mutations that alter the $5' \rightarrow 3'$ exonuclease activity of DNA polymerase I are lethal to the cell under conditions where the exonuclease is non-functional.

10.4 (a)

$3'$ ⓅP-TGCGAATTAGCGACAT-Ⓟ $5'$
$5'$ Ⓟ-ATCGGTACGACGCTTAATCGCTGTA-OH $3'$;

note that DNA synthesis will *not* occur on the left end since the 3'-terminus of the potential primer strand is blocked with a phosphate group—all DNA polymerases require a free 3'-OH terminus.

(b) The first step will be the removal of the mismatched C (exiting as dCMP) from the 3'-OH primer terminus by the $3' \rightarrow 5'$ exonuclease ("proofreading") activity.

10.5 If nascent DNA is labeled by exposure to ^{3}H-thymidine for very short periods of time, continuous replication predicts that the label would be incorporated into chromosome-sized DNA molecules, whereas discontinuous replication predicts that the label would first appear in small pieces of nascent DNA (prior to covalent joining, catalyzed by polynucleotide ligase).

10.6

Proteins	Functions
1. DNA polymerase III	1. (a) Catalyzes polymerization (covalent extension) of new DNA chains. (b) The $3' \to 5'$ exonuclease activity "proofreads" the product, removing any mismatched base-pairs at the $3'$ end of the primer strand.
2. DNA polymerase I	2. Removes the RNA primers ($5' \to 3'$ exonuclease activity) and replaces them with new DNA strands ($5' \to 3'$ polymerase activity).
3. DNA ligase	3. Catalyzes covalent joining of "Okazaki fragments."
4. Primase (*dnaG* protein)	4. Catalyzes RNA primer synthesis.
5. DNA gyrase	5. Catalyzes the formation of negative supercoils; facilitates unwinding?
6. DNA helicase (*dnaB* protein)	6. Catalyzes unwinding.
7. DNA single-strand binding protein	7. Maintains an "extended" single-stranded template; aids unwinding?
8. Proteins i, n, n', *dnaC* protein	8. "Prepriming"— required prior to initiation of primer synthesis.
9. *dnaI, dnaJ*, etc., proteins	9. Required; but functions unknown.

10.7

Two  Plus two For both the large and small chromosomes

10.8

Two [ ] Plus two [ ] For both the large and small chromosomes

10.9 That DNA replication was unidirectional rather than bidirectional. As the intracellular pools of radioactive ^{3}H-thymidine are gradually diluted after transfer to nonradioactive medium, less and less ^{3}H-thymidine will be incorporated into DNA at each replicating fork. This will produce autoradiograms with tails of decreasing grain density at each growing point. Since such tails appear at only one end of each track, replication must be unidirectional. Bidirectional replication would produce such tails at both ends of an autoradiographic track (see Figure 10.29).

10.10 The correct sequence of action is 4, 5, 3, 2, 1.

10.11 DNA polymerases α, β, δ, $\in$, and η are located in the nuclei of cells; polymerase γ is located in mitochondria and chloroplasts. Current evidence suggests that polymerases α, δ, and/or $\in$ are both required for the replication of nuclear DNA. Polymerase δ and/or $\in$ is thought to catalyze the continuous synthesis of the leading strand, and polymerase α is believed to function as a primase in the discontinuous synthesis of the lagging strand. Polymerases β, ζ, and η function in DNA repair pathways.

10.12 (a) Given bidirectional replication of a single replicon, each replication fork must traverse 2×10^6 nucleotide pairs in *E. coli* and 3×10^7 nucleotide pairs in the largest *Drosophila* chromosome. If the rates were the same in both species, it would take 15 times $\dfrac{3 \times 10^7}{2 \times 10^6}$ as long to replicate the *Drosophila* chromosome or 10 hours (40 minutes × 15 = 600 minutes).

(b) If replication forks in *E. coli* move 20 times as fast as replication forks in *Drosophila* (100,000 nucleotide pairs per minute/5,000 nucleotide pairs per minute), the largest *Drosophila* chromosome would require 8.3 days (10 hours × 20 = 200 hours) to complete one round of replication.

(c) Each *Drosophila* chromosome must contain many replicons in order to complete replication in less than 10 minutes.

10.13 Sucrose velocity density gradient centrifugation is the standard technique for separating DNA molecules in this size range. Pulsed-field gel electrophoresis (Chapter 10) could also be used.

10.14 (a) and (b)

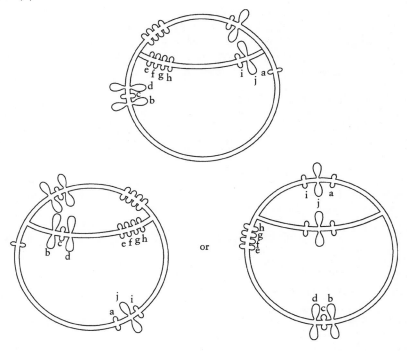

10.15 (a) DNA gyrase; (b) primase; (c) the $5' \rightarrow 3'$ exonuclease activity of DNA polymerase I; (d) the $5' \rightarrow 3'$ polymerase activity of DNA polymerase III; (e) the $3' \rightarrow 5'$ exonuclease activity of DNA polymerase III.

10.16 (a) (34 nm/100 bp)$(2 \times 10^{10}$ bp$) = 6.8 \times 10^9 = 6.8$ meters.

(b) $2 \times 10^{10}/10$ chromosomes $= 2 \times 10^9$ bp

4×10^4 bp/min (bidirectional); $2 \times 10^9/4 \times 10^4 = 5 \times 10^4$ min = 50,000 min.

(c) (50,000 min)(1 ori) = (300 min)(X ori); X = 50,000/300 = 167 ori.

(d) $(2 \times 10^9$ bp/chrom.)/(167 ori/chrom.) = 1.2×10^7 bp/ori.

10.17 In eukaryotes, the rate of DNA synthesis at each replication fork is about 2,500 to 3,000 nucleotide pairs per minute. Large eukaryotic chromosomes often contain 10^7 to 10^8 nucleotide pairs. A single replication fork could not replicate the giant DNA in one of these large chromosomes fast enough to permit the observed cell generation times.

10.18 The $5' \rightarrow 3'$ exonuclease activity of DNA polymerase I is essential to the survival of the bacterium, whereas the $5' \rightarrow 3'$ polymerase activity of the enzyme is not essential.

10.19 No. *E. coli* strains carrying *polA* mutations that eliminate the $3' \rightarrow 5'$ exonuclease activity of DNA polymerase I will exhibit unusually high mutation rates.

10.20 Because A:T base pairs are held together by only two hydrogen bonds instead of the three hydrogen bonds present in G:C base pairs, the two strands of A:T-rich regions of double helices are separated more easily, providing the single-stranded template regions required for DNA replication.

10.21 Rolling-circle replication begins when an endonuclease cleaves one strand of a circular DNA double helix. This cleavage produces a free 3'-OH on one end of the cut strand, allowing it to function as a primer. The discontinuous synthesis of the lagging strand requires the *de novo* initiation of each Okazaki fragment, which requires DNA primase activity.

10.22 DNA polymerase III does not have a $5' \rightarrow 3'$ exonuclease activity that acts on double-stranded nucleic acids. Thus it cannot excise RNA primer strands from replicating DNA molecules. DNA polymerase I is present in cells at much higher concentrations and functions as a monomer. Thus DNA polymerase I is able to catalyze the removal of RNA primers from the vast number of Okazaki fragments formed during the discontinuous replication of the lagging strand.

10.23 DNA helicase unwinds the DNA double helix, and single-strand DNA-binding protein coats the unwound strands, keeping them in an extended state. DNA gyrase catalyzes the formation of negative supercoiling in *E. coli* DNA, and this negative supercoiling behind the replication forks is thought to drive the unwinding process because superhelical tension is reduced by unwinding the complementary strands.

10.24 DNA polymerase I is a single polypeptide of molecular weight 109,000, whereas DNA polymerase III is a complex multimeric protein. The DNA polymerase holoenzyme has a molecular mass of about 900,000 daltons and is composed of at least 20 different polypeptides. The *dnaN* gene product, the β subunit of DNA polymerase III, forms a dimeric clamp that encircles the DNA molecule and prevents the enzyme from dissociating from the template DNA during replication.

10.25 DnaA protein initiates the formation of the replication bubble by binding to the 9-bp repeats of *OriC*. DnaA protein is known to be required for the initiation process because bacteria with temperature-sensitive mutations in the *dnaA* gene cannot initiate DNA replication at restrictive temperatures.

10.26 The primosome is a protein complex that initiates the synthesis of Okazaki fragments during lagging strand synthesis. The major components of the *E. coli* DNA primosome are DNA primase and DNA helicase. Geneticists have been able to show that both DNA primase and DNA helicase are required for DNA replication by demonstrating that mutations in the genes encoding these enzymes result in the arrest of DNA synthesis in mutant cells under conditions where the altered proteins are inactive.

10.27 Nucleosomes and replisomes are both large macromolecular structures, and the packaging of eukaryotic DNA into nucleosomes raises the question of how a replisome can move past a nucleosome and replicate the DNA in the nucleosome in the process. The most obvious solution to this problem would be to completely or partially disassemble the nucleosome to allow the replisome to pass. The nucleosome would then reassemble after the replisome had passed (see Figure 10.32).

10.28 The product of the first gene is required for DNA chain extension, whereas the product of the second gene is only required for the initiation of DNA synthesis.

10.29 (1) DNA replication usually occurs continuously in rapidly growing prokaryotic cells but is restricted to the S phase of the cell cycle in eukaryotes.

(2) Most eukaryotic chromosomes contain multiple origins of replication, whereas most prokaryotic chromosomes contain a single origin of replication.

(3) Prokaryotes utilize two catalytic complexes that contain the same DNA polymerase to replicate the leading and lagging strands, whereas eukaryotes utilize two or three distinct DNA polymerases for leading and lagging strand synthesis.

(4) Replication of eukaryotic chromosomes requires the partial disassembly and reassembly of nucleosomes as replisomes move along parental DNA molecules. In prokaryotes, replication probably involves a similar partial disassembly/reassembly of nucleosome-like structures.

(5) Most prokaryotic chromosomes are circular and thus have no ends. Most eukaryotic chromosomes are linear and have unique termini called telomeres that are added to replicating DNA molecules by a unique, RNA-containing enzyme called telomerase.

10.30   (a) 2,000 to 4,000 Okazaki fragments. (b) 300,000 to 600,000 Okazaki fragments.

10.31   The chromosomes of haploid yeast cells that carry the *est1* mutation become shorter during each cell division. Eventually, chromosome instability results from the complete loss of telomeres, and cell death occurs because of the deletion of essential genes near the ends of chromosomes.

10.32   Without telomerase, the 5′ end of the newly replicated strand will be missing some bases. The exact number of missing bases doesn't matter:

5′-GATTCCCCGGGAAGCTTGGGGGGCCCATCT T CGT ACGTCTTTGCA-3′
3′-CTAAGGGGCCCTTCGAACCCCCCGCGGGTAGAAGCAT G-5′
5′-AAGCTT GGGGGGCCCATCT T CGT ACGTCT T T G CA-3′
3′-CTAAGGGGCCCTTCGAA CCCCCCGGGTAGAAGCATG CAGAAACG T-5′

CHAPTER 11

11.1   (a) RNA contains the sugar ribose, which has an hydroxyl (OH) group on the 2-carbon; DNA contains the sugar 2-deoxyribose, with only hydrogens on the 2-carbon. RNA usually contains the base uracil at positions where thymine is present in DNA. However, some DNAs contain uracil, and some RNAs contain thymine. DNA exists most frequently as a double helix (double-stranded molecule); RNA exists more frequently as a single-stranded molecule; but some DNAs are single-stranded and some RNAs are double-stranded. (b) The main function of DNA is to store genetic information and to transmit that information from cell to cell and from generation to generation. RNA stores and transmits genetic information in some viruses that contain no DNA. In cells with both DNA and RNA: (1) mRNA acts as in intermediary in protein synthesis, carrying the information from DNA in the chromosomes to the ribosomes (sites at which proteins are synthesized). (2) tRNAs carry amino acids to the ribosomes and function in codon recognition during the synthesis of polypeptides. (3) rRNA molecules are essential components of the ribosomes. (c) DNA is located primarily in the chromosomes (with some in cytoplasmic organelles, such as mitochondria and chloroplasts), whereas RNA is located throughout cells.

11.2   3′—ACGUCUGU—5′.

11.3   3′—GACTA—5′.

11.4   The genetic information of cells is stored in DNA, which is located predominantly in the chromosomes. The gene products (polypeptides) are synthesized primarily in the cytoplasm on ribosomes. Some intermediate must therefore carry the genetic information from the chromosomes to the ribosomes. RNA molecules (mRNAs) were shown to perform this function by means of RNA pulse-labeling and pulsechase experiments combined with autoradiography (see Figure 11.6). The enzyme RNA polymerase was subsequently shown to catalyze the synthesis of mRNA using chromosomal DNA as a template. Finally, the mRNA molecules synthesized by RNA polymerase were shown to faithfully direct the synthesis of specific polypeptides when used in *in vitro* protein synthesis systems.

11.5   Protein synthesis occurs on ribosomes. In eukaryotes, most of the ribosomes are located in the cytoplasm and are attached to the extensive membranous network of endoplasmic reticulum. Some protein synthesis also occurs in cytoplasmic organelles such as chloroplasts and mitochondria.

11.6   (1) Eukaryotes have a 5′ cap on their mRNAs, prokaryotes do not. (2) Messenger RNAs of eukaryotes generally have a 3′ poly-A tail, prokaryotic mRNAs never do. (3) Messenger RNA formation in eukaryotes involves removal of introns (when present) and the splicing together of exons. Prokaryotic genes (with very rare exceptions) do not have introns.

11.7   Both prokaryotic and eukaryotic organisms contain messenger RNAs, transfer RNAs, and ribosomal RNAs. In addition, eukaryotes contain small nuclear RNAs and micro RNAs. Messenger RNA molecules carry genetic information from the chromosomes (where the information is stored) to the ribosomes in the cytoplasm (where the information is expressed during protein synthesis). The linear sequence of triplet codons in an mRNA molecule specifies the linear sequence of amino acids in the polypeptides produced during translation of that mRNA. Transfer RNA molecules are small (about 80 nucleotides long) molecules that carry amino acids to the ribosomes and provide the codon-recognition specificity during translation. Ribosomal RNA molecules provide part of the structure and function of ribosomes; they represent an important part of the machinery required for the synthesis of polypeptides. Small nuclear RNAs are structural components of spliceosomes, which excise noncoding intron sequences from nuclear gene transcripts. Micro RNAs play important roles in regulating gene expression (see chapter 21).

11.8 The entire nucleotide-pair sequences—including the introns—of the genes are transcribed by RNA polymerase to produce primary transcripts that still contain the intron sequences. The intron sequences are then spliced out of the primary transcripts to produce the mature, functional RNA molecules. In the case of protein-encoding nuclear genes of higher eukaryotes, the introns are spliced out by complex macromolecular structures called spliceosomes (see Figure 11.27).

11.9 "Self-splicing" of RNA precursors demonstrates that RNA molecules can also contain catalytic activity; this property is not restricted to proteins.

11.10 Spliceosomes excise intron sequences from nuclear gene transcripts to produce the mature mRNA molecules that are translated on ribosomes in the cytoplasm. Spliceosomes are complex macromolecular structures composed of snRNA and protein molecules (see Figure 11.28).

11.11 The introns of protein-encoding nuclear genes of higher eukaryotes almost invariably begin (5′) with GT and end (3′) with AG. In addition, the 3′ subterminal A in the "TACTAAC box" is completely conserved; this A is involved in bond formation during intron excision.

11.12 (a) 7; (b) 5; (c) 10; (d) 3; (e) 11; (f) 1; (g) 12; (h) 6; (i) 2; (j) 8; (k) 13; (l) 4; (m) 9.

11.13 (a) Sequence 5. It contains the conserved intron sequences: a 5′ GU, a 3′ AG, and a UACUAAC internal sequence providing a potential bonding site for intron excision. Sequence 4 has a 5′ GU and a 3′ AG, but contains no internal A for the bonding site during intron excision. (b) 5′—UAGUCUCAA—3′; the putative intron from the 5′ GU through the 3′ AG has been removed.

11.14 This is a wide-open question at present! There is much speculation, but little hard evidence. One popular hypothesis is that introns enhance exon shuffling by increasing recombination events between sequences encoding adjacent domains of a polypeptide. Also, in one yeast mitochondrial gene, the introns contain open reading frames that encode "maturases" that splice out these introns—a neat negative feedback control. Other introns may be merely relics of evolution.

11.15

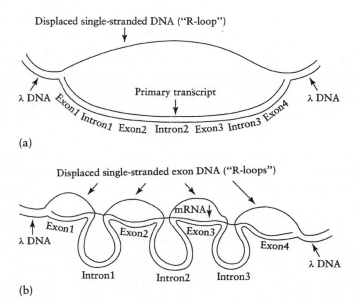

11.16 If there is a promoter located upstream from this DNA segment, the nucleotide sequence of this portion of the RNA transcript will be 5'-UACGAUGACGAUAAGCGACAUAGC-3'. If there is no upstream promoter, this segment of DNA will not be transcribed.

11.17 Assuming that there is a −35 sequence upstream from the consensus −10 sequence in this segment of the DNA molecule, the nucleotide sequence of the transcript will be 5'-ACCCGACAUAGCUAC-GAUGACGAUAAGC-GACAUAGC-3'.

11.18 Given the consensus −35 and −10 sequences in this segment of DNA and the fact that transcripts almost always start with a purine, the predicted nucleotide sequence of the transcript is 5'-ACCCGACAUAGCUACGAUG-ACGAUA-3'.

11.19 Assuming that there is a CAAT box located upstream from the TATA box shown in this segment of DNA, the nucleotide sequence of the transcript will be 5'-ACCCGACAUAGCUACGAUGACGAUA-3'.

11.20 Given the vast amount of information that can be stored on a small computer chip by using a binary code, it is clear that large quantities of genetic information can be stored in the genomes of organisms by using the four-letter alphabet of the genetic code.

11.21 Although in theory it would be possible to produce six different enzyme activities by alternative splicing of one gene transcript or by combining the polypeptide products of two or three genes into different combinations with distinct enzyme activities, in reality these possibilities are unlikely. In living organisms, six enzymes are usually specified by at least six genes—with each gene encoding a single polypeptide. However, many enzymes are composed of two or more distinct polypeptides. Thus the synthesis of six enzymes may require more than six genes.

11.22 According to the central dogma, genetic information is stored in DNA and is transferred from DNA to RNA to protein during gene expression. RNA tumor viruses store their genetic information in RNA, and that information is copied into DNA by the enzyme reverse transcriptase after a virus infects a host cell. Thus the discovery of RNA tumor viruses or retroviruses—retro for backwards flow of genetic information—provided an exception to the central dogma.

11.23 DNA, RNA, and protein synthesis all involve the synthesis of long chains of repeating subunits. All three processes can be divided into three stages: chain initiation, chain elongation, and chain termination.

11.24 The two stages of gene expression are:

(1) transcription—the transfer of genetic information from DNA to RNA.

(2) translation—the transfer of genetic information from RNA to protein. In eukaryotes, transcription occurs in the nucleus and translation occurs in the cytoplasm on complex macromolecular structures called ribosomes. In prokaryotes, transcription and translation are often coupled with mRNA molecules often being translated by ribosomes while still being synthesized during transcription.

11.25 The primary transcripts of eukaryotes undergo more extensive post-transcriptional processing than those of prokaryotes. Thus the largest differences between mRNAs and primary transcripts occur in eukaryotes. Transcript processing is usually restricted to the excision of terminal sequences in prokaryotes. In contrast, eukaryotic transcripts are usually modified by: (1) the excision of intron sequences; (2) the addition of 7-methyl guanosine caps to the 5' termini; (3) the addition of poly(A) tails to the 3' termini. In addition, the sequences of some eukaryotic transcripts are modified by RNA editing processes.

11.26 The five types of RNA molecules that are involved in gene expression are mRNAs, rRNAs, tRNAs micro RNAs, and snRNAs. mRNA molecules carry genetic information from genes to the sites of protein synthesis and specify the amino acid sequences of polypeptides. rRNAs are major structural components of the ribosomes and provide functions required for translation. tRNA molecules are the adapters that provide amino acid-codon specificity during translation; each tRNA is activated by a specific amino acid and contains an anticodon sequence that is complementary or partially complementary to one, two, or three codons in mRNAs. Micro RNAs are involved in regulative gene expression (see chapter 21). snRNAs are structural components of the spliceosomes that excise introns from gene transcripts in eukaryotes. snRNAs perform their splicing functions in the nucleus. mRNAs carry information from the nucleus to the cytoplasm, so they function in both compartments of the cell. However, their most prominent function is to direct the synthesis of polypeptides during translation, which occurs in the cytoplasm. rRNAs and tRNAs perform their functions during translation in the cytoplasm. Micro RNAs become incorporated into ribonucleoprotein complexes in the cytoplasm.

11.27 The gene that contains a single exon will be transcribed in the least time. The rate of RNA chain extension is the same—about 30 nucleotides per second—for both intron and exon sequences. Thus the synthesis of the long intron sequence will take about 23 minutes. On the other hand, the time required to translate the two mRNAs will be the same because intron sequences are spliced out of transcripts prior to their translation.

11.28 In eukaryotes, the genetic information is stored in DNA in the nucleus, whereas proteins are synthesized on ribosomes in the cytoplasm. How could the genes, which are separated from the sites of protein synthesis by a double membrane—the nuclear envelope, direct the synthesis of polypeptides without some kind of intermediary to carry the specifications for the polypeptides from the nucleus to the cytoplasm? Researchers first used labeled RNA and protein precursors and autoradiography (see Figure 11.6) to demonstrate that RNA synthesis and protein synthesis occurred in the nucleus and the cytoplasm, respectively.

11.29 A simple pulse-and pulse/chase-labeling experiment will demonstrate that RNA is synthesized in the nucleus and is subsequently transported to the cytoplasm. This experiment has two parts: (1) Pulse label eukaryotic culture cells by growing them in ^{3}H-uridine for a few minutes, and localize the incorporated radioactivity by autoradiography. (2) Repeat the experiment, but this time add a large excess of nonradioactive uridine to the medium in which the cells are growing after the labeling period, and allow the cells to grow in the nonradioactive medium for about an hour. Then localize the incorporated radioactivity by autoradiography. The expected results are shown in Figure 11.6.

11.30 RNA-DNA duplexes will be formed when the template strand is used, but not when the nontemplate strand is used. Only one strand—the template strand—of most genes is transcribed. Thus RNA will contain nucleotide sequences complimentary to the template strand, but not to the nontemplate strand.

11.31 The first preparation of RNA polymerase is probably lacking the sigma subunit and, as a result, initiates the synthesis of RNA chains at random sites along both strands of the *argH* DNA. The second preparation probably contains the sigma subunit and initiates RNA chains only at the site used *in vivo*, which is governed by the position of the −10 and −35 sequences of the promoter.

11.32 In eukaryotes, transcription occurs in the nucleus and translation occurs in the cytoplasm. Because these processes occur in different compartments of the cell, they cannot be coupled as they are in prokaryotes.

11.33 TATA and CAAT boxes. The TATA and CAAT boxes are usually centered at positions −30 and −80, respectively, relative to the starpoint (+1) of transcription. The TATA box is responsible for positioning the transcription starpoint; it is the binding site for the first basal transcription factor that interacts with the promoter. The CAAT box enhances the efficiency of transcriptional initiation.

11.34 (1) Intron sequences are spliced out of gene transcripts to provide contiguous coding sequences for translation. (2) The 7-methyl guanosine caps added to the 5′ termini of most eukaryotic mRNAs help protect them from degradation by nucleases and are recognized by proteins involved in the initiation of translation. (3) The poly(A) tails at the 3′ termini of mRNAs play an important role in their transport from the nucleus to the cytoplasm and enhance their stability.

11.35 RNA editing sometimes leads to the synthesis of two or more distinct polypeptides from a single mRNA. Guide RNAs are partially complementary to the pre-mRNAs that are edited and serve as templates for the editing process. During editing, uracils are inserted in pre-mRNAs opposite adenines in the guide RNAs.

11.36 The introns of tRNA precursors, *Tetrahymena* rRNA precursors, and nuclear pre-mRNAs are excised by completely different mechanisms. (1) Introns in tRNA precursors are excised by cleavage and joining events catalyzed by splicing nucleases and ligases, respectively. (2) Introns in *Tetrahymena* rRNA precursors are excised autocatalytically. (3) Introns of nuclear pre-mRNAs are excised by spliceosomes. snRNAs are involved in nuclear pre-mRNA splicing as structural components of spliceosomes. In addition, snRNA U1 is required for the cleavage events at the 5′ termini of introns; U1 is thought to base-pair with a partially complementary consensus sequence at this position in pre-mRNAs.

11.37 Some individuals with systemic lupus erythematosus produce antibodies that react with proteins in snRNPs. These antibodies have been used to immunoprecipitate snRNPs, facilitating their purification and subsequent characterization.

11.38 This zygote will probably be inviable because the gene product is essential and the elimination of the 5′ splice site will almost certainly result in the production of a nonfunctional gene product.

11.39 RNA-DNA duplexes would be expected when RNA isolated from nuclei is used in the renaturation experiment, but not when cytoplasmic RNA is used. The introns of genes are transcribed along with the exons. Thus the primary transcripts present in nuclei will contain both exon and intron sequences. However, intron sequences are excised by nuclear spliceosomes before the transcripts are exported to the cytoplasm. Thus cytoplasmic RNAs does not contain intron sequences and will not form RNA-DNA duplexes when incubated with single-stranded intron sequences.

CHAPTER 12

12.1 Proteins are long chainlike molecules made up of amino acids linked together by peptide bonds. Proteins are composed of carbon, hydrogen, nitrogen, oxygen, and usually sulfur. They provide the enzymatic capacity and much of the structure of living organisms. DNA is composed of phosphate, the pentose sugar 2-deoxyribose, and four nitrogen-containing organic bases (adenine, cytosine, guanine, and thymine). DNA stores and transmits the genetic information in most living organisms. Protein synthesis is of particular interest to geneticists because proteins are the primary gene products — the key intermediates through which genes control the phenotypes of living organisms.

12.2 Protein synthesis occurs on ribosomes. In eukaryotes, most of the ribosomes are located in the cytoplasm and are attached to the extensive membranous network of endoplasmic reticulum. Some protein synthesis also occurs in cytoplasmic organelles such as chloroplasts and mitochondria.

12.3 Ribosomes are from 10 to 20 nm in diameter. They are located primarily in the cytoplasm of cells. In bacteria, they are largely free in the cytoplasm. In eukaryotes, many of the ribosomes are attached to the endoplasmic reticulum. Ribosomes are complex structures composed of over 50 different polypeptides and three to five different RNA molecules.

12.4 (a) The nucleus, specifically the nucleoli.

   (b) The cytoplasm.

12.5 Messenger RNA molecules carry genetic information from the chromosomes (where the information is stored) to the ribosomes in the cytoplasm (where the information is expressed during protein synthesis). The linear sequence of triplet codons in an mRNA molecule specifies the linear sequence of amino acids in the polypeptide(s) produced during translation of that mRNA. Transfer RNA molecules are small (about 80 nucleotides long) molecules that carry amino acids to the ribosomes and provide the codon-recognition specificity during translation. Ribosomal RNA molecules provide part of the structure and function of ribosomes; they represent an important part of the machinery required for the synthesis of polypeptides.

12.6 (a) Polysomes are formed when two or more ribosomes are simultaneously translating the same mRNA molecule. Ribosomes are usually spaced about 90 nucleotides apart on an mRNA molecule. Thus, polysome size is determined by mRNA size.

   (b) A ribosome, which contains rRNA molecules, can participate in the synthesis of any polypeptide specified by the ribosome-associated mRNA. In that sense, rRNA is *nonspecific*. Messenger RNAs and tRNAs, in contrast, are *specific*, in directing the synthesis of a particular polypeptide or set of polypeptides (mRNA) or in attaching to a particular amino acid (tRNA).

   (c) Transfer RNA molecules are much smaller (about 80 nucleotides) than DNA or mRNA molecules. They are single-stranded molecules but have complex secondary structures because of the base pairing between different segments of the molecules.

12.7 A specific aminoacyl-tRNA synthetase catalyzes the formation of an amino acid-AMP complex from the appropriate amino acid and ATP (with the release of pyrophosphate). The same enzyme then catalyzes the formation of the aminoacyl-tRNA complex, with the release of AMP. The amino acid-AMP and amino-acyl-tRNA linkages are both high-energy phosphate bonds.

12.8 Synthetic RNA molecules (polyuridylic acid molecules) containing only the base uracil were prepared. When these synthetic molecules were used to activate *in vitro* protein synthesis systems, small polypeptide containing only the amino acid phenylalanine (polyphenylalanine molecules) were synthesized. Codons composed only of uracil were therefore shown to specify phenylalanine. Similar experiments were carried out using synthetic RNA molecules with different base compositions. Later, *in vitro* systems activated with synthetic RNA molecules with known repeating base sequences were developed. Ultimately, *in vitro* system in which specific aminoacyl-tRNAs where shown to bind to ribosomes activated with specific mini-mRNA which were trinucleotides of known base sequence, were developed and used in codon identification.

12.9 (a) The genetic code is degenerate in that all but 2 of the 20 amino acids are specified by two or more codons. Some amino acids are specified by six different codons. The degeneracy occurs largely at the third or 3′ base of the codons. "Partial degeneracy" occurs where the third base of the codon may be either of the two purines or either of the two pyrimidines and the codon still specifies the same amino acid. "Complete degeneracy" occurs where the third base of the codon may be any one of the four bases and the codon still specifies the same amino acid. (b) The code is ordered in the sense that related codons (codons that differ by a single base change) specify chemically similar amino acids. For example, the codons CUU, AUU, and GUU specify the structurally related amino acids, leucine, isoleucine, and valine, respectively. (c) The code appears to be almost completely universal. Known exceptions to universality include strains carrying suppressor mutations that alter the reading of certain codons (with low efficiencies in most cases) and the use of UGA as a tryptophan codon in yeast and human mitochondria.

12.10 (a) Met → Val. This substitution occurs as a result of a transition. All other amino acid substitutions listed would require transversions.

12.11 His → Arg results from a transition; His → Pro would require a transversion (not induced by 5-bromouracil).

12.12 (a) By a complex reaction involving mRNA, ribosomes, initiation factors (IF-1, IF-2, and IF-3), GTP, the initiator codon AUG, and a special initiator tRNA (tRNA$_f^{Met}$). It also appears to involve a base-pairing interaction between a base sequence near the 3′-end of the 16S rRNA and a base sequence in the "leader sequence" of the mRNA. (b) By recognition of one or more of the chain-termination codons (UAG, UAA, and UGA) by the appropriate protein release factor (RF-1 or RF-2).

12.13 At least 813 nucleotides.

12.14 Crick's wobble hypothesis explains how the anticodon of a given tRNA can base-pair with two or three different mRNA codons. Crick proposed that the base-pairing between the 5′ base of the anticodon in tRNA and the 3′ base of the codon in mRNA was less stringent than normal and thus allowed some "wobble" at this site. As a result, a single tRNA often recognizes two or three of the related codons specifying a given amino acid (see Table 12.2).

12.15 (a) Inosine.

(b) Two.

12.16 Because of wobble (Table 12.2), one tRNA can recognize both UUA and UUG codons for leucine. However, it takes two more tRNAs to recognize CUU, CUC, CUA, and CUG. Therefore, a minimum of three tRNAs are required to recognize the six codons for leucine.

12.17 (a) Two nucleotides in all combinations of four (2^4) would produce sixteen codons. Therefore, the minimum number of nucleotides comprising the Martian genetic code must be four.

(b) Sixteen codons would allow code words for 14 amino acids, one initiation codon, and a translational termination codon. The Martian genetic code could not be degenerate.

12.18 (a) Singlet and doublet codes provide a maximum of 4 and $(4)^2$ or 16 codons, respectively. Thus neither code would be able to specify all 20 amino acids. (b) 20. (c) $(20)^{146}$.

12.19 Translation occurs by very similar mechanisms in prokaryotes and eukaryotes; however, there are some differences.

(1) In prokaryotes, the initiation of translation involves base pairing between a conserved sequence (AGGAGG) —the Shine-Dalgarno box — in mRNA and a complementary sequence near the 5′ end of the 16S rRNA. In eukaryotes, the initiation complex forms at the 5′ end of the transcript when a cap-binding protein interacts with the 7-methyl guanosine on mRNA. The complex then scans the mRNA processively and initiates translation (with a few exceptions) at the AUG closest to the 5′ terminus.

(2) In prokaryotes, the amino group of the initiator methionyl-tRNA$_f^{Met}$ is formylated; in eukaryotes, the amino group of methionyl-tRNA$_i^{Met}$ is not formylated.

(3) In prokaryotes, two soluble protein release factors (RFs) are required for chain termination. RF-1 terminates polypeptides in response to UAA and UAG codons; RF-2 terminates chains in response to UAA and UGA codons. In eukaryotes, one release factor responds to all three termination codons.

12.20 (a) Attachment of an amino acid to the correct tRNA. (b) Recognition of termination codons UAA and UAG and release of the nascent polypeptide from the tRNA in the *P* site of the ribosome. (c) Formation of a peptide bond between the amino group of the aminoacyl-tRNA in the *A* site and the carboxyl group of the growing polypeptide on the tRNA in the *P* site. (d) Formation of the initiation complex required for translation; all steps leading up to peptide bond formation. (e) Translocation of the peptidyl-tRNA from the *A* site on the ribosome to the *P* site.

12.21 Assuming 0.34 nm per nucleotide pair in B-DNA, a gene 68 nm long would contain 200 nucleotide pairs. Given the triplet code, this gene would contain 200/3 = 66.7 triplets, one of which must specify chain termination. Disregarding the partial triplet, this gene could encode a maximum of 65 amino acids.

12.22 (a) A nonsense mutation changes a codon specifying an amino acid to a chain-termination codon, whereas a missense mutation changes a codon specifying one amino acid to a codon specifying a different amino acid. (b) Missense mutations are more frequent. (c) Of the 64 codons, only three specify chain termination. Thus the number of possible missense mutations is much larger than the number of possible nonsense mutations. Moreover, nonsense mutations almost always produce nonfunctional gene products. As a result, nonsense mutations in essential genes are usually lethal in the homozygous state.

12.23 426 nucleotides—3 × 121 = 423 specifying amino acids plus three (one codon) specifying chain termination.

12.24 The incoming aminoacyl-tRNA enters the *A* site of the ribosome, the nascent polypeptide-tRNA occupies the *P* site, and the uncharged exiting tRNA occupies the E site.

12.25 (a) Related codons often specify the same or very similar amino acids. As a result, single base-pair substitutions frequently result in the synthesis of identical proteins (degeneracy) or proteins with amino acid substitutions involving very similar amino acids.

(b) Leucine and valine have very similar structures and chemical properties; both have nonpolar side groups and fold into essentially the same three-dimensional structures when present in polypeptides. Thus, substitutions of leucine for valine or valine for leucine seldom alter the function of a protein.

12.26 (a) The Shine-Dalgarno sequence is a conserved polypurine tract, consensus AGGAGG, that is located about seven nucleotides upstream from the AUG initiation codon in mRNAs of prokaryotes. It is complementary to, and is believed to base-pair with, a sequence near the 5′ terminus of the 16S ribosomal RNA. (b) Prokaryotic mRNAs with the Shine-Dalgarno sequence deleted are either not translated or are translated inefficiently.

12.27 (a) Ribosomes and spliceosomes both play essential roles in gene expression, and both are complex macromolecular structures composed of RNA and protein molecules.

(b) Ribosomes are located in the cytoplasm; spliceosomes in the nucleus. Ribosomes are larger and more complex than spliceosomes.

12.28 Incorporation of alanine into polypeptide chains.

12.29 Met-Ser-Ile-Cys-Leu-Phe-Gln-Ser-Leu-Ala-Ala-Gln-Asp-Arg-Pro-Gly.

12.30 NH$_2$-Met-Ala-Ile-Cys-Leu-Phe-Gln-Ser-Leu-Ala-Ala-Gln-Asp-Arg-Pro-Gly-COOH.

12.31 *Amber* (UAG). This is the only nonsense codon that is related to tryptophan, serine, tyrosine, leucine, glutamic acid, glutamine, and lysine codons by a single base-pair substitution in each case.

12.32 (a) 5′-CAAUCAUGGACUGCCAUGCUUCAUAUGAAUAGUUGACAU-3′.

(b) NH$_2$ -fMet-Asp-Cys-His-Ala-Ser-Tyr-Glu-COOH.

(c) NH$_2$-fMet-Asp-Cys-Met-Leu-His-Met-Asn-Ser-COOH.

12.33 (a) 5′-GAAUGUCAGAACUGCCAUGCUUCAUAUGAAUAGACCUCUAG-3′.

(b) NH₂ - fMet-Ser-Glu-Leu-Pro-Cys-Phe-Ile-COOH.

(c) NH₂ - fMet-Ser-Glu-Leu-Pro-Cys-Phe-Ile-Arg-Ile-Asp-Leu-COOH.

CHAPTER 13

13.1 (a) Transition, (b) transition, (c) transversion, (d) transversion, (e) frameshift, (f) transition.

13.2 6:1.

UGG transitions → UGA (nonsense), UAG (nonsense), CGG (Arg).

UGG transversions → UGC (Cys), UGU (Cys), UCG (Ser), UUG (Leu), AGG (Arg), GGG (Gly).

13.3 (a) *CIB* method, (b) attached-X method (see Chapter 6).

13.4 Bacteria treated with a mutagen or expected to carry mutations may be introduced into media with particular drugs in appropriate concentrations. Colonies that appear have originated from cells carrying preexisting mutations for resistance. This may be verified by the replica-plating technique (see Figure 14.3). The frequency of mutations of wild-type (drug-sensitive) cells to drug resistance can be measured in the presence or absence of the drug.

13.5 Probably not. A human is larger than a bacterium, with more cells and a longer life span. If mutation frequencies are calculated in terms of cell generations, the rates for human cells and bacterial cells are similar.

13.6 A dominant mutation presumably occurred in the woman in whom the condition was first known.

13.7 The X-linked gene is carried by mothers, and the disease is expressed in half of their sons. Such a disease is difficult to follow in pedigree studies because of the recessive nature of the gene, the tendency for the expression to skip generations in a family line, and the loss of the males who carry the gene. One explanation for the sporadic occurrence and tendency for the gene to persist is that, by mutation, new defective genes are constantly being added to the load already present in the population.

13.8 Plants can be propagated vegetatively, but no such methods are available for widespread use in animals.

13.9 The sheep with short legs could be mated to unrelated animals with long legs. If the trait is expressed in the first generation, it could be presumed to be inherited and to depend on a dominant gene. On the other hand, if it does not appear in the first generation, F_1 sheep could be crossed back to the short-legged parent. If the trait is expressed in one-half of the backcross progeny, it might be presumed to be inherited as a simple recessive. If two short-legged sheep of different sex could be obtained, they could be mated repeatedly to test the hypothesis of dominance. In the event that the trait is not transmitted to the progeny that result from these matings, it might be considered to be environmental or dependent on some complex genetic mechanism that could not be identified by the simple test used in the experiments.

13.10 Enzymes may discriminate among the different nucleotides that are being incorporated. Mutator enzymes may utilize a higher proportion of incorrect nucleotides, whereas antimutator enzymes may select fewer incorrect bases in DNA replication. In the case of the phage T4 DNA polymerase, the relative efficiencies of polymerization and proofreading by the polymerase's $3′ → 5′$ exonuclease activity play key roles in determining the mutation rate.

13.11 If both mutators and antimutators operate in the same living system, an optimum mutation rate for a particular organism in a given environment may result from natural selection.

13.12 *Dt* is a mutator gene that induces somatic mutations in developing kernels.

13.13 (a) Yes. (b) A block would result in the accumulation of phenylalanine and a decrease in the amount of tyrosine, which would be expected to result in several different phenotypic expressions.

13.14 These hemoglobins can be distinguished by mobility of molecules in an electric field (electrophoretic mobility) and by the amino acid sequences of their β polypeptides.

13.15

Amino Acid	mRNA	DNA
Glumatic acid	— GAA →	— GAA →
		← C̈T̈T̈ — ← Transcribed strand
		↓ Mutation
Valine	— GUA →	— GTA →
		← CAT —
		↓ Mutation
Lysine	— AAA →	— AAA →
		← T̈T̈T̈ —

13.16 The label "molecular disease" became common in speaking of sickle-cell anemia because its molecular basis (the substitution of a valine residue for the glutamic acid residue at amino acid position number 6 in the β chain) was recognized quite early during the emergence of the science of molecular biology. Actually, most if not all inherited diseases probably have very similar molecular bases. We just don't know what the molecular defects are in most instances.

13.17 Mutations; transitions, transversions, and frameshifts.

13.18 Irradiate the nonresistant strain and plate the irradiated organisms on a medium containing streptomycin. Those that survive and produce colonies are resistant. They could then be replicated to a medium without streptomycin. Those that survive would be of the first type; those that can live with streptomycin but not without it would be the second type.

13.19 3%; 4%; 6%.

13.20 Each quantum of energy from the X rays that is absorbed in a cell has a certain probability of hitting and breaking a chromosome. Hence, the greater the number of quanta of energy or dosage, the more likely breaks are to occur. The rate at which this dosage is delivered does not change the probability of each quantum inducing a break.

13.21 Radioactive iodine is concentrated by living organisms and food chains.

13.22 The person receiving a total of 100 r would be expected to have twice as many mutations as the one receiving 50 r.

13.23 During the replicating process, untraviolet light produces mispairing alterations mostly in pyrimidines (for example, cytosine to thymine transitions). Thymine may be altered to cytosine (or a modified pyrimidine with the base-pairing potential of cytosine), which pairs with guanine. A reverse mutation may occur when cytosine is changed to thymine (or a derivative of cytosine with the hydrogen-bonding potential of thymine), which pairs with adenine. The T-A base pair may thus be changed to a C-G, and the reverse mutation may occur from C-G to T-A.

13.24 Nitrous acid brings about a substitution of an OH group for an NH_2 group in those bases (A, C, and G) having NH_2 side groups. In so doing, adenine is converted to hypoxanthine, which base-pairs with cytosine, and cytosine is converted to uracil, which base-pairs with adenine. The net effects are GC ⇔ AT base-pair substitutions (see Figure 13.19).

13.25 Transitions.

13.26 P #1 - Causes transition mutation. Liver enzymes convert it into non-mutagen. Does not cause frameshift mutations. P #2 - Does not cause transition mutations. Liver enzymes convert it into a frameshift mutagen. P #3 - Causes transition mutations. Liver enzymes have no effect on mutagenicity. Does not cause frameshift mutations.

13.27 Nitrous acid acts as a mutagen on either replicating or nonreplicating DNA and produces transitions from A to G or C to T, whereas 5-bromouracil does not affect nonreplicating DNA but acts during the replication process causing GC ⇔ AT transitions. 5-Bromouracil must be incorporated into DNA during the replication process in order to induce mispairing of bases and thus mutations.

13.28 The reading frame will be shifted between the two frameshift mutations. This shift in reading frame does not read the normal nonsense codon and termination does not occur.

13.29 5-BU causes GC ⇔ AT transitions. 5-BU can, therefore, revert almost all of its induces by enhancing the transition event that is the reverse of the one that produced the mutation. In contrast, the spontaneous mutations will include transversions, frameshifts, deletions, etc., as well as transitions. Only the spontaneous transitions will show enhanced reversion by 5-BU.

13.30 Mutations induced by acridine dyes are primarily insertions or deletions of single base-pairs. Such mutations alter the reading frame (the in-phase triplets specifying mRNA codons) for that portion of the gene distal (relative to the direction of transcription and translation) to the mutation (see Figure 13.15b). This would be expected to totally change the amino acid sequences of polypeptides distal to the mutation site and produce inactive polypeptides. In addition, such frameshift mutations frequently produce inframe termination codons that result in truncated proteins.

13.31 (a) Frameshift due to the insertion of C at the 9th, 10th, or 11th nucleotide from the 5′ end.

(b) normal: 5′-AUGCCGUACUGCCAGCUAACUGCUAAAGAACAAUUA-3′.
mutant: 5′-AUGCCCGUACUGCCAGCUAACUGCUAAAGAACAAUUA-3′.

(c) normal: NH_2-fMet-Pro-Tyr-Cys-Gln-Leu-Thr-Ala-Lys-Glu-Gln-Leu.
mutant: NH_2-fMet-Pro-Val-Leu-Pro-Ala-Asn-Cys.

13.32 Proline and serine.

13.33 No. Leucine ⟶ proline would occur more frequently. Leu (CUA) $\xrightarrow{\text{5-BU}}$ Pro (CCA) occurs by a single base-pair transition, whereas Leu (CUA) $\xrightarrow{\text{5-BU}}$ Ser (UCA) requires two base-pair transitions. Recall that 5-bromouracil (5-BU) induces only transitions (see Figure 13.18).

13.34 No. 5-Bromouracil is mutagenic only to replicating nucleic acids.

13.35 Yes:

```
DNA:      ◄—GGX—            ◄—GGX—
          —CCX′—► HNO₂ ► —UCX′—►
               ↓                 ↓
mRNA:        GGX               AGX
               ↓                 ↓
Polypeptide: Gly            Ser or Arg
                            (depending on X)
```

or

```
DNA:      ◄—GGX—            ◄—GGX—
          —CCX′—► HNO₂ ► —CUX′—►
               ↓                 ↓
mRNA:        GGX               GAX
               ↓                 ↓
Polypeptide: Gly            Asp or Glu
                            (depending on X)
```

or

```
DNA:      ◄—GGX—            ◄—GGX—
          —CCX′—► HNO₂ ► —UUX′—►
               ↓                 ↓
mRNA:        GGX               AAX
               ↓                 ↓
Polypeptide: Gly            Asn or Lys
                            (depending on X)
```

Note: The X at the third position in each codon in mRNA and in each triplet of base pairs in DNA refers to the fact that there is complete degeneracy at the third base in the glycine codon. Any base may be present in the codon, and it will still specify glycine.

13.36 No. The glycine codon is GGX, where X can be any one of the four bases. Because of this complete degeneracy at the third position of the glycine codon, changing X to any other base will have no effect (that is, the codon will still specify glycine). Nitrous acid deaminates guanine (G) to xanthine, but xanthine still base-pairs with cytosine. Thus guanine is not a target for mutagenesis by nitrous acid.

13.37 Tyr $\longrightarrow$ Cys substitutions; Tyr to Cys requires a transition, which is induced by nitrous acid. Tyr to Ser would require a transversion, and nitrous acid is not expected to induce transversions.

13.38 (b) Met → Thr. 5-Bromouracil induces transitions, not transversions. All other changes listed require transversions.

13.39 5'-UGG-UGG-UGG-AUG-CGA or AGA-GAA or GAG-UGG-ACG-AUG-3'.

13.40 5'-AUGCCCUUUGGGGAAAGGUUUCCCUAA-3'.

CHAPTER 14

14.1 Prior to 1940, the gene was considered a "bead-on-a-string," not subdivisible by recombination or mutation. Today, the gene is considered to be the unit of genetic material that codes for one polypeptide. The unit of structure, not subdivisible by recombination or mutation, is known to be the single nucleotide pair.

14.2 The recombination observed between lz^s and lz^g, two functionally allelic mutations at the *lozenge* locus of *Drosophila*.

14.3 The *cis-trans* test, which defines the unit of genetic material specifying the amino acid sequence of one polypeptide.

14.4 (a) Nucleotide. (b) Between adjacent nucleotide base-pairs.

14.5 Precursor $\xrightarrow{2}$ B $\xrightarrow{4}$ A $\xrightarrow{3}$ C $\xrightarrow{1}$ D

14.6

14.7 They provide powerful selective sieves for identifying rare recombinants. This is accomplished by using the restrictive environmental conditions to select wild-type recombinant progeny from crosses between pairs of conditional lethal mutants.

14.8 Two genes; mutations 1, 2, 3, 4, 5, 6, and 8 are in one gene; mutation 7 is in a second gene.

14.9 Four genes; mutations 1 and 2 in one gene; mutations 3 and 4 in a second gene; mutations 5 and 6 in a third gene; mutations 7 and 8 in a fourth gene.

14.10 m4   m7   m1   m6   m5   m2   m3.

14.11 (a) (A, B, F, G, H, I) and (C, D) and (E); (b) Blue colony requires at least three genes; (c) The maximum number of genes for blue color cannot be estimated from the above data.

14.12 The size of the gene (assuming that all nucleotide pairs in the gene are capable of undergoing base-pair substitutions, as seems highly probable). Dominant lethal alleles and recessive lethal alleles in haploids will (under normal conditions) exist only transiently, of course.

14.13 Homoalleles are structurally and functionally allelic; they are not separable by recombination. Heteroalleles are functionally allelic (based on *cis-trans* tests) but are structurally nonallelic (based on recombination tests). Heteroalleles thus result from mutations occurring at different sites within a gene.

14.14 (1) Cross the two white-flowered varieties. The F_1 plants will be *trans*-heterozygotes. If the F_1 plants have white flowers, the two varieties probably carry mutations in the same gene, causing white flowers. (2) Cross white-flowered varieties with red-flowered varieties and self-pollinate or intercross the F_1 plants. If alleles of a single gene are involved, monohybrid F_2 ratios should be observed in all cases.

14.15 (a) Five genes. (b) Mutations 1, 3, and 5 are in one gene; mutations 7 and 8 are in a second gene; mutations 2, 4, and 6 identify genes 3, 4, and 5, respectively.

14.16 A maximum of three mutant homoalleles in addition to the wild-type base pair at any one site.

14.17 (a) True. (b) False. (c) True. (d) True. (e) True.

14.18 The observed complementation between ry^2 and ry^{42} is *intra*genic complementation. Xanthine dehydrogenase is a dimeric protein, and dimers that contain one polypeptide encoded by the ry^2 allele and one polypeptide encoded by the ry^{42} allele are partially active. Presumably, the wild-type segment of the ry^2 polypeptide somehow stabilizes the mutant segment of the ry^{42} polypeptide, and vice versa, yielding a dimer with enzymatic activity.

14.19 *Am* mutations result in UAG chain-termination codons within the coding sequence of the mRNA product of a gene; they thus produce truncated polypeptide gene-products. Since all *am* mutant alleles of a gene will produce polypeptides lacking the COOH-terminus, they would not be expected to exhibit intragenic complementation except in very rare cases. In contrast, most *ts* mutations are caused by missense mutations that change the amino acid sequence of the polypeptide gene-product making it more heat-labile. However, most *ts* mutant alleles produce a complete, although altered, gene-product. As a result, *ts* mutant alleles often exhibit intragenic complementation when the active form of the protein gene-product is a homomultimer. For this reason, *cis-trans* tests carried out with *am* mutants, not *ts* mutants, have been used whenever possible to define the genes of phage T4.

14.20 (a) The seven *sus* mutants are located in three different genes. (b) Mutant strains 1, 3, and 7 contain mutations in one gene; 4, 5, and 6 carry mutations in a second gene; 2 has a mutation in a third gene.

14.21 It depends on how you define alleles. If every variation in nucleotide sequence is considered to be a different allele, even if the gene product and the phenotype of the organism carrying the mutation are unchanged, then the number of alleles will be directly related to gene size. However, if the nucleotide sequence change must produce an altered gene product or phenotype before it is considered a distinct allele, then there will be a positive correlation, but not a direct relationship, between the number of alleles of a gene and its size in nucleotide pairs. The relationship is more likely to occur in prokaryotes where most genes lack introns. In eukaryotic genes, nucleotide sequence changes within introns usually are neutral; that is, they do not affect the activity of the gene product or the phenotype of the organism. Thus, in the case of eukaryotic genes with introns, there may be no correlation between gene size and number of alleles producing altered phenotypes.

14.22 *White* and *eosin* are located in the same gene; *carnation* is located in a different gene.

14.23 (a) Four genes. (b) Mutations 1, 3, and 7 are in one gene; mutations 4 and 6 are in a second gene; mutation 2 is in a third gene; and mutation 5 is in a fourth gene.

14.24 One gene; all seven mutations are in the same gene.

14.25 No. Because the two mutations map to different chromosomes, they could not be located within the same gene, at least, based on our current concept of the gene. However, if two transcripts are spliced in *trans*, it is possible for two parts of a gene—a nucleotide sequence encoding one polypeptide—to be located on two different chromosomes. Remember that our concept of the gene has evolved considerably since it was introduced by Mendel in 1865, and it will undoubtedly continue to evolve in the future.

14.26 Several enzymes were shown to contain two or more different polypeptides, and these polypeptides were sometimes controlled by genes that mapped to different chromosomes. Thus the mutations clearly were not in the same gene.

14.27 *Neurospora* has many advantages over humans as an experimental organism. The most important advantages are the ability to grow organisms under carefully controlled conditions, to enhance mutation frequency by treatment with mutagenic agents, and to perform controlled crosses for genetic analysis.

14.28 (a) Homoalleles. (b) Heteroalleles.

14.29 One. The *trpA58* and *trpA78* mutations alter the same codon. If they alter the 5′ and 3′ nucleotide pairs of the triplet, they will be separated by the middle nucleotide pair of the triplet specifying one mRNA codon.

14.30 All four mutations are in the same gene. However, ry^{42} and ry^{406} exhibit intragenic complementation with each other, whereas ry^{5} and ry^{41} do not.

14.31

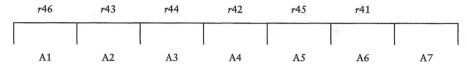

14.32 (a) A.

(b) Either a double mutation or a deletion covering both genes.

(c) It is either not an rII mutant or a mutant allele of T that shows intragenic complementation.

(d) S.

(e) No, unless this mutant and V exhibit intragenic complementation.

(f) B2.

(g) B4.

14.33 The most reliable way to determine if mutation is a deletion is to perform a reverse mutation analysis. Deletions do not revert to wild type, but point mutations do.

14.34 (a) The linear order of the six mutations that can be ordered based on the data given is 1 - 2 - 3 - 4 - 5 - 6. (b) Mutation 7 can not be ordered based on the data provided.

14.35 (a) The linear order of the six mutations that can be ordered based on the data given is 3 - 8 - 7 - 4 - 2 - 5. (b) Mutations 1 and 6 can not be ordered based on the data provided.

14.36 (a) The amino acid sequences that are specified by overlapping genes or genes-within-genes can not evolve independently. However, the degeneracy of the genetic code (see Chapter 12) will allow for some evolutionary flexibility.

(b) 3.

(c) 6.

(d) A single nucleotide change within the coding sequences of overlapping genes will often cause amino acid changes in both polypeptide gene products. Mutations that produce beneficial changes in one gene product may be deleterious to the other gene product. Thus, the two genes will have to evolve together, with natural selection acting on the effects of mutations on both gene products.

14.37 They are usually referred to as gene segments because they are joined together by somatic recombination to produce a sequence of nucleotide pairs that encodes a single polypeptide.

14.38 Several different polypeptides can be produced from a single gene by alternate pathways of transcript splicing. In the case of the tropomyosins, the exons of the transcripts are known to be spliced together in different combinations to produce overlapping, but distinct polypeptides.

14.39 No. These mutations are located far apart on the X chromosome. They are separated by millions of nucleotide pairs and many other genes, and it is virtually impossible for mutations located that far apart to be part of the same gene. For these mutations to be located in the same gene, all of the intervening genes would have to be part of a huge intron or some type of *trans* splicing would have to occur.

14.40 Mutations 1, 2, and 4 do not complement one another and thus appear to be located in the same gene, as do mutations 3 and 5. The anomaly is that mutation 7 does not complement mutations 3, 5, and 6, even though mutation 6 does complement mutations 3 and 5. (a) There are three simple explanations of the seemingly anomalous complementation behavior of mutation 7. (1) It is a deletion spanning all or parts of two genes. (2) It is a double mutation with defects in two genes. (3) It is a polar mutation in the promoter-proximal gene of a multigenic transcription unit. (b) Three simple genetic operations will distinguish between these three possibilities.

(1) Reversion. Plate a large number of mutant 7 phage on *E. coli* strain Z and look for wild-type revertants. (2) Backcross mutant 7 to wild-type phage and test the mutant progeny for the ability to complement mutations 3 and 6. (3) Introduce F's carrying tRNA nonsense suppressor genes into *E. coli* strain Z and determine whether any of them suppress the *loz7* mutation. (c) If mutation 7 is a deletion, it will not revert, and, if it is a double mutation, the reversion rate will probably be below the level of detection in your experiment. On the other hand, if it is a polar nonsense mutation, *loz*+ revertants will be obtained. If mutation 7 is a deletion, no new genotypes will be produced in the backcross to wild-type. However, if it is a double mutation, some recombinant single-mutant progeny will be produced in the backcross to wild-type, and these single mutations will complement either mutation 3 or mutation 6. If mutation 7 is a polar nonsense mutation, it should be suppressed by one or more of the tRNA suppressor genes introduced into *E. coli* strain Z.

CHAPTER 15

15.1 (a) Both introduce new genetic variability into the cell. In both cases, only one gene or a small segment of DNA representing a small fraction of the total genome is changed or added to the genome. The vast majority of the genes of the organism remain the same. (b) The introduction of recombinant DNA molecules, if they come from a very different species, is more likely to result in a novel, functional gene product in the cell, if the introduced gene (or genes) is capable of being expressed in the foreign protoplasm. The introduction of recombinant DNA molecules is more analogous to duplication mutations (see Chapter 7) than to other types of mutations.

15.2 Restriction endonucleases cleave at palindromes in double-stranded DNA. A palindrome (indicated in bold-faced letters) can be found in all the double-stranded sequences except d. Therefore, sequence d would not be cleaved by a restriction endonuclease, whereas a, b and c could be cleaved by the appropriate enzyme.

(a) ACTCC**AGAATTC**ACTCCG
    TGCGGTC**TTAAG**TGAGGC

(b) GCCTC**ATTCGAAG**CCTGA
    CGGAGT**AAGCTT**CGGACT

(c) CTCGCC**AATTG**ACTCGTC
    GAGCG**GTTAACT**GAGCAG

(d) ACTCCACTCCCGACTCCA
    TGAGGAGAGGGCTGAGGT

15.3 (a) $(1/4)^4 = 1/256$; (b) $(1/4)^6 = 1/4096$.

15.4 Restriction endonucleases recognize and cut specific nucleotide sequences in DNA. Most other endonucleases are not sequence-specific; many cut DNA sequences at random.

15.5 Recombinant DNA and gene-cloning techniques allow geneticists to isolate essentially any gene or DNA sequence of interest and to characterize it structurally and functionally. Large quantities of a given gene can be obtained in pure form, which permits one to determine its nucleotide-pair sequence (to "sequence it" in common lab jargon). From the nucleotide sequence and our knowledge of the genetic code, geneticists can predict the amino acid sequence of any polypeptide encoded by the gene. By using an appropriate subclone of the gene as a hybridization probe in northern blot analyses, geneticists can identify the tissues in which the gene is expressed. Based on the predicted amino acid sequence of a polypeptide encoded by a gene, geneticists can synthesize oligopeptides and use these to raise antibodies that, in turn, can be used to identify the actual product of the gene and localize it within cells or tissues of the organism. Thus, recombinant DNA and gene-cloning technologies provide very powerful tools with which to study the genetic control of essentially all biological processes. These tools have played major roles in the explosive progress in the field of biology during the last two decades.

15.6 The nucleotide-pair sequence. Restriction endonucleases recognize a specific nucleotide-pair sequence in DNA regardless of the source of the DNA. In most cases, this is a 4 or 6 nucleotide-pair sequence; in a few cases, the recognition sequence is longer (for example, 8 nucleotide pairs). Most restriction enzymes cleave the two strands of the DNA at a specific position (between the same 2 adjacent nucleotides in each strand) within the recognition sequence. A few restriction enzymes bind at a specific recognition sequence, but cut the DNA at a nearby site outside of the recognition sequence. Some restriction endonucleases cut both strands between the same two nucleotide pairs ("blunt end" cutters), whereas others cut the two strands at different positions and yield complementary single-stranded ends ("sticky or staggered end" cutters). See Table 15.1 for examples.

15.7 Restriction endonucleases are believed to provide a kind of primitive immune system to the microorganisms that produce them—protecting their genetic material from "invasion" by foreign DNAs from viruses or other pathogens or just DNA in the environment that might be taken up by the microorganism. Obviously, these microorganisms do not have a sophisticated immune system like that of higher animals (Chapter 23).

15.8 Microorganisms that produce restriction endonucleases also produce enzymes that modify one or more bases in the recognition sequence for that endonuclease so that it can no longer cleave the DNA at that site. In most cases, the modifying enzyme is a methylase that attaches a methyl group to one or more of the bases in the recognition sequence. For example, *E. coli* strains that produce the restriction endonuclease *Eco*RI, also produce *Eco*RI methylase, an enzyme that transfers a methyl group from S-adenosylmethionine to the 3′ adenine residue in each strand of the recognition sequence (5′-GAATCC-3′) producing N^6-methyladenines at these positions. *Eco*RI cannot cleave DNA that contains N^6-methyladenine at these positions even if the *Eco*RI recognition sequence is present in this DNA. Thus, if one wishes to digest DNA with *Eco*RI, that DNA must not be isolated from an *E. coli* strain that is producing *Eco*RI methylase.

15.9 A foreign DNA cloned using an enzyme that produces single-stranded complementary ends can always be excised from the cloning vector by cleavage with the same restriction enzyme that was originally used to clone it. For example, if a *Hin*dIII fragment from the human genome is cloned into *Hin*dIII-cleaved pUC119, the human *Hin*dIII fragment can be excised from a plasmid DNA preparation of this clone by cleavage with restriction endonuclease *Hin*dIII. The human *Hin*dIII fragment will be flanked in the recombinant plasmid DNA clone by *Hin*dIII cleavage sites. When terminal transferase is used to add complementary single-stranded ends during cloning, the original restriction endonuclease cleavage sites are destroyed. Thus, the restriction enzyme used to generate the fragment for cloning cannot be used to excise the original fragments from the cloning vector.

15.10 Step 1: Digest genomic DNA isolated from your research organism with *Eco*RI. Step 2: Treat pUC118 DNA with *Eco*RI. Step 3: Mix the *Eco*RI-digested genomic and vector DNAs under annealing conditions and incubate with DNA ligase. Step 4: Transform *amp^s* *E. coli* cells carrying the *lacZ*Δ*M15* gene with the resulting ligation products. Step 5: Plate the transformed cells on nutrient agar medium containing Xgal and ampicillin. Only transformed cells will produce colonies in the presence of ampicillin. Step 6: Prepare your genomic DNA library by using bacteria from white colonies; these bacteria will contain pUC118 DNA with genomic DNA inserts. Bacteria harboring pUC118 plasmids with no insert will produce blue colonies.

15.11 Most genes of higher plants and animals contain noncoding intron sequences. These intron sequences will be present in genomic clones, but not in cDNA clones, because cDNAs are synthesized using mRNA templates and intron sequences are removed during the processing of the primary transcripts to produce mature mRNAs.

15.12 Higher eukaryotes have very large genomes; for example, the genomes of mammals contain approximately 3×10^9 nucleotide pairs. Thus, trying to identify a particular single-copy gene from a clone library is like looking for the proverbial "needle-in-a-haystack." To accomplish this, one needs a nucleic acid hybridization probe specific for the gene or an antibody probe specific for the gene product. Given a specific cell or tissue type producing the mRNA and/or the protein gene product in large amounts, it was relatively easy to obtain pure mRNA or pure protein to use in making a hybridization or antibody probe, respectively, with which to screen a library for the gene or cDNA of interest. These approaches are much more difficult for the majority of the genes that encode products that represent only a small proportion of the total gene products in any given cell type.

15.13 The maize *gln2* gene contains many introns, and one of the introns contains a *Hin*dIII cleavage site. The intron sequences (and thus the *Hin*dIII cleavage site) are not present in mRNA sequences and thus are also not present in full-length *gln2* cDNA clones.

15.14 (a) 1.

 (b) 2.

 (c) 2.

15.15 By oligonucleotide-directed site-specific mutagenesis, you can change each of the eight nucleotide pairs in the protein binding site to each of the other three possible nucleotide pairs. You can then examine the binding of the regulatory protein to each of the mutant sequences (24 single nucleotide-pair changes are possible at the eight positions), and you can examine the ability of each of the mutant sequences to mediate induction of transcription of the gene at 45°C *in vivo*. You could study the binding of the protein using synthetic oligonucleotide sequences more easily, but that approach would not let you study induction of transcription of the gene *in vivo*. The site-specific mutagenesis approach will allow you to carry out a saturation mutational analysis of the octameric regulatory protein binding sequence.

15.16 The PCR technique has much greater sensitivity than any other method available for analyzing nucleic acids. Thus PCR procedures permits analysis of nucleic acid structure given extremely minute amounts of starting material. DNA sequences can be amplified and structurally analyzed from very small amounts of tissue like blood or sperm in assault and rape cases. In addition, PCR methods permit investigators to detect the presence of rare gene transcripts (for example, in specific types of cells) that could not be detected by less sensitive procedures such as northern blot analyses or *in situ* hybridization studies.

15.17 Some bacterial viruses package single-stranded DNA rather than double-stranded DNA. These bacteriophages employ a double-stranded DNA intermediate for replication, but then switch to the production of single-stranded DNAs for packaging during phage maturation. When a foreign DNA segment is cloned into the replicative form of the chromosome of one of these phages, only one of the two strands gets packaged during the subsequent maturation processes. In the case of the filamentous single-stranded DNA phages such as M13 and f2, the progeny phage particles are simply extruded through the cell membrane and wall without lysing the host bacterium. Because the phage are so small relative to the size of the bacteria, the phage particles can be separated from the bacteria by a simple low-speed centrifugation step. The bacteria form a pellet at the bottom of the centrifuge tube; the phage remain in the supernatant suspension. The bacterial pellets are discarded, and the phage particles can be collected by high-speed centrifugation or by precipitation with polyethylene glycol and low-speed centrifugation. Pure single-stranded DNA can then be separated from the phage proteins by phenol-chloroform extractions. Many cloning vectors like pUC118 and pUC119 (see Figure 15.17) are phage-plasmid hybrids. They have both plasmid and M13 phage origins of replication as well as M13 phage packaging signals so that they can replicate either as plasmids or as phages and can package single-stranded phage DNA when in the phage mode of replication. The switch from the plasmid mode of replication to the phage mode is accomplished by superinfecting the host bacteria with a mutant "helper" phage.

15.18 All modern cloning vectors contain a "polycloning site"— a cluster of cleavage sites for a number of different restriction endonucleases in a nonessential region of the vector into which foreign DNAs can be inserted. In general, the greater the complexity of the polycloning site—that is, the more restriction endonuclease cleavage sites that are present—the greater the utility of the vector for cloning a wide variety of different restriction fragments. For example, see the polycloning site present in pUC118 and pUC119 shown in Figure 15.17.

15.19 (a) 7000 bp + 7000 bp; (b) 14,000 bp; (c) 7000 bp + 7000 bp; (d) 4000 bp + 2000 bp + 5000 bp; (e) 4000 bp + 7000 bp.

15.20 The name "western blot" has no literal significance; it is pure laboratory jargon. In 1975, E. M. Southern published a procedure for the transfer of DNA molecules that had been separated by agarose gel electrophoresis to nitrocellulose membranes and their subsequent detection by hybridization to radioactive probes and autoradiography. The resulting autoradiographs showing the positions of the DNA bands that had hybridized to the probes were called Southern blots in reference to E. M. Southern who had developed the technique. When similar procedures were developed for analyzing RNA molecules that had been separated by gel electrophoresis, researchers started calling them northern blots because of their similarity to Southern blots. When the methodology was extended to proteins, the term western blot supposedly was the logical extension of the Southern, northern laboratory jargon. Or, was it? Why wasn't the western blot called an eastern blot?

15.21 Because the nucleotide-pair sequences of both the normal *CF* gene and the *CFΔ508* mutant gene are known, labeled oligonucleotides can be synthesized and used as hybridization probes to detect the presence of each allele (normal and Δ508). Under high-stringency hybridization conditions, each probe will hybridize only with the *CF* allele that exhibits perfect complementarity to itself. Since the sequences of the *CF* gene flanking the Δ508 site are known, oligonucleotide PCR primers can be synthesized and used to amplify this segment of the DNA obtained from small tissue explants of putative CF patients and their relatives by PCR. The amplified DNAs can then be separated by agarose gel electrophoresis, transferred to nylon membranes, and hybridized to the respective labeled oligonucleotide probes, and the presence of each *CF* allele can be detected by autoradiography. For a demonstration of the utility of this procedure, see B. Kerem et al. "Identification of the cystic fibrosis gene: Genetic analysis." *Science 245*: 1073–1080, 1989. Kerem and coworkers used two synthetic oligonucleotide probes (oligo-N = 3'-CTTTTATAGTAGAAACCAC-5' and oligo-ΔF = 3'-TTCTTTTATAGTA—ACCACAA-5'; the dash indicates the deleted nucleotides in the *CFΔ508* mutant allele) to analyze the DNA of CF patients and their parents. For confirmed CF families, the results of these Southern blot hybridizations with the oligo-N (normal) and oligo-ΔF (*CFΔ508*) labeled probes were often as follows:

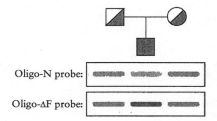

Both parents were heterozygous for the normal *CF* allele and the mutant *CFΔ508* allele as would be expected for a rare recessive trait, and the CF patient was homozygous for the *CFΔ508* allele. In such families, one-fourth of the children would be expected to be homozygous for the *Δ508* mutant allele and exhibit the symptoms of CF, whereas three-fourths would be normal (not have CF). However, two-thirds of these normal children would be expected to be heterozygous and transmit the allele to their children. Only one-fourth of the children of this family would be homozygous for the normal *CF* allele and have no chance of transmitting the mutant *CF* gene to their offspring. Note that the screening procedure described here can be used to determine which of the normal children are carriers of the *CFΔ508* allele: that is, the mutant gene can be detected in heterozygotes as well as homozygotes.

15.22 You could attempt to isolate either a dihydropicolinate synthase (DHPS) cDNA clone or a DHPS genomic clone. Once you have isolated either DHPS clone, it can be used as a hybridization probe to isolate the other (genomic or cDNA) by screening an appropriate library by *in situ* colony hybridization. Four approaches that have proven effective in isolating other eukaryotic coding sequences of interest are the following:

(1) You could obtain a clone of the DHPS gene of a lower eukaryote (a clone of the DHPS gene of *Saccharomyces cerevisiae* is available) or even a prokaryote and use it as a heterologous hybridization probe to screen a maize cDNA library using low stringency conditions. Sometimes this approach is successful; sometimes it is not successful. Whether or not this approach works depends on how similar the coding sequences of the specific gene of interest are in the two species.

(2) You could purify the DHPS enzyme from corn and use the purified protein to produce an antibody to DHPS. This DHPS-specific antibody could then be used to screen a maize cDNA expression library by a protocol analogous to the western blot procedure. (An expression library contains the cDNA coding sequences fused to appropriate transcription and translation signals so that they are expressed in *E. coli* or other host cells in which the cDNA library is prepared.)

(3) You could purify the DHPS enzyme from maize and determine the amino acid sequence of its NH_2-terminus by microsequencing techniques. From the amino acid sequence and the known genetic code, you could predict the possible nucleotide sequences encoding this segment of the protein. Because of the degeneracy in the code, there would be a set of nucleotide sequences that would all specify the same amino acid sequence, and you would not know which one was present in the maize DHPS gene. However, the synthesis of oligonucleotides is now routine and quite inexpensive. Thus, you could synthesize a mixture of oligonucleotides containing all possible coding sequences and use this mixture as a set of hybridization probes to screen an appropriate library by *in situ* colony hybridization.

(4) Finally, you might try a simple and very quick genetic approach based on the ability of cDNAs in an expression library to rescue DHPS mutants of *E. coli* or other species that can be transformed at high frequencies. You would obtain a DHPS-deficient mutant of *E. coli* (available from the *E. coli* Genetics Stock Center at Yale University), transform it with your cDNA expression library, and plate the transformed cells on medium lacking diaminopimelic acid (the product of DHPS). DHPS-deficient *E. coli* mutants cannot grow in the absence of diaminopimelic acid; thus, any colonies that grow on your selection plates should be the result of rescue of the DHPS mutant bacteria by corn DHPS encoded by cDNAs in the library. This entire screening procedure can be carried out in three or four days; thus, it is much simpler than the preceding approaches. In fact, David A. Frisch first isolated a maize DHPS cDNA by this simple, but powerful genetic approach. However, note that this approach would only be expected to work in the case of enzymes that are active as monomers or homomultimers; it is not applicable when the active form of the enzyme is a heteromultimer.

15.23 (a) Southern, northern, and western blot procedures all share one common step, namely, the transfer of macromolecules (DNAs, RNAs, and proteins, respectively) that have been separated by gel electrophoresis to a solid support—usually a nitrocellulose or nylon membrane—for further analysis. (b) The major difference between these techniques is the class of macromolecules that are separated during the electrophoresis step: DNA for Southern blots, RNA for northern blots, and protein for western blots.

15.24 No. The genome of the species in question may contain one, two, or three copies of the gene (or family of closely related genes) encoding this protein. The possibilities are:

(1) one copy of the gene with two *Eco*RI cleavage sites located within intron sequences.

(2) two copies of the gene with one *Eco*RI cleavage site located within an intron sequence of one of the copies.

(3) three copies of the gene with no *Eco*RI cleavage site in any of the copies, that is, each copy present on a single *Eco*RI restriction fragment.

15.25

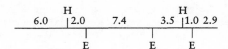

15.26

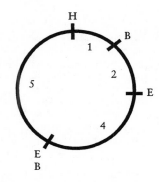

15.27 There are two possible restriction maps for these data as shown below:

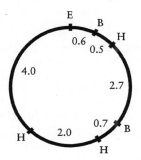 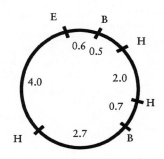

Restriction enzyme cleavage sites for *Bam*HI, *Eco*RI, and *Hin*dIII are denoted by B, E, and H, respectively. The numbers give distances in kilobase pairs.

15.28 3′ CCTACTGGAATTGCCAGATCTGGTTGAGGT 5′.

15.29

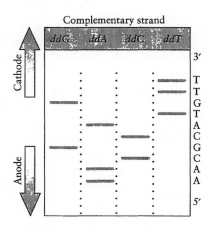

15.30 Nascent strand: 5′-AGTTCTAGAGCGGCCGCCACCGCGTGGAGCTCCAGCTTTTGTTCCCTTT-3′.
Template strand: 3′-TCAAGATCTCGCCGGCGGTGGCGCACCTCGAGGTCGAAAACAAGGGAAA-5′.

CHAPTER 16

16.1   Genetic map distances are determined by crossover frequencies. Cytogenetic maps are based on chromosome morphology or physical features of chromosomes. Physical maps are based on actual physical distances—the number of nucleotide pairs (0.34 nm per bp)— separating genetic markers. If a gene or other DNA sequence of interest is shown to be located near a mutant gene, a specific knob on a chromosome, or a particular DNA restriction fragment, that genetic or physical marker (mutation, knob, or restriction fragment) can be used to initiate a chromosome walk (see Figure 16.9).

16.2   Chromosome walks are performed by isolating overlapping genomic clones and using them to "walk" along a chromosome or, more accurately, along a chromosomal DNA molecule (see Figure 16.9). Chromosome jumps use large genomic DNA fragments produced by partial digestion of genomic DNA with a restriction endonuclease to "jump" to a new position 100 kb or more away from the original site in a DNA molecule (see Figure 16.10). If there is no molecular marker near a gene of interest, isolating the gene by chromosome walking would take too long. In such cases, chromosome jumps must be used to speed up the process.

16.3   A contig (*contig*uous clones) is a physical map of a chromosome or part of a chromosome prepared from a set of overlapping genomic DNA clones. An RFLP (*r*estriction *f*ragment *l*ength *p*olymorphism) is a variation in the length of a specific restriction fragment excised from a chromosome by digestion with one or more restriction endonucleases. A VNTR (*v*ariable *n*umber *t*andem *r*epeat) is a short DNA sequence that is present in the genome as tandem repeats and in highly variable copy number. An STS (*s*equence *t*agged *s*ite) is a unique DNA sequence that has been mapped to a specific site on a chromosome. An EST (*e*xpressed *s*equence *t*ag) is a cDNA sequence—a genomic sequence that is transcribed. Contig maps permit researchers to obtain clones harboring genes of interest directly from DNA Stock Centers—to "clone by phone." RFLPs are used to construct the high density genetic maps that are needed for positional cloning. VNTRs are especially valuable RFLPs that are used to identify multiple sites in genomes. STSs and ESTs provide molecular probes that can be used to initiate chromosome walks to nearby genes of interest.

16.4   (a) a1, a2, a3, a4, b1, b2, and b3. (b) Band a1 represents a locus whose DNA is homologous to cDNA1. Since the marker is not polymorphic in the parents used, it cannot be mapped in this cross. (c) The cDNA1 probe detects one RFLP locus with alleles that are visualized as band a4 and bands a2/a3. The cDNA2 detects a second RFLP locus with alleles that are visualized as band b3 and bands b1/b2. The two loci are linked with 20% recombination observed in this cross.

16.5   (a)

10 cM	25 cM	15 cM	1 cM	14 cM
STS1	STS5	STS3	C	STS4	STS2

(b) 3.3×10^9 bp / 3.3×10^3 cM = 1×10^6 bp/cM. The total map length is 65 cM, which equates to about 65×10^6 or 65 million bp. (c) The cancer gene (C) and STS4 are separated by 1 cM or about one million base pairs.

16.6   Microsatellites are polymorphic tandem repeats of sequences only two to four nucleotide pairs long. They are called microsatellites because they short in length and are components of the highly repetitive satellite DNAs of eukaryotes (see Chapter 9).

16.7   With a clone of the gene available, fluorescent *in situ* hybridization (FISH) can be used to determine which human chromosome carries the gene and to localize the gene on the chromosome. Single-stranded copies of the clone are coupled to a fluorescent probe and hybridized to denatured DNA in chromosomes spread on a slide (see "Focus on *In Situ* Hybridization" in Chapter 9). After hybridization and removal on nonhybridized probe, autoradiography is performed to determine the location of the fluorescent probe (see Figure 16.6b).

16.8   An EST must be placed on the physical map of a chromosome before it can be called an STS. If is also positioned on the genetic map of the chromosome, then it is called an anchor marker.

16.9   VNTRs and microsatellites are both polymorphic tandem repeats of relatively short nucleotide sequences. Microsatellites, however, are repeats of sequences only 2–5 nucleotide pairs in length and are a subset of the satellite sequences (sequences that separate from the main band of DNA and form satellite bands in density gradients) in the highly repetitive DNAs of eukaryotes. Most VNTRs are present in mainband DNA—not satellite bands—when eukaryotic DNAs are analyzed by equilibrium density gradient centrifugation (see Chapter 10).

16.10 The resolution of genetic mapping in humans is quite low—in the range of one to ten million base pairs. Radiation hybrid mapping provides higher resolution—to about 50 kb. However, even with 50-kb resolution there could be several genes separating the RFLP including the mutation responsible for albinism.

16.11 You can start a chromosome walk using the hybridization probe that detects the RFLP (see Figure 16.9). However, if a physical map of this region of the chromosome already exists (see Figure 16.7), a chromosome walk might not be necessary. cDNAs can be used to locate candidate genes in the region covered by the chromosome walk, and the sequences of genes in individuals with this form of inherited deafness can be compared with the sequences of homologous genes of individuals with normal hearing, looking for changes that would be expected to cause a loss of gene function. The overall process is illustrated in Figure 16.8.

16.12 The resolution of radiation hybrid mapping is higher than that of standard somatic-cell hybrid mapping because the frequency of recombination is greatly increased in radiation hybrids by using X rays to fragment the human chromosomes prior to cell fusion. The rationale of radiation hybrid mapping is that the probability of breaking a DNA molecule in the region between two genes and thus separating them is directly proportional to the physical distance (number of base pairs) between them.

16.13 The goals of the Human Genome Project are to prepare genetic and physical maps showing the locations of all the genes in the human genome and to determine the nucleotide sequences of all 24 chromosomes in the human genome. These maps and nucleotide sequences of the human chromosomes will help scientists identify mutant genes that result in inherited diseases and, hopefully, will lead to successful treatments, including gene therapies, for at least some of these diseases. Potential misuses of these data include invasions of privacy by governments and businesses—especially employment agencies and insurance companies. Individuals must not be denied educational opportunities, employment, or insurance because of inherited diseases or mutant genes that result in a predisposition to mental or physical abnormalities.

16.14 If the RFLP is on the same homolog as a mutant gene causing an inherited abnormality, the presence of the RFLP usually will indicate that the mutant gene is also present. If the marker is on the other homolog, the RFLP and the mutant gene will segregate from one another during gametogenesis. In that case, the presence of the RFLP can no longer be used as an indication that the mutant gene is present.

16.15 An EST is more likely than an RFLP to occur in a disease-causing human gene. ESTs all correspond to expressed sequences in a genome. RFLPs occur throughout a genome, in both expressed and unexpressed sequences. Because less than 5 percent of the human genome encodes proteins, most RFLPs occur in non-coding DNA.

16.16 The bacteriophage ΦX174 genome has the highest gene density—one gene per 490 nucleotide pairs. (Recall that ΦX174 contains overlapping genes; see Chapter 14.) The human genome has the lowest gene density—about one gene per 100 kb. In the species mentioned in this question, there is a striking correlation between genome size and developmental complexity. With some exceptions, including species with polyploid genomes, there does appear to be a rough correlation between genome size and developmental complexity.

16.17 (a) Segment 5; (b) segment 4; (c) segment 1, 6, or 10.

16.18 EST markers D and E appear to be closely linked. The eight human-Chinese hamster radiation hybrids contain either both D and E or neither marker. More radiation hybrids would need to be tested for the presence of these ESTs to obtain convincing evidence of this linkage.

16.19 The advantage that gene chips have over traditional dot blot hybridization methods is that a single gene chip can be used to quantify thousands of distinct nucleotide sequences simultaneously. The hybridization probes are arranged on gene chips in microarrays on a solid surface, whereas traditional dot blots were macroarrays of a few to, at most, a few hundred sequences. The gene-chip technology allows researchers to investigate the levels of expression of large numbers of genes more efficiently than was possible using traditional dot blot techniques.

16.20 The green fluorescent protein (GFP) can be used to study protein localization and movement over time in living cells. Most other procedures for studying protein localization require that cells be permeabilized and/or fixed and exposed to antibodies coupled to radioactive or fluorescent compounds prior to visualization. As a result, these procedures only provide information about the location of a protein at a single time point. In contrast, GFP-tagged proteins can be used to study the synthesis and movement of proteins in living cells over time (hours to days).

16.21 The DNA sequences in human chromosome-specific cDNA libraries can be coupled to fluorescent dyes and hybridized *in situ* to the chromosomes of other primates. The hybridization patterns can be used to detect changes in genome structure that have occurred during the evolution of the various species of primates from common ancestors (see Figure 16.26). Such comparisons are especially effective in detecting new linkage relationships resulting from translocations and centric fusions.

16.22 The presence of two copies of each block of genes in the corn genome indicates that maize has evolved fron a tetraploid ancestor. The presence of one set of genes primarily in the large chromosomes and the second set largely in the small chromosomes suggests that maize has evolved from an allotetraploid (see Chapter 6) produced by combining the diploid genomes of two ancestral cereal grass species.

16.23 (a) Order of STS sites: 2-5-1-4-3-6.

(b) STS markers: 2   5   1   4   3   6

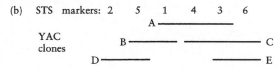

16.24 Scientists at deCODE have found that specific alleles of the gene encoding 5-lipoxygenase activating protein result in an approximately two-fold greater risk of both myocardial infarction (heart attack) and stroke.

16.25 All of the sequences identified by the megablast search encode histone H2a proteins. The query sequence is identical to the coding sequence of the *Drosophila melanogaster* histone *H2aV* gene (a member of the gene family encoding histone H2a proteins). The query sequence encodes a *Drosophila* histone H2a polypeptide designated variant V. The same databank sequences are identified when one-half or one-fourth of the given nucleotide sequence is used as the query in the megablast search. Query sequences as short as 15 to 20 nucleotides can be used to identify the *Drosophila* gene encoding the histone H2a variant. However, the results will vary depending on the specific nucleotide sequence used as the query sequence.

16.26 Your query sequence is a portion of histone H1.2 of the mouse (*Mus musculus*). This portion of the mouse histone H1.2 polypeptide differs by one amino acid from the corresponding part of the human H1 histone (aligned sequence 2).

16.27 Reading frame 5′ → 3′ number 1 has a large open reading frame with a methionine codon near the 5′ end. You can verify that this is the correct reading frame by using the predicted translation product as a query to search one of the protein databases (see Question 16.26).

CHAPTER 17

17.1 CpG islands are clusters of cytosines and guanines that are often located just upstream (5′) from the coding regions of human genes. Their presence in nucleotide sequences can provide hints as to the location of genes in human chromosomes.

17.2 The gene was named *huntingtin* after the disease that it causes when defective. The gene will probably be renamed after the function of its gene product has been determined.

17.3 The *CF* gene was identified by map position-based cloning, and the nucleotide sequences of *CF* cDNAs were used to predict the amino acid sequence of the *CF* gene product. A computer search of the protein data banks revealed that the *CF* gene product was similar to several ion channel proteins. This result focused the attention of scientists studying cystic fibrosis on proteins involved in the transport of salts between cells and led to the discovery that the *CF* gene product was a transmembrane conductance regulator—now called the CFTR protein.

17.4 Once the function of the CF gene product has been established, scientists should be able to develop procedures for introducing wild-type copies of the *CF* gene into the appropriate cells of cystic fibrosis patients to alleviate the devastating effects of the mutant gene. A major obstacle to somatic-cell gene-therapy treatment of cystic fibrosis is the size of the *CF* gene—about 250 kb, which is too large to fit in the standard gene transfer vectors. Perhaps a shortened version of the gene constructed from the *CF* cDNA—about 6.5 kb—can be used in place of the wild-type gene. A second major obstacle is getting the transgene into enough of the target cells of the cystic fibrosis patient to alleviate the symptoms of the disease. A third challenge is to develop an expression vector containing the gene that will result in long-term expression of the introduced gene in transgenic cells. Another concern is how to avoid possible side effects caused by overexpression or inappropriate expression of the transgene in cystic fibrosis patients. Despite these obstacles, many scientists are optimistic that cystic fibrosis will be effectively treated by somatic-cell gene therapy in the future.

17.5   Oligonucleotide primers complementary to DNA sequences on both sides (upstream and downstream) of the CAG repeat region in the *MD* gene can be synthesized and used to amplify the repeat region by PCR. One primer must be complementary to an upstream region of the template strand, and the other primer must be complementary to a downstream region of the nontemplate strand. After amplification, the size(s) of the CAG repeat regions can be determined by gel electrophoresis (see Figure 17.2). Trinucleotide repeat lengths can be measured by including repeat regions of known length on the gel. If less than 30 copies of the trinucleotide repeat are present on each chromosome, the newborn, fetus, or pre-embryo is homozygous for a wild-type *MD* allele or heterozygous for two different wild-type *MD* alleles. If more than 50 copies of the repeat are present on each of the homologous chromosomes, the individual, fetus, or cell is homozygous for a dominant mutant *MD* allele or heterozygous for two different mutant alleles. If one chromosome contains less than 30 copies of the CAG repeat and the homologous chromosome contains more than 50 copies, the newborn, fetus, or pre-embryo is heterozygous, carrying one wild-type *MD* allele and one mutant *MD* allele.

17.6   Yes. A somatic-cell gene therapy procedure similar to that used for ADA-SCID (see Figure 17.8) might be effective in treating purine nucleoside phosphorylase (PNP) deficiency. White blood cells could be isolated from the patient, transfected with a vector carrying a wild-type *PNP* gene, grown in culture and assayed for the expression of the *PNP* transgene, and then infused back into the patient after the expression of the transgene had been verified.

17.7   The transcription initiation and termination and translation initiation signals of eukaryotes differ from those of prokaryotes such as *E. coli*. Therefore, to produce a human protein in *E. coli*, the coding sequence of the human gene must be joined to appropriate *E. coli* regulatory signals—promoter, transcription terminator, and translation initiator sequences. Moreover, if the gene contains introns, they must be removed or the coding sequence of a cDNA must be used, because *E. coli* does not possess the spliceosomes required for the excision of introns from nuclear gene transcripts. In addition, many eukaryotic proteins undergo post-translational processing events that are not carried out in prokaryotic cells. Such proteins are more easily produced in transgenic eukaryotic cells growing in culture.

17.8   You would first construct a chimeric gene containing your synthetic gene fused to a plant promoter such as the 35S promoter of cauliflower mosaic virus and a plant transcription termination and polyadenylation signal such as the one from the *nos* gene of the Ti plasmid. This chimeric gene would then be inserted into the T-DNA of a Ti plasmid carrying a dominant selectable marker gene (for example, 35S/NPTII/*nos*, which confers resistance to kanamycin to host cells) and introduced into *Agrobacterium tumefaciens* cells by transformation. Tissue explants from *Arabidopsis* plants would be co-cultivated with *A. tumefaciens* cells harboring the recombinant Ti plasmid, and plant cells that carry T-DNAs inserted into their chromosomes would be selected by growth on medium containing the appropriate selective agent (for example, kanamycin). Transgenic plants would then be regenerated from the transformed cells and tested for resistance to glyphosate.

17.9   Eleven, ranging, in multiples of 3, from 15 to 45 nucleotides long.

17.10   DNA fingerprints are the specific patterns of bands present on Southern blots of genomic DNAs that have been digested with particular restriction enzymes and hybridized to appropriate DNA probes such as VNTR sequences. DNA fingerprints, like epidermal fingerprints, are used as evidence for identity or nonidentity in forensic cases. Geneticists have expressed concerns about the statistical uses of DNA fingerprint data. In particular, they have questioned some of the methods used to calculate the probability that DNA from someone other than the suspect could have produced the observed DNA fingerprints. This concern is based in part on the lack of adequate databases for various human subpopulations and the lack of precise information about the amount of variability in DNA fingerprints for individuals of different ethnic backgrounds. These concerns can be best addressed by the acquisition of data on fingerprint variability in different subpopulations and ethnic groups.

17.11   Contamination of blood samples would introduce more variability into the fingerprint. This would lead to a lack of allelic matching of fingerprints obtained from the blood samples and obtained from the defendant. Mixing errors would be expected only to lead to the acquittal of a guilty person and not to the conviction of an innocent person. Only the mislabeling of samples could implicate someone who is innocent.

17.12   The woman could be the mother of both children, having passed alleles *A* 8 and *B* 5 to child 1 and *A* 8 and *B* 3 to child 2. Male 3 could have been the father of child 1. If so, he contributed alleles *A* 7 and *B* 7.

17.13 Neither F1 nor F2 could be the girl's biological father; only individual F3 could be the child's father.

17.14 The T in T-DNA is an abbreviation for "transferred." The T-DNA region of the Ti plasmid is the segment that is transferred from the Ti plasmid of the bacterium to the chromosomes of the plant cells during *Agrobacterium tumefaciens*-mediated transformation.

17.15 Probing Southern blots of restriction enzyme-digested DNA of the transgenic plants with ^{32}P-labeled transgene may provide evidence of multiple insertions, but would not reveal the genomic location of the inserts. Fluorescence *in situ* hybridization (FISH) is a powerful procedure for determining the genomic location of gene inserts. FISH is used to visualize the location of transgenes in chromosomes.

17.16 Disarmed retroviruses are lacking genes essential for reproduction in host cells, but can still integrate into the DNA of the host cell in the proviral state. The retroviral genomes are small enough to allow them to be manipulated easily *in vitro* and yet will accept foreign DNA inserts of average gene size (see Figure 17.7). The retroviruses contain strong promoters in their long terminal repeats that can be used to drive high levels of transcription of the foreign gene insert.

17.17 Transgenic mice are usually produced by microinjecting the genes of interest into pronuclei of fertilized eggs or by infecting preimplantation embryos with retroviral vectors containing the genes of interest. Transgenic mice provide invaluable tools for studies of gene expression, mammalian development, and the immune system of mammals. Transgenic mice are of major importance in medicine; they provide the model system most closely related to humans. They have been, and undoubtedly will continue to be, of great value in developing the tools and technology that will be used for human gene therapy in the future.

17.18 An antisense RNA molecule is an RNA that is complementary to the pre-mRNA or mRNA ("sense" sequence) of a gene. In some cases, antisense RNAs can be used to block gene expression and study the effect(s) of the absence of a particular gene product on the growth and development of an organism. In agriculture, antisense RNAs have been used to block the synthesis of specific gene products to achieve desired changes in the phenotype of an organism. The best known example is the FlavrSavr™ tomato, which remains firm longer during the ripening process and allows the tomatos to remain on the vines longer before picking. The FlavrSavr™ tomato was produced by using antisense RNA to decrease the rate of expression of the gene encoding the enzyme polygalacturonase, which breaks down cell walls and causes softening during the ripening process. An organism that synthesizes an antisense RNA is usually produced by inverting the coding sequence of a gene relative to the promoter and transcription termination signal and reintroducing the altered gene into the host organism by transformation. With the coding sequence inverted, the nontemplate strand will be transcribed — rather than the template strand — producing antisense RNAs (see Figure 17.19).

17.19 Posttranslationally modified proteins can be produced in transgenic eukaryotic cells growing in culture or in transgenic plants and animals. Indeed, transgenic sheep have been produced that secrete human blood-clotting factor IX and α1-antitrypsin in their milk. These sheep were produced by fusing the coding sequences of the respective genes to a DNA sequence that encodes the signal peptide required for secretion and introducing this chimeric gene into fertilized eggs that were then implanted and allowed to develop into transgenic animals. In principle, this approach could be used to produce any protein of interest.

17.20 The noncoding regions of the ten actin genes can be subcloned and used as gene-specific hybridization probes to measure individual gene transcript levels in various organs and tissues of developing plants by either northern blot or *in situ* hybridization experiments. Alternatively, the gene-specific noncoding regions can be used to design PCR primers, and PCR can be used to measure individual gene transcript levels in organs and tissues. Perhaps the most powerful approach would be

(1) to fuse the promoter regions of the individual genes to the coding region of a reporter gene that encodes a product that can be detected by histochemical assays of plant organs and tissues.

(2) to introduce these chimeric reporter genes into *Arabidopsis* plants by *Agrobacterium tumefaciens*-mediated transformations.

(3) to analyze the temporal and spatial patterns of expression of the reporter genes in the transgenic plants. Indeed, Meagher and colleagues have already used these approaches to document the striking temporal and tissue-specific patterns of actin gene expression in *Arabidopsis*.

17.21 The vector described contains the HGH gene; however, it does not contain a mammalian HGH-promoter that will regulate the expression of the transgene in the appropriate tissues. Construction of vectors containing a properly positioned mammalian HGH-promoter sequence should result in transgenic mice in which HGH synthesis is restricted to the pituitary gland.

17.22 Classical genetic approaches use mutational dissection to probe the functions of genes. Mutant alleles that produce altered phenotypes are identified and used to investigate the functions of the wild-type alleles. Comparative molecular analyses of mutant and wild-type organisms sometimes allow researchers to determine the precise function of a gene. Reverse genetics approaches use the known nucleotide sequences of genes to design procedures to either produce null mutations in them or to inhibit their expression. Antisense RNA and RNA interference protocols are used to block the expression of specific genes. T-DNA and transposon insertion protocols are used to produce null mutations (to "knock out" the function) of specific genes.

17.23 Both antisense RNA and RNAi make use of RNA strands that are complementary to the mRNA target strand to block its translation (compare Figures 17.19 and 17.22). Antisense RNA uses single stranded RNA that is complementary to the mRNA. RNAi involves the use of double-stranded RNA, where one strand is complementary to the mRNA and the other strand is equivalent to the mRNA, to silence the expression of the target gene. RNAi makes use of the RNA-induced silencing complex (RISC) to block gene expression.

17.24 Insertional mutagenesis approaches produce mutant alleles—usually null alleles, or "knockouts,"— by the insertion of foreign DNA into genes. Other reverse genetic approaches, for example, antisense RNA and RNA interference, inhibit the expression of the genes but leave them structurally intact.

17.25 Plants have an advantage over animals in that once insertional mutations are induced they can be stored for long periods of time and distributed to researchers as dormant seeds.

17.26 The vectors used in somatic-cell gene therapy must not insert themselves preferentially into important genes, especially protooncogenes. Ideally, the vectors should insert themselves into unessential regions of the human genome. However, such vectors may not exist. At the minimum, however, vectors should be used that insert into the genome at random sites so that their chance of inserting into an important regulatory gene such as a protooncogene is very low. Gene therapy will probably never be completely risk free, because anomalous insertion events will always occur at some low frequency. No biological process is 100 percent accurate. However, the potential benefits of gene therapy must greatly outweigh the potential risks before the procedure will become an accepted tool for treating inherited diseases.

17.27 (a) You can invert the coding sequence of a gene, introduce it into *Arabidopsis* plants by transformation with *Agrobacterium tumefaciens*, and examine the effects of the antisense RNA on the phenotypes of the transgenic plants. (b) You would first want to check the Salk Institute's Genome Analysis Laboratory web site to see if a T-DNA of transposon insertion has already been identified in this gene (see Question 17.28). If so, you can simply order seeds of the transgenic line from the Arabidopsis Biological Resource Center at Ohio State University. If no insertion is available in the gene, you can determine where it maps in the genome and use transposons that preferentially jump to nearby sites to identify a new insertional mutation (see http.//www.arabidopsis.org/abrc/-ima/jsp). (c) You can construct a gene that has sense and antisense sequences transcribed to a single mRNA molecule (see Figure 17.22*b*). The transcript will form a partially base-paired hairpin that will enter the RISC silencing pathway and block the expression of the gene.

17.28 A search of the Salk Institute's Genome Analysis Laboratory (SIGnAL) web site reveals that there are numerous T-DNA insertions (Salk T-DNA insertions and several others) located in the promoter region of this gene. Moreover, there are two insertion lines (FLAG_177C02 and FLAG_216D01) in the collection at the Institute of Agronomic Research in Versailles, France with inserts in the coding region of this gene. All of these insertion lines in these collections are available to researchers on request. Thus, the function of this gene can be studied by using these insertion lines.

CHAPTER 18

18.1 The pair in (d) are inverted repeats and could therefore qualify.

18.2 The pair in (a) are direct repeats and could therefore qualify.

18.3 Resistance for the second antibiotic was acquired by conjugative gene transfer between the two types of cells.

18.4 Tn3 elements carry a gene that is not essential for transposition.

18.5   In the first strain, the F factor integrated into the chromosome by recombination with the IS element between genes *C* and *D*. In the second strain, it integrated by recombination with the IS element between genes *D* and *E*. The two strains transfer their genes in different orders because the two chromosomal IS elements are in opposite orientation.

18.6   No. IS*1* and IS*2* are mobilized by different transposases.

18.7   Both IS*50* elements should be able to excise from the transposon and insert elsewhere in the chromosome.

18.8   IS*50L* inserted on each side of the cluster of antibiotic resistance genes.

18.9   The *tnpA* mutation: no; the *tnpR* mutation: yes.

18.10   Two enzymes, transposase and resolvase, are needed for replicative transposition (see Figure 18.6). These enzymes are encoded by genes of Tn3. Transposase catalyzes formation of a cointegrate between donor and recipient plasmids. During this process, Tn3 is replicated so that there is a copy of it at each junction in the cointegrate. Resolvase catalyzes the site-specific recombination between the two Tn3 elements and thereby resolves the cointegrate, generating two molecules each with a copy of the transposon. Resolvase also represses the synthesis of both the transposase and resolvase enzymes.

18.11   Many bacterial transposons carry genes for antibiotic resistance and it is relatively simple for these genes to move from one DNA molecule to another. DNA molecules that acquire resistance genes can be passed to other cells in a bacterial population, both vertically (by descent) and horizontally (by conjugative transfer). Over time, continued exposure to an antibiotic will select for cells that have acquired a gene for resistance to that antibiotic. The antibiotic will therefore no longer be useful in combating these bacteria.

18.12   The *Ac* element consists of 4563 nucleotide pairs bounded by inverted repeats that are 11 nucleotide pairs long (see Figure 18.10a). The *Ac* element is flanked by direct repeats 8 nucleotide pairs long; however, these repeats are created at the time the element is inserted into a chromosome (target site duplications), and are therefore not considered to be integral parts of the element itself. *Ds* elements possess the same inverted terminal repeats as *Ac*, but their internal sequences vary. Some residue of the *Ac* sequence may be present, or non-*Ac* sequences may be present; sometimes one *Ds* element is contained within another *Ds* element.

18.13   The *c^m* mutation is due to a *Ds* or an *Ac* insertion.

18.14   The *Ac* element must be tightly linked to the *O2* allele.

18.15   The paternally inherited *Bz* allele was inactivated by a transposable element insertion.

18.16   M cytotype and P transposase.

18.17   Cross dysgenic (highly mutable) males carrying a wild type X chromosome to females homozygous for a balancer X chromosome; then cross the heterozygous F₁ daughters individually to their brothers and screen the F₂ males that lack the balancer chromosome for mutant phenotypes, including failure to survive (lethality). Mutations identified in this screen are probably due to *P* element insertions in X-linked genes.

18.18

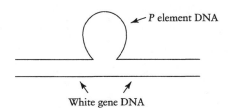

White gene DNA

18.19   Factors made by the host's genome are required.

18.20   (a) Retroviruslike elements resemble integrated retroviruses in overall structure and behavior. (b) Ty*1* element in yeast, and the copia element in *Drosophila*. (c) Retroviruslike elements transpose using an RNA intermediate. The element DNA is transcribed into single-stranded RNA, which is reverse-transcribed into double-stranded DNA (cDNA). The double-stranded cDNA is then inserted into a site in the genome. (d) No. However, the LTRs could pair and recombine to excise all but one LTR.

18.21 Through crossing over between the LTRs of a *gypsy* element.

18.22 No. The intron sequences would be removed by RNA processing prior to reverse transcription into DNA.

18.23 *In situ* hybridization to polytene chromosomes using a *TART* probe (see Chapter 10).

18.24 Same orientation: a deletion; opposite orientation: an inversion.

18.25 *TART* and *HeT-A* replenish the ends of *Drosophila* chromosomes.

18.26 The transposition rate in humans may be very much less than it is in *Drosophila*.

18.27 The *Sleeping Beauty* element could be used as a transformation vector in vertebrates much like the *P* element has been used in *Drosophila*. The *gfp* gene could be inserted between the ends of the *Sleeping Beauty* element and injected into eggs or embryos along with an intact *Sleeping Beauty* element capable of encoding the element's transposase. If the transposase that is produced in the injected egg or embryo acts on the element that contains the *gfp* gene, it might cause the latter to be inserted into genomic DNA. Then, if the egg or embryo develops into an adult, that adult can be bred to determine if a *Sleeping Beauty/gfp* transgene is transmitted to the next generation. In this way, it would be possible to obtain strains of mice or zebrafish that express the *gfp* gene.

18.28 The processed pseudogenes will, in the best of cases; contain sequences from the transcription start site to the poly-A tail of the transcript (if there is one); however, they will not contain the gene's promoter or any of its introns. Because these pseudogenes will not have a promoter, they are not likely to found new retrotransposon families. Without a promoter, they will not be transcribed; hence, they will not produce RNA to be reverse transcribed into DNA for insertion into other sites in the genome. Most likely, the copy number of each of these processed pseudogenes will not increase as the copy number of the *Alu* element has. The *Alu* element contains an "internal" promoter recognized by RNA polymerase III. Each insertion of the *Alu* element contains this promoter and can therefore be transcribed into RNA, which can subsequently be reverse transcribed into DNA. The *L1* element also contains an "internal" promoter, but this promoter is recognized by RNA polymerase II. Most protein-coding genes contain an "external" promoter—that is, one that is not transcribed—and this promoter is recognized by RNA polymerase II.

CHAPTER 19

19.1 The white phenotype is lethal in whole plants.

19.2 In *Mirabilis*, the chloroplasts are not inherited through the pollen, whereas in *Pelargonium*, they are.

19.3 Color depends on the proportion of wild-type and mutant chloroplasts in the tissue. Female gametes from green sectors should transmit the green color, female gametes from white sectors should transmit the white color, and female gametes from pale green sectors should transmit all three colors. Male gametes should have no effect on the trait.

19.4 Organelle heredity should affect both sexes identically, whereas sex chromosome heredity should affect them differently.

19.5 If *O. hookeri* had yellow plastids and *O. muricata* had green plastids.

19.6 Spectinomycin resistance is encoded by a chloroplast gene.

19.7 The results show biparental inheritance of chloroplast genes—a phenomenon made possible by irradiating the mt^+ parent prior to mating. Without irradiation, all the progeny would be resistant to spectinomycin and sensitive to streptomycin.

19.8 Very rarely. Mutations would have to occur in two different cpDNA molecules within the same cell, and then these would have to recombine to produce a molecule that carried both mutations—an extremely unlikely series of events.

19.9 Neutral petite.

19.10 Suppressive petite.

19.11 Plant mtDNA molecules are much larger than animal mtDNA molecules and sometimes, two or more different mtDNA molecules are present within a single plant cell.

19.12 No. The genetic code for mitochondrial genes is not the same as that for *E. coli* genes.

19.13 Probably not. The polypeptide could be shorter or longer because of differences between the mitochondrial and bacterial genetic codes.

19.14 More than one gene encodes a polypeptide.

19.15 Yes. Although the disease is inherited maternally, it can affect either sex.

19.16 The results indicate that: (1) the enzyme contains a polypeptide encoded by a nuclear gene; (2) this polypeptide either is modified by a cytoplasmically inherited factor—probably mitochondrial—or forms a dimer with a mitochondrial-encoded polypeptide; (3) the polypeptide encoded by the nuclear gene exists in two forms, each specified by a different allele; and (4) differences in the mitochondrial genotype create two different cytoplasms. Nuclear allele 1 in cytoplasm I produces enzyme form A; nuclear allele 2 in cytoplasm I produces enzyme form B; nuclear allele 1 in cytoplasm II produces enzyme form C, and nuclear allele 2 in cytoplasm II produces enzyme form D.

19.17 The expression of genes in two different genetic systems must be coordinated.

19.18 Bacterial genomes do not have introns. Thus, the presence of introns in some mitochondrial and chloroplast DNAs would seem to argue that these DNAs were not derived from bacterial genomes. However, it is possible that the introns were acquired by the mitochondrial and chloroplast DNAs secondarily, that is, after they became endosymbiotic genomes within eukaryotic cells. Some of these introns might have been acquired by the transfer of nuclear genes to the organellar genomes.

19.19 (a) Cardiomyopathy is transmitted from mother to both male and female offspring, but not from father to offspring. This pattern of maternal inheritance suggests that the trait is due to a mutation in the mitochondrial genome. (b) The varying degrees of severity of cardiomyopathy may be due to varying numbers of mutant mitochondria that result from cytoplasmic segregation during cell division. Other factors that could influence the severity of the condition are (1) interactions with nuclear genes, and (2) environmental variables.

CHAPTER 20

20.1 By studying the synthesis or lack of synthesis of the enzyme in cells grown on chemically defined media. If the enzyme is synthesized only in the presence of a certain metabolite or a particular set of metabolites, it is probably inducible. If it is synthesized in the absence but not in the presence of a particular metabolite or group of metabolites, it is probably repressible.

20.2 Repression occurs at the level of transcription during enzyme synthesis. The end-product, or a derivative of the end-product, of a repressible system acts as an effector molecule that usually, if not always, combines with the product of one or more regulator genes to turn off the *synthesis* of the enzymes in the biosynthetic pathway. Feedback inhibition occurs at the level of enzyme activity; it usually involves the first enzyme of the biosynthetic pathway. Feedback inhibition thus brings about an immediate arrest of the biosynthesis of the end-product. Together, feedback inhibition and repression rapidly and efficiently turn off the synthesis of both the enzymes and the end-products that no longer need to be synthesized by the cell.

20.3

	Gene or Regulatory Element	Function
(a)	Regulator gene	Codes for repressor
(b)	Operator	Binding site of repressor
(c)	Promoter	Binding site of RNA polymerase and CAP-cAMP complex
(d)	Structural gene z	Encodes β-galactosidase
(e)	Structural gene y	Encodes β-galactoside permease

20.4   (1) Constitutive synthesis of the *lac* enzymes.

(2) Constitutive synthesis of the *lac* enzymes.

(3) Uninducibility of the *lac* enzymes.

(4) No β-galactosidase activity.

(5) No β-galactoside permease activity.

20.5   (a) 1, 2, 3, and 5; (b) 2, 3, and 5.

20.6   The O^c mutant prevents the repressor from binding to the operator. The I^s mutant repressor cannot bind to O^c. The I^s mutant protein has a defect in the allosteric site that binds allolactose, but has a normal operator binding site.

20.7   (a) β-Galactosidase will be produced only when lactose is present. Permease will not be produced at all. (b) *Cis* dominance.

20.8   (a) Inducible, this is the wild-type genotype and phenotype.

(b) constitutive, the O^c mutation produces an operator that is not recognized by the lac repressor.

(c) constitutive, same as for **(b)**.

(d) inducible, I^+ is dominant to I^-.

(e) constitutive, no active repressor is synthesized in this bacterium.

20.9   (a) $\dfrac{I^+O^cZ^+Y^-}{I^+O^+Z^-Y^+}$;

(b) $\dfrac{I^sO^cZ^+Y^-}{I^sO^+Z^-Y^+}$

20.10   (a) The O^c mutations map very close to the Z structural gene; I^- mutations map slightly farther from the structural gene (but still very close by; see Figure 20.5).

(b) An $I^+O^cZ^+Y/^+I^+O^+Z^+Y^+$ partial diploid would exhibit constitutive synthesis of β-galactosidase and β-galactoside permease an $I^+O^+Z^+Y^+/I^+O^+Z^+Y^+$ partial diploid would be inducible for the synthesis of these enzymes.

(c) The O^c mutation is cis-dominant; I^- is *trans*-recessive.

20.11   (a) 1, 5.

(b) 2, 5.

(c) 5.

(d) 1, 2, 5.

(e) 3, 4.

20.12   The system could have developed from a series of tandem duplications of a single ancestral gene. Mutational changes that make the system more efficient and, therefore, favored in selection could have brought the system to its present level of efficiency.

20.13 Catabolite repression has apparently evolved to assure the use of glucose as a carbon source when this carbohydrate is available, rather than less efficient energy sources.

20.14 Possibly by directly or indirectly inhibiting the enzyme adenylcyclase, which catalyzes the synthesis of cyclic AMP from ATP.

20.15 Positive regulation; the CAP-cAMP complex has a positive effect on the expression of the *lac* operon. It functions in turning on the transcription of the structural genes in the operon.

20.16 Yes; in the gene coding for CAP. Some mutations in this gene might result in a CAP that binds to the promoter in the absence of cAMP. Also, mutations in the gene (or genes) coding for the protein (or proteins) that regulate the cAMP level as a function of glucose concentration.

20.17 Negative regulatory mechanisms such as that involving the repressor in the lactose operon block the transcription of the structural genes of the operon, whereas positive mechanisms such as the CAP-cAMP complex in the *lac* operon promote the transcription of the structural genes of the operon.

20.18 0.1; 100; 25.1; 200.

20.19 Repression/derepression of the *trp* operon occurs at the level of transcription initiation, modulating the frequency at which RNA polymerase initiates transcription from the *trp* operon promoters. Attenuation modulates *trp* transcript levels by altering the frequency of termination of transcription within the *trp* operon leader region (*trpL*).

20.20 Deletion of the *trpL* region would result in the levels of the tryptophan biosynthetic enzymes in cells growing in the presence of tryptophan being increased about tenfold because attenuation would no longer occur if this region were absent.

20.21 First, remember that transcription and translation are coupled in prokaryotes. When tryptophan is present in cells, tryptophan-charged tRNA^{Trp} is produced. This allows translation of the *trp* leader sequence through the two UGG Trp codons to the *trp* leader sequence UGA termination codon. This translation of the *trp* leader region prevents base-pairing between the partially complementary mRNA leader sequences 75–83 and 110–121 (see Figure 20.13*b*), which in turn permits formation of the transcription-termination "hairpin" involving leader sequences 110–121 and 126–134 (see Figure 20.13*c*).

20.22 No. Attenuation of the *trp* operon would now be controlled by the presence or absence of Gly-tRNA^{Gly}.

20.23 A prophage is a chromosome of a bacteriophage after it has become integrated into the chromosome of the host bacterium. The lambda prophage is present as a transcriptionally inactive linear structure in the *E. coli* chromosome. During lytic infections, the lambda chromosome begins its reproductive cycle by replicating as a transcriptionally active circular DNA molecule.

20.24 The autogenously maintained repression of λ lytic genes during lysogeny occurs at the level of transcriptional initiation, whereas the λ lytic cascade is regulated at the level of transcriptional termination. In the first case, a repressor protein binds to the promoter and blocks the transcription of the λ lytic genes. In the second case, two key antiterminator proteins act to prevent the termination of transcription at sites distal to the immediate early and delayed early coding sequences. These antitermination events result in the transcription of delayed early and late λ genes.

20.25 The operator region of the *lac* operon in *E. coli* contains a single repressor binding site, whereas the lambda operators O_L and O_R each contain three repressor binding sites that are differentially occupied under different conditions. The differential affinity of lambda repressor for the three binding sites in these operators plays an important role in the maintenance of the lysogenic state of a lambda prophage in *E. coli*.

20.26 Autogenous means self-generated. The λ repressor regulates its own synthesis; the presence of λ repressor stimulates the synthesis of more repressor. This autoregulatory mechanism enhances the stability of the lysogenic state by virtually assuring that once repressor has been synthesized in an infected bacterium, it will continue to be synthesized in amounts that are sufficient to keep the λ lytic genes repressed.

20.27 Lysogeny will continue, because $C_1{}^+$ acts in *trans* and is dominant.

20.28 The precise temporal pattern of gene expression—immediate early genes followed by delayed early genes and then late genes—during the lytic cycle of phage lambda is controlled at the level of transcriptional termination. The two immediate early genes, *cro* and *N*, are adjacent to the origins of transcription at P_L and P_R (see Figure 20.17). The product of gene *N* is a transcriptional antiterminator that causes transcription to continue past the termination signals t_{L1} and t_{R1} and on through the delayed early genes (see Figure 20.19). Similarly, the product of one of the delayed early genes, *Q*, functions as a transcriptional antiterminator at site t_{R3}, resulting in the continuation of transcription through the late genes.

20.29 The N protein, the product of the first gene transcribed from promoter P_L, prevents the termination of transcription at two sites (t_{R1} and t_{L1}), which results in the transcription of the delayed early genes located downstream from these sites. N protein binds to *nut* sites just upstream from the termination sites, and, together with the Nus protein and ribosomal protein S10, modifies the specificity of RNA polymerase so that termination does not occur at sites t_{R1} and t_{L1} (see Figure 20.19).

20.30 This mutation will produce the so-called clear-plaque phenotype. In the absence of functional lambda repressor, lambda phage can only reproduce lytically. Lysogeny requires the lambda repressor to be present to keep the lytic genes of the prophage repressed. Thus all infected cells will be killed and lysed during the lambda lytic cycle, resulting in clear plaques rather than the turbid plaques formed by wild-type lambda.

20.31 No. Since transcription (nucleus) and translation (cytoplasm) are not coupled in eurkaryotes, attenuation of the type occurring in prokaryotes would not be possible.

CHAPTER 21

21.1 In multicellular eukaryotes, the environment of an individual cell is relatively stable. There is no need to respond quickly to changes in the external environment. In addition, the development of a multicellular organism involves complex regulatory hierarchies composed of hundreds of different genes. The expression of these genes is regulated spatially and temporally, often through intricate intercellular signaling processes.

21.2 The spatial dimension of gene expression is that not all genes are expressed in all tissues. An example is the expression of the β tubulin genes of *Arabidopsis*. The different β tubulin genes show tissue-specific expression. The temporal dimension of gene expression is that not all genes are expressed at all times. An example is the expression of different globin genes during human development. Some of these genes are expressed in the early embryo, some during fetal development, and some exclusively after birth.

21.3 The concentration of mRNA is determined by the rate of transcription relative to the rate of degradation. The longevity of an mRNA is influenced by the length of its poly(A) tail, the structure of its 3' untranslated region, and the metabolic state of the cell.

21.4 Steroid hormones are small, lipid-soluble molecules that have little difficulty passing through the cell membrane. Peptide hormones are typically too large to pass through the cell membrane freely; rather, their influence must be mediated by a signaling system that begins with a membrane-bound receptor protein that binds the hormone.

21.5 By monitoring puffs in response to environmental signals, such as heat shock, or to hormonal signals.

21.6 An enhancer can be located upstream, downstream, or within a gene and it functions independently of its orientation. A promoter is almost always immediately upstream of a gene and it functions only in one direction with respect to the gene.

21.7 By alternate splicing of the transcript.

21.8 The two genes are controlled by different enhancers, one active in roots and the other in vascular tissues.

21.9 Northern blotting of RNA extracted from plants grown with and without light, or PCR amplification of cDNA made by reverse transcribing these same RNA extracts.

21.10 It may make an RNA but not a functional polypeptide.

21.11 The green fluorescent protein will be made after the flies are heat shocked.

21.12 Probably not unless the promoter of the *gfp* gene is recognized and transcribed by the *Drosophila* RNA polymerase independently of the heat shock response elements.

21.13 This enzyme plays an important role in photosynthesis, a light-dependent process. Thus, it makes sense that its production should be triggered by exposure to light.

21.14 A zinc finger is a short peptide loop that forms when two cysteines in one part of a polypeptide and two histidines in another part nearby jointly bind a zinc ion. A leucine zipper is a peptide sequence with a leucine at every seventh position.

21.15 The flies would be phenotypically normal males.

21.16 Both XX and XY animals would develop as intersexes.

21.17 Yes. Enhancers are able to function in different positions in and around a gene.

21.18 Polytene chromosome puffs.

21.19 (a) HATs, histone acetyl transferases, transfer acetyl groups to the amino acid lysine in histones; (b) HDACs, histone deacetylases, remove acetyl groups from these lysines; (c) HMTs, histone methyl transferases, transfer methyl groups to lysine, arginine, and histidine in histones.

21.20 Yes.

21.21 H4 is acetylated.

21.22 Short interfering RNAs target messenger RNA molecules, which are devoid of introns. Thus, if siRNA were made from double-stranded RNA derived from an intron, it would be ineffective against an mRNA target.

21.23 The Dicer enzymes cleave double-stranded RNA to produce siRNA and miRNA molecules.

21.24 Some cytosines in CG dinucleotides are methylated in mammalian DNA.

21.25 Imprinting is a condition in which the expression of a gene depends on its parental origin.

21.26 The paternally contributed allele (*b*) will be expressed in the F_1 progeny.

21.27 The ribosomal RNA genes are extrachromosomally amplified in oocytes from the toad *Xenopus*.

21.28 A silent *vsg* gene located at a nontelomeric location can be copied into the expression site near the telomere of a chromosome and then become the expressed *vsg* gene, or a silent telomeric *vsg* gene can be recombined into an active expression site near the telomere of another chromosome and then become the expressed *vsg* gene.

21.29 RNA could be isolated from liver and brain tissue. Northern blotting or RT-PCR with this RNA could then establish which of the genes (A or B) is transcribed in which tissue. For northern blotting, the RNA samples would be fractionated in a denaturing gel and blotted to a membrane, and then the RNA on the membrane would be hybridized with gene-specific probes, first for one gene, then for the other (or the researcher could prepare two separate blots and hybridize each one with a different probe). For RT-PCR, the RNA samples would be reverse transcribed into cDNA using primers specific for each gene; then the cDNA molecules would be amplified by standard PCR and the products of the amplifications would be fractionated by gel electrophoresis to determine which gene's RNA was present in the original samples.

21.30 The sample of chromatin from brain tissue would be expected to show the greater signal on a Southern blot hybridized with a probe specific for gene A. The reason is that this gene is not so well transcribed in brain cells; consequently, it will be more resistant to digestion with DNase I in chromatin derived from brain cells than in chromatin derived from liver cells in which it is actively transcribed (and therefore more open to digestion with DNase I).

21.31 The *mle* gene is not functional in females.

21.32 X chromosome inactivation begins at the X inactivation center (XIC) which contains the *XIST* gene. During early embryogenesis in XX mammals, the *XIST* genes produce unstable 17-kb-long RNAs that remain closely associated with the XIC region. Transcripts from one of the two *XIST* genes stabilize and eventually coat the entire X chromosome containing that gene; transcripts from the other *XIST* gene disintegrate and further transcription from that gene is repressed by methylation of nucleotides in the gene's promoter. The chromosome that becomes coated with *XIST* RNA becomes the inactive X chromosome; it condenses and forms a Barr body.

CHAPTER 22

22.1 Cell division forms the blastula; cell movement forms the gastrula.

22.2 Unequal division of the cytoplasm during the meiotic divisions; transport of substances into the oocyte from surrounding cells such as the nurse cells in *Drosophila*.

22.3 Determination establishes cell fates; differentiation realizes those fates.

22.4 No. The primary germ layers are ectoderm, endoderm and mesoderm.

22.5 Collect mutations with diagnostic phenotypes; map the mutations and test them for allelism with one another; perform epistasis tests with mutations in different genes; clone individual genes and analyze their function at the molecular level.

22.6 Mitotic division is so rapid that there is not enough time for membranes to form between cells.

22.7 Imaginal discs.

22.8 1/4.

22.9 Mate homozygous *unc* hermaphrodites that carry one of the *dumpy* mutations with males that carry the other *dumpy* mutation and score the non-uncoordinated (*unc/+*) F_1 hermaphrodites for the dumpy phenotype. If these worms are dumpy, the two mutations are allelic; if they are wild-type, the two mutations are not allelic.

22.10 The mutation would kill males but not females.

22.11 Cross heterozygous carriers separately with *tra/+* and *tra2/+* flies. If any of the XX progeny of these matings are transformed into sterile males, the new mutation is an allele of the *tra* or *tra2* tester.

22.12 X:A ratio --> *Sxl* --> *tra* --> *dsx* --> differentiation as male or female.
  tra2

22.13 Intersex. The *dsx* mutation is epistatic to the *tra* mutation.

22.14 (a) Male with abnormal Muscle of Lawrence (MOL) anatomy and abnormal courting behavior; (b) male with abnormal MOL anatomy and abnormal courting behavior; (c) intersex, probably with abnormal MOL anatomy and abnormal courting behavior; (d) intersex, probably with abnormal MOL anatomy and abnormal courting behavior.

22.15 Examine *fru* mRNAs synthesized in wild-type flies and in homozygous *tra*, *tra2*, and *dsx* mutant flies. If *fru* expression is controlled by the *tra* and *tra2* gene products (both splicing factors), then XX flies homozygous for mutations in either of these genes should produce the male-specific form of *fru* mRNA; however, these mutations should have no effect on *fru* expression in XY animals because the *tra* and *tra2* gene products are nonfunctional in XY animals. If *fru* expression is not controlled by the *dsx* gene products, then XY *dsx* mutants should produce the male-specific *fru* mRNA and XX *dsx* mutants should produce the female-specific *fru* mRNA.

22.16 The X:A ratio is 2:3--between the 1:2 ratio needed for normal male development and the 1:1 ratio needed for normal female development. By chance, the autosomal denominator proteins will be in excess in some cells of the triploid, causing the *Sxl* gene to be inactivated; these cells will develop male characteristics. In other cells, the

X-linked numerator proteins will be in excess, causing the *Sxl* gene will be activated; these cells will develop female characteristics. Thus, the triploid fly will be a mosaic of male and female cells--that is, an intersex.

22.17 XX, male phenotype; XO, female phenotype.

22.18 The *xol-1* gene is needed in *Caenorhabditis* males but not in hermaphrodites; the *Sxl* gene is needed in *Drosophila* females but not in males.

22.19 Between positions 5 and 8.

22.20 (a) Wild-type; (b) embryonic lethal; (c) embryonic lethal; (d) wild-type; (e) wild-type.

22.21 Some structures fail to develop in the posterior portion of the embryo.

22.22 Female sterility.

22.23 Screen for lethal mutations that prevent regions of the embryo from developing normally.

22.24 In homozygous condition, the mutation that causes phenylketonuria has a maternal effect. Women homozygous for this mutation influence the development of their children *in utero*.

22.25 *ey --> boss --> sev -->* R7 differentiation.

22.26 Extra mouse eyes or eye primorida.

22.27 Northern blotting of RNA extracted from the tissues at different times during development. Hybridize the blot with gene-specific probes.

22.28 Methylation of DNA, acetylation of histones, and packaging of DNA into chromatin by certain kinds of proteins all portend difficulties for therapeutic cloning as well as for reproductive cloning. These epigenetic modifications of somatic cell DNA would have to be "reprogrammed" in the oocyte or they could affect how the stem cells derived from that oocyte would develop.

CHAPTER 23

23.1 The genetic information specifying antibody chains is stored in sequences of nucleotide pairs encoding sequences of amino acids, just like the genetic information specifying other polypeptides. However, the information specifying an antibody chain is stored in bits and pieces that are assembled into functional genes encoding antibody chains by genome rearrangements (somatic-cell recombination events) occurring during the development of the B lymphocytes (the antibody-producing cells). See Figures 23.10 and 23.11.

23.2 4; 2; 1.

23.3 (1) The joining of different *V*, *D*, and *J* gene segments by somatic recombination during B lymphocyte development (see Figures 23.10 and 23.11); (2) variability in the exact location of the joining reaction during *V-D-J* joining events (see Figure 23.13); (3) somatic mutation.

23.4 Antibody class is determined by the the type of heavy chain that is present. Antibodies of classes IgA, IgD, IgE, IgG, and IgM have heavy chains of type α, δ, ε, γ, and μ, respectively. The different types of heavy chains are specified by different gene segments encoding the heavy-chain constant regions.

23.5 IgM is the initial antibody produced by a B cell. Later, B cells may switch to the production of other classes of antibodies.

23.6 The predominant class of immunoglobin molecules in human blood is IgG.

23.7 IgE.

23.8 The mammalian genome consist of about 3.0×10^9 nucleotide pairs. If all of this DNA were in the form of uninterrupted coding sequences of genes, each about 1000 nucleotide pairs, the genome would contain about 3 million genes. However, the mammalian immune system consists of about a trillion white blood cells, individually carrying one of billions of different antibody molecules.

23.9 At the DNA level; class switching occurs by somatic recombination during B lymphocyte differentiation (see Figure 23.11, steps 3a and 3b).

23.10 (a) B lymphocytes, T lymphocytes, and macrophages.

(b) B lymphocytes differentiate into the plasma cells that synthesize the antibodies responsible for the humoral immune response; T lymphocytes synthesize the T-cell antigen receptors responsible for the cellular immune response; and macrophages ingest and degrade antigen-antibody complexes.

23.11 The *MHC* locus encodes the transplantation antigens that play a major role in the rejection of foreign tissues after organ and tissue transplant operations.

23.12 (a) The *MHC* genes are said to be highly polymorphic because each is represented by a large number of alleles segregating in human populations.

(b) The polymorphic nature of the *MHC* genes makes it difficult to match the tissue types of potential organ donors and organ recipients so as to avoid rejection of the foreign organ following transplant surgery.

23.13 Antibody-antigen complexes are eliminated by (1) ingestion and degradation by phagocytes or (2) the proteolytic activity of complement proteins, which degrades the protein coats of viruses and lyses cellular pathogens.

23.14 The epitope is a short region of the antigen, sometimes containing as few as six amino acids, that stimulates the production of an antibody. The antibody recognizes and binds to the epitopic region of the antigen.

23.15 Cytokines, lymphokines, and interleukens are secreted by helper T cells. These molecules stimulate the development of immature B and T lymphocytes. The immature T lymphocytes develop into killer T cells and memory T cells.

23.16 This response occurs by a process called clonal selection during which antibodies on the surface of the B lymphocyte producing them bind to the antigen. This in turn stimulates the cell to divide and produce a population (clone) of plasma cells all producing the same antibody.

23.17 Yes, the T lymphocyte antigen receptor proteins.

23.18 The somatic recombination events are apparently controlled by short nucleotide-pair sequences that provide *V-J*, *V-D*, and *D-J* joining signals. All *V* gene segments have the same adjacent signals. All *J* gene segments have their own specific signal sequences located adjacent to the coding sequences. *D* and *C* gene segments also have their own signal sequences.

23.19 Both contain polypeptides with variable regions that form the antigen-binding domains and constant regions that facilitate interactions with cell membranes and other components of the immune response.

23.20 After the exposure of virgin B and T cells to an antigen, they develop into activated cells and memory cells. The activated cells differentiate into antibody-producing plasma cells and receptor-producing T cells, respectively, that are responsible for the primary immune response. The long-lived memory cells are also activated and remain in the activated state until a subsequent exposure to the same antigen triggers their rapid differentiation into antibody-producing plasma cells and receptor-producing T cells. Thus a secondary immune response is faster than a primary immune response because the memory B and memory T cells do not have to go through the activation processes required with virgin B cells and T cells.

23.21 Nonspecific immune responses include inflammation of infected tissues resulting in increased blood flow to the tissues and recruitment of phagocytes to ingest and destroy the infecting agent(s). The two major specific immune responses are the production of antigen-specific antibodies and killer T cells.

23.22 Most of the cells of the immune system have short life spans—usually a few days to a week. In contrast, memory cells have long life spans—months to years.

23.23 (a) Monoclonal antibodies are produced by fusing B lymphocytes from mammals that have been exposed to a specific antigen with cancer cells called myelomas to produce immortal hybrid cells called hybridomas. Each hybridoma clone produces a single antibody. The most difficult step in monoclonal antibody production is screening the resulting population of hybridoma cells for the clone producing the desired antibody. Once identified, the antibody-producing cells can be cultured indefinitely as a result of their cancerous heritage. (b) Monoclonal antibodies are invaluable for use in identifying individual polypeptides separated by gel electrophoresis (see Figure 15.26) and in determining the location of specific macromolecules in tissues and cells by immunolocalization techniques. They are also used to immunoprecipitate proteins during purification processes and to diagnose human diseases.

23.24 One. Mammals are diploids. Thus, in the absence of allelic exclusion, each plasma cell could produce two different heavy chains because heavy-chain antibody genes could be assembled from the gene segments on both homologous chromosomes. Allelic exclusion assures that only heavy-chain gene segments on one chromosome — the maternal or the paternal chromosome — are assembled into a functional antibody heavy-chain gene.

23.25 When helper T cells encounter antigens on the surfaces of macrophages, they secrete cytokines, lymphokines, and interleukins that stimulate B and T lymphocytes to differentiate into antibody-producing plasma cells and killer T cells, respectively. Thus, helper T cells are involved in both the antibody-mediated and T cell-mediated immune responses.

23.26 One type of phagocyte, the macrophage, plays a key role in initiating the immune response in mammals. Macrophages ingest bacteria, viruses, and other foreign substances and partially degrade them. Antigens produced by the degradation process become anchored on the surfaces of the macrophages, where they are recognized by the helper T cells that mediate both the humoral and cellular immune responses. Other phagocytes ingest and destroy foreign cells, viruses, and antibody-antigen complexes. The complement proteins also lyse foreign cells and degrade antigens that have IgG or IgM antibodies bound to them. Although phagocytes do ingest and destroy viruses, bacteria, and other pathogenic cells in a nonspecific manner, they are most effective in destroying foreign substances that have been complexed with antibodies.

23.27 Autoimmune diseases are disorders in which the immune system of an individual attacks specific tissues or cells of her or his own body ("self" cells). The combining term *auto* means self; they are called autoimmune diseases because of the immune system's attack on the patient's own cells.

23.28 1,088,000,000 different antibodies.

23.29 One hypothesis regarding the cause of autoimmune diseases is that an immune response is induced against a foreign antigen—for example, one found on an invading pathogen—that is similar to an epitope of a human protein. If the immune response is not suppressed after the antigen has been eliminated, then the antibodies may attack the epitope-containing protein of the individual producing them and cause an autoimmune disease.

CHAPTER 24

24.1 Cancer has been called a genetic disease because it results from mutations of genes that regulate cell growth and division. Nonhereditary forms of cancer result from mutations in somatic cells. These mutations, however, can be induced by environmental factors including tobacco smoke, chemical pollutants, ionizing radiation, and UV light. Hereditary forms of cancer also frequently involve the occurrence of environmentally-induced somatic mutations.

24.2 Cancer cells do not display contact inhibition—they pile up on top of each other—whereas embryonic cells spread out in flat sheets. Cancer cells are frequently aneuploid; embryonic cells are euploid.

24.3 Aneuploidy might involve the loss of functional copies of tumor suppressor genes, or it might involve the inappropriate duplication of proto-oncogenes. Loss of tumor suppressor genes would remove natural brakes on cell division, and duplication of proto-oncogenes would increase the abundance of factors that promote cell division.

24.4 No. A virus that carried a processed copy of a tumor suppressor gene would not be expected to induce tumor formation because the product of the tumor suppressor gene would help to restrain cell growth and division.

24.5 They possess introns.

24.6 The absence of introns might speed up the expression of the gene's protein product because there would be no need for splicing. In addition, some introns contain sequences called silencers that negatively regulate transcription. Removal of these sequences might cause transcription to occur when it otherwise would not.

24.7 The products of these genes play important roles in cell activities.

24.8 Mutations in these codons cause amino acid changes that activate the Ras protein.

24.9 The cultured NIH 3T3 cells probably carry other mutations that predispose them to become cancerous; transfection of such cells with a mutant *c-H-ras* oncogene may be the last step in the process of transforming the cells into cancer cells. Cultured embryonic cells probably do not carry the predisposing mutations needed for them to become cancerous; thus, when they are transfected with the mutant *c-H-ras* oncogene, they continue to divide normally.

24.10 Ras protein is an activator of cell metabolism; RB protein is a suppressor of cell metabolism.

24.11 Retinoblastoma results from homozygosity for a loss-of-function (recessive) allele. The sporadic occurrence of retinoblastoma requires two mutations of this gene in the same cell or cell lineage. Therefore retinoblastoma is rare among individuals who, at conception, are homozygous for the wild-type allele of the *RB* gene. For such individuals, we would expect the frequency of tumors in both eyes to be the square of the frequency of tumors in one eye. Individuals who are heterozygous for a mutant *RB* allele require only one somatic mutation to occur for them to develop retinoblastoma. Because there are millions of cells in each retina, there is a high probability that this somatic mutation will occur in at least one cell in each eye, causing both eyes to develop tumors.

24.12 At the cellular level, loss-of-function mutations in the *RB* gene are recessive; a cell that is heterozygous for such a mutation divides normally. However, when a second mutation occurs, that cell becomes cancerous. If the first *RB* mutation was inherited, there is a high probability that the individual carrying this mutation will develop retinoblastoma because a second mutation can occur any time during the formation of the retinas in either eye. Thus, the individual is predisposed to develop retinoblastoma, and it is this predispostion that shows a dominant pattern of inheritance in pedigrees.

24.13 The probability that a carrier does not develop retinoblastoma is 0.05, which is equal to the probability that the wild-type *RB* allele is not mutationally inactivated during the cell divisions that form the retinas of the eyes. If the rate of mutational inactivation is $u = 10^{-6}$ per cell division, and n is the number of cell divisions, then $0.05 = (1 - u)^n$. If we take logarithms of both sides, $\log (0.05) = n \log (1 - u)$, which to a good approximation, is equal to $n \log (u)$. After substituting 10^{-6} for u and solving, we find that $n = 1.3 \times 10^6$.

24.14 II-1 should be tested for the *BRCA1* mutation that apparently was involved in the ovarian cancer that developed in her mother and sister. If she is found to carry this mutation, a prophylactic oophorectomy can be prescribed to reduce the chance that she will develop cancer. If she is found to be free of the mutation carried by her mother and sister, then she is not more likely to develop ovarian cancer than a woman in the general population.

24.15 By binding to E2F transcription factors, pRB prevents those transcription factors from activating their target genes—which encode proteins involved in progression of the cell cycle; pRB is therefore a negative regulator of transcription factors that stimulate cell division.

24.16 At 25°, cell division should be normal—the same as for cells homozygous for a wild-type allele of the *E2F* gene. At 35°, division would be expected to be impaired, either because the mutant E2F protein binds unproductively to the sequence in its target gene or because a mutant E2F polypeptide dimerizes with a wild-type E2F polypeptide and abolishes the activation function of the wild-type polypeptide.

24.17 Cells homozygous for a loss-of-function mutation in the *p16* gene might be expected to divide in an uncontrolled fashion because the p16 protein would not be able to inhibit cyclin-CDK activity during the cell cycle. The *p16* gene would therefore be classified as a tumor suppressor gene.

24.18 Cells heterozygous for a dominant activating mutation in the *BCL-2* gene would be expected to be unable to execute the programmed cell death pathway in response to DNA damage induced by radiation treatment. Such cells would continue to divide and accumulate mutations; ultimately they would have a good chance of becoming cancerous. The *BCL-2* gene would therefore be classified as a proto-oncogene.

24.19 Cells homozygous for a loss-of-function mutation in the *BAX* gene would be unable to prevent repression of the programmed cell death pathway by the BCL-2 gene product. Consequently, these cells would be unable to execute that pathway in repsonse to DNA damage induced by radiation treatment. Such cells would continue to divide and accumulate mutations; ultimately, they would have a good chance of becoming cancerous. The *BAX* gene would therefore be classified as a tumor suppressor gene.

24.20 Loss-of-function mutations in the DNA-binding domain of p53 abolish the ability of that protein to activate transcription of target genes whose products are involved in the restraint of cell division or in the promotion of programmed cell death. Without restraint of cell division or promotion of programmed cell death, cells accumulate damage to their DNA and ultimately become cancerous.

24.21 If a cell were heterozygous for a mutation that caused p53 to bind tightly and constitutively to the DNA of its target genes, its growth and division might be retarded, or it might be induced to undergo apoptosis. Such a cell would be expected to be more sensitive to the effects of ionizing radiation because radiation increases the expression of p53, and in this case, the p53 would be predisposed to activate its target genes, causing the cell to respond vigorously to the radiation treatment.

24.22 Homozygous *TP53* knock-out mice might actually be less sensitive to the killing effects of ionizing radiation because p53 would be unable to mediate the apoptotic response to the radiation treatment.

24.23 They would probably decrease the ability of pAPC to bind β-catenin.

24.24 The increased irritation to the intestinal epithelium caused by a fiber-poor, fat-rich diet would be expected to increase the need for cell division in this tissue (to replace the cells that were lost because of the irritation), with a corresponding increase in the opportunity for the occurrence of cancer-causing mutations.

24.25 All three genes (*RB*, *p130*, and *p107*) are essential for embryonic development, although by themselves, *p130* and *p107* are dispensable, possibly because their products are functionally redundant. (Both gene products must be inactivated before any deleterious effect is seen.) In adults, only pRB appears to play a role in suppressing tumor formation.

24.26 The *KAI1* gene is a prostate tumor suppressor gene. Functional inactivation of this gene allows prostate tumors to develop.

24.27 No. Apparently there is another pathway—one not mediated by p53—that leads to the activation of the *p21* gene.

CHAPTER 25

25.1 The phenotypes are determined by the number of primed alleles in the genotypes. This is a classic dihybrid cross with no dominance; the phenotypic frequencies in the F_2 are: 1/16 white (0 primed alleles), 4/16 light red (1 primed allele), 6/16 intermediate red (2 primed alleles), 4/16 darker red (3 primed alleles), 1/16 darkest red (4 primed alleles).

25.2 Some of the genes implicated in heart disease are listed in Table 25.2. Environmental factors might include diet, amount of exercise, and whether or not the person smokes.

25.3 The concordance for monozygotic twins is almost twice as great as that for dizygotic twins. Monozygotic twins share twice as many genes as dizygotic twins. The data strongly suggest that alcoholism has a genetic basis.

25.4 Because 8/2012 is approximately $1/256 = (1/4)^4$, it appears that four size-determining genes were segregating in the crosses.

25.5 (a) 4; (b) 3 inches; (c) frequency of 1/4.

25.6 (a) The mean is 20.45 inches. (b) The variance is 2.37 inches^2. (c) The standard deviation is 1.54 inches.

25.7 Because $\Sigma(X_i - \text{mean}) = 0$.

25.8 For the F_1, $V_g = 0$ because they are all genetically identical and heterozygous; for the F_2, $V_g > 0$ because genetic differences result from the segregation and independent assortment of genes.

25.9 $3.17/6.08 = 0.52$.

25.10 The broad-sense heritability within a highly inbred strain is expected to be zero because there is no genetic variability.

25.11 V_e is estimated by the average of the variances of the inbreds: 9.4 cm^2. V_g is estimated by the difference between the variances of the randomly pollinated population and the inbreds: $(26.4 - 9.4) = 7.4$ cm^2. The broad-sense heritability is $H^2 = V_g/V_t = 7.4/26.4 = 0.64$.

25.12 Because the F_1 plants have maturation times midway between those of the parental strains, there seems to be little or no dominance for this trait.

25.13 Broad-sense heritability must be greater than narrow-sense heritability because $H^2 = V_g/V_t > V_a/V_t = h^2$.

25.14 $(125 - 100)(0.4) + 100 = 110$ units.

25.15 $(15 - 12)(0.3) + 12 = 12.9$ bristles.

25.16 $(90 - 100)(0.2) + 100 = 98$ days.

25.17 $h^2 = R/S = (12.5 - 10)/(15 - 10) = 0.5$; selection for increased growth rate should be effective.

25.18 $(98 - 100)(0.4) + 100 = 99.2$.

25.19 $(40)(0.3)(10) + 2000 = 2120$ µg.

25.20 $F_I = (1/2)^3 (1 + F_C) = 0.25$; thus, $F_C = 1$.

25.21 The pedigree is shown below:

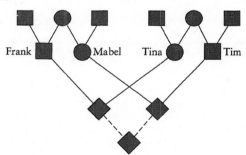

The coefficient of relationship between the offspring of the two couples is obtained by calculating the inbreeding coefficient of the imaginary child from a mating between these offspring and multiplying by 2; $[(1/2)^5 \times 2] \times 2 = 1/8$.

25.22 From the pedigree below, the inbreeding coefficient is $(1/2)^3 (1 + F_C) + (1/2)^4 (1 + F_B) = 3/8$ because $F_B = F_C = 1$.

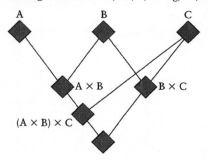

25.23 The mean ear length for randomly mated maize is 24 cm and that for maize from one generation of self-fertilization is 20 cm. The inbreeding coefficient of the offspring of one generation of self-fertilization is 1/2, and the inbreeding coefficient of the offspring of two generations of self-fertilization is $(1/2)(1 + 1/2) = 3/4$. Mean ear length (Y) is expected to decline linearly with inbreeding according to the equation $Y = 24 - b\ F$, where b is the slope of the line. The value of b can be determined from the two values of Y that are given. The difference between these two values (4 cm) corresponds to an increase in F from 0 to 1/2. Thus, $b = 4/(1/2) = 8$ cm, and for $F = 3/4$, the predicted mean ear length is $Y = 24 - 8 \times (3/4) = 18$.

25.24 $F_A = (1/2)^5 = 1/32$; $F_B = 2 \times (1/2)^6 = 1/32$; $F_C = 2 \times (1/2)^7 = 1/64$.

25.25 To estimate the narrow-sense heritability, divide the observed correlation between the relatives by the fraction of genes shared by the relatives; in this case, $h^2 = 0.14/0.25 = 0.56$.

25.26 The correlations for MZT are not much different from those for MZA. Evidently, for these personality traits, the environmentality (C^2 in Table 25.3) is negligible.

CHAPTER 26

26.1 Frequency of L^M in Central American population: $p = (2 \times 53 + 29)/(2 \times 86) = 0.78$; $q = 0.22$. Frequency of L^M in North American population: $p = (2 \times 78 + 61)/(2 \times 278) = 0.39$; $q = 0.61$.

26.2 $2pq = 2(0.2)(0.8) = 0.32$.

26.3 $q^2 = 0.0004$; $q = 0.02$.

26.4

Phenotype	Hardy-Weinberg Frequency
M	$(0.78)^2 = 0.61$
MN	$2(0.78)(0.22) = 0.34$
N	$(0.22)^2 = 0.05$

26.5 Frequency of heterozygotes combined = $2[(0.6)(0.3) + (0.3)(0.1) + ((0.6)(0.1)] = 0.54$.

26.6 Frequency of tasters (genotypes TT and Tt): $(0.4)^2 + 2(0.4)(0.6) = 0.64$. Frequency of TT tasters among all tasters: $(0.4)^2/(0.64) = 0.25$.

26.7 $(0.00025)^2 = 6.25 \times 10^{-8}$.

26.8 In females, the frequency of the dominant phenotype is 0.36. The frequency of the recessive phenotype is $0.64 = q^2$; thus, $q = 0.8$ and $p = 0.2$. The frequency of the dominant phenotype in males is therefore $p = 0.2$.

26.9 (a) Half the males will be ruby-eyed and 5 percent $(0.50 \times 0.10 \times 100$ percent) of the females will be ruby-eyed. (b) Among males, the frequency of the recessive allele will be 0.5, which was its frequency among females in the previous generation; among females, the frequency of the recessive allele will be $(0.5 + 0.1)/2 = 0.3$, which is the average of the frequencies of this allele in males and females in the previous generation.

26.10 The probability that the unaffected child of the couple is a carrier of the mutant allele for cystic fibrosis is 2/3. The probability that the mate of this individual is a carrier can be determined by using the population incidence of the disease. The mutant allele frequency is the square root of the incidence—0.045—and the frequency of heterozygotes under the assumption of random mating is $2 \times 0.045 \times (1 - 0.045) = 0.086$. If both individuals are carriers, the chance that they will have an affected child is 1/4. Putting all this analysis together, the risk for the child to have cystic fibrosis is therefore $2/3 \times 0.086 \times 1/4 = 0.014$, which is seven times the incidence in the population at large.

26.11 Frequency of heterozygotes = $H = 2pq = 2p(1 - p)$. Using calculus, take the derivative of H and set the result to zero to solve for the value of p that maximizes H: $dH/dp = 2 - 4p = 0$ implies that $p = 2/4 = 0.5$.

26.12 In a Hardy-Weinberg population the frequency of $I^A I^A$ homozygotes is $(0.15)^2 = 0.0225$ and the frequency of $I^A i$ heterozygotes is $2 \times 0.15 \times 0.60 = 0.18$. The sum of these frequencies—0.2025—is the frequency of individuals with type A blood. Thus, the frequency of $I^A I^A$ homozygotes among all individuals with type A blood is $0.0225/0.2025 = 0.1111$.

26.13 Under the assumption that the population is in Hardy-Weinberg equilibrium, the frequency of the allele for light coloration is the square root of the frequency of recessive homozygotes. Thus, $q = \sqrt{0.49} = 0.7$, and the frequency of the allele for dark color is $1 - q = p = 0.3$. From $p^2 = 0.09$, we estimate that $0.09 \times 100 = 9$ of the dark moths in the sample are homozygous for the dominant allele.

26.14 (a) $P^1 = 0.6$ and $P^2 = 0.4$. (b) Assuming random mating, the Hardy-Weinberg expectations are 36 ($P^1 P^1$), 48 ($P^1 P^2$), and 16 ($P^2 P^2$). c2 = $(30 - 36)^2/36 + (60 - 48)^2/48 + (10 - 16)^2/16 = 6.25$; with one degree of freedom, this value of χ^2 warrants rejection of the hypothesis that the population is in Hardy-Weinberg genotype proportions. (c) Hardy-Weinberg proportions will be established after one generation of random mating.

26.15 $2pq(1 - F) = 2(0.6)(0.4)(1 - 0.5) = 0.24$.

26.16 Ultimate frequency of GG is 0.2; ultimate frequency of gg is 0.8.

26.17 (a) Frequency of A in merged population is 0.5, and that of a is also 0.5;

(b) 0.25 (AA), 0.50 (Aa), and 0.25 (aa);

(c) Frequencies in (b) will persist.

26.18 Ultimately, all members of the population will either be TT or tt, each 50% of the total.

26.19 The relative fitnesses can be obtained by dividing each of the survival probabilities by the largest probability (0.92). Thus, the relative fitnesses are 1 for GG, $0.98 = 1 - 0.02$ for Gg, and $0.61 = 1 - 0.39$ for gg. The selection coefficients are $s_1 = 0.02$ for Gg and $s_2 = 0.39$ for gg.

26.20 Frequency of a is $q = 0.4$; (a) thus $p = 1 - q = 0.6$; (b) Use the following scheme:

Genotype	AA	Aa	aa
Hardy-Weinberg Frequency	0.36	0.48	0.16
Relative Fitness	1	1	0.95
Relative Contribution to next generation	0.36	0.48	0.152
Proportional Contribution to next generation	0.363	0.484	0.153

26.21 Use the following scheme:

Genotype	CC	Cc	cc
Hardy-Weinberg frequency	$(0.98)^2 = 0.9604$	$2(0.98)(0.02) = 0.0392$	$(0.02)^2 = 0.0004$
Relative fitness	1	1	0
Relative contribution to next generation	$(0.9604) \times 1$	$(0.0392) \times 1$	0
Proportional contribution	$0.9604/0.9996 = 0.9608$	$0.0392/0.9996$	$= 0.0392$

New frequency of the allele for cystic fibrosis = $(0.5)(0.0392) = 0.0196$; thus, the incidence of the disease will be $(0.0196)^2 = 0.00038$, which is very slightly less than the incidence in the previous generation.

26.22 Case 1: selection is operating against a deleterious recessive allele. Case 2: selection is operating against a deleterious dominant allele. Case 3: selection is operating against a deleterious allele that has some expression in heterozygotes, that is, it is partially dominant. Case 4: selection is operating against both alleles in homozygous condition; this is a case of balancing selection.

26.23 $q^2 = 4 \times 10^{-5}$; thus $q = 6.3 \times 10^{-3}$ and $2pq = 0.0126$.

26.24 $1/(2N) = 1/100 = 0.01$.

26.25 Probability of ultimate fixation of A_2 is 0.5; probability of ultimate loss of A_3 is $1 - 0.3 = 0.7$.

26.26 0.25.

26.27 $p = 0.2$; at equilibrium, $p = t/(s + t)$. Because $s = 1$, we can solve for t; $t = 0.25$.

26.28 The relative fitnesses of the genotypes *HH*, *Hh* and *hh* are 0.5, 1, and 0.5, respectively. At equilibrium, the frequency of *h* will be $s/(t + s) = (0.5)/(0.5 + 0.5) = 0.5$.

26.29 At mutation-selection equilibrium $q = \sqrt{u/s} = \sqrt{10^{-6}/1} = 0.001$.

26.30 $q = \sqrt{u/s} = \sqrt{10^{-6}/(0.2)} = 2.2 \times 10^{-3}$.

CHAPTER 27

27.1 Among other things, Darwin observed species on islands that were different from each other and from continental species, but that were still similar enough to indicate that they were related. He also observed variation within species, especially within domesticated breeds, and saw how the characteristics of an organism could be changed by selective breeding. His observations of fossilized organisms indicated that some species have become extinct.

27.2 Darwin did not understand the mechanism of inheritance; he did not know of Mendel's principles.

27.3 The frequency of the *a* allele is 0.06 in the South African population and 0.42 in the English population. The predicted genotype frequencies under the assumption of random mating are

Genotype	South Africa	England
aa	$(0.06)^2 = 0.004$	$(0.42)^2 = 0.18$
ab	$2(0.06)(0.94) = 0.11$	$2(0.42)(0.58) = 0.49$
bb	$(0.94)^2 = 0.88$	$(0.58)^2 = 0.33$

27.4 The frequency of the ST banding pattern declines with increasing altitude. Thus, the data indicate that this chromosomal polymorphism exhibits an altitudinal cline.

27.5 In the sample, the frequency of the *F* allele is $(2 \times 32 + 46)/(2 \times 100) = 0.55$ and the frequency of the *S* allele is $1 - 0.55 = 0.45$. The predicted and observed genotype frequencies are

Genotype	Observed	Hardy-Weinberg Predicted
FF	32	$100 \times (0.55)^2 = 30.25$
FS	46	$100 \times 2(0.55)(0.45) = 49.5$
SS	22	$100 \times (0.45)^2 = 20.25$

To test for agreement between the observed and predicted values, we compute a chi-square statistic with 1 degree of freedom: $\chi^2 = \Sigma(\text{obs.} - \text{pred.})^2/\text{pred.} = 0.50$, which is not significant at the 5 percent level. Thus, the population appears to be in Hardy-Weinberg equilibrium for the alcohol dehydrogenase locus.

27.6 There are four alleles.

27.7 In the third position of some of the codons. Due to the degeneracy of the genetic code, different codons can specify the same amino acid. The degeneracy is most pronounced in the third position of many codons, where different nucleotides can be present without changing the amino acid that is specified.

27.8 Introns do not encode amino acids; most exons do. Many—perhaps most—of the nucleotides within an intron can be changed without impairing the expression of a gene or the integrity of its polypeptide product. By contrast, many of the nucleotides within exons—especially the first and second positions of codons—are functionally constrained by the amino acids specified.

27.9 Complex carbohydrates are not "documents of evolutionary history" because although they are polymers, they are typically made of one subunit incorporated repetitiously into a chain. Such a polymer has little or no "information content." Thus, there is little or no opportunity to distinguish a complex carbohydrate obtained from two different organisms. Moreover, complex carbohydrates are not part of the genetic machinery; their formation is ultimately specified by the action of enzymes, which are gene products, but they themselves are not genetic material or the products of the genetic material.

27.10 The tree on the left would require 9 mutations. In this tree, we regard the gap in the intron as ancestral—which means that sequence 4 has acquired an "insertion" at the position of the gap. The other mutations are a TE insertion in the branch leading to the common ancestor of sequences 1 and 2, and seven base-pair changes: one leading to sequence 1, another to sequence 2, two leading to sequence 3, and three leading to sequence 4. The tree on the right would also require 9 mutations. Here again we regard the gap in the intron as ancestral; evidently, an insertion occurred at the position of the gap in the branch leading to sequence 4. The TE insertion can also be regarded as ancestral, with loss occurring in the branch leading to the common ancestor of sequences 3 and 4. There are also seven base-pair changes in this tree: one leading to sequence 1, another leading to sequence 2, two leading to sequence 3 and three leading to sequence 4. These interpretations of the data assume that insertions and deletions (gaps) are reversible and that there is no way of telling which way the sequence evolved—that is, through insertion or deletion. However, if we know, for example, that the TE is a retrotransposon incapable of excision, then we would need more than 9 mutations to explain the tree on the right. The TE insertions must have occurred independently in the branches leading to sequences 1 and 2, and there is no need to postulate excision of the TE in the branch leading to the common ancestor of sequences 3 and 4. Thus, on the assumption that the TE is not excisable, 10 mutations are needed to explain the tree on the right.

27.11 The total elapsed evolutionary time is 0.800/0.9 = 880 million years, which must be apportioned equally to the α and β gene lineages by dividing by 2; thus, the time since the duplication event is estimated to be 440 million years.

27.12 The histidines are rigorously conserved because they perform an important function—anchoring the heme group in hemoglobin. Because these amino acids are strongly constrained by natural selection, they do not evolve by mutation and random genetic drift.

27.13 With n alleles having equal frequency, the frequency of any one allele is $1/n$. Under random mating, the frequency of homozygotes for a particular allele is $(1/n)^2$. The frequency of all the homozygotes is therefore $\Sigma(1/n)^2 = n(1/n)^2 = 1/n$.

27.14 Estimate the average number of substitutions per site in the ribonuclease molecule as $-\ln(S)$, where $S = (124 - 40)/124 = 0.68$, the proportion of amino acids that are the same in the rat and cow molecules. The average number of substitutions per site since the cow and rat lineages diverged from a common ancestor is therefore 0.39. The evolutionary rate in the cow and rat lineages is $0.39/(2 \times 80$ million years$) = 2.4$ substitutions per site every billion years.

27.15 The fraction of polymorphic sites among the silent nucleotides is 13/192 = 0.068. If the same level of polymorphism existed among the nonsilent sites, the number of amino acid polymorphisms would be $255 \times 0.067 = 17.2$. Kreitman actually observed only one amino acid polymorphism—evidence that amino acid changes in alcohol dehydrogenase are deleterious.

27.16 The reciprocal of the rate, that is, $1/K$.

27.17 The protein with the higher evolutionary rate is not as constrained by natural selection as the protein with the lower evolutionary rate.

27.18 The difference in the NS:S ratios suggests that in at least one lineage, positive selection has been operating to change nucleotides in the gene.

27.19 Repetitive sequences that are near each other can mediate displaced pairing during meiosis. Exchange involving the displaced sequences can duplicate the region between them. See Figures 18.20.

27.20 Exon shuffling is a way of creating genes that have pieces from disparate sources. Through exon shuffling, the number of genes in a genome could be increased without having to evolve the genes "from scratch." However, the genome sequencing projects indicate that the gene number in complex, multicellular vertebrates is not much different from the gene number in simple, multicellular invertebrates or in plants. If, judging from their phenotypes, multicellular vertebrates need more gene products than phenotypically simpler invertebrates like *C. elegans*, alternate splicing could provide some of these gene products without increasing the number of genes.

27.21 Cross *D. mauritiana* with *D. simulans* and determine if these two species are reproductively isolated. For instance, can they produce offspring; if they can, are the offspring fertile?

27.22 Allopatric speciation occurs when populations diverge genetically while they are geographically separated. Sympatric speciation occurs when populations diverge genetically while they inhabit the same territory.

27.23   The *Kpn-pn* interaction is an example of the kind of negative epistasis that might prevent populations that have evolved separately from merging into one panmictic population. The *Kpn* mutation would have evolved in one population and the *pn* mutation in another, geographically separate population. When the populations merge, the two mutations can be brought into the same fly by interbreeding. If the combination of these mutations is lethal, then the previously separate populations will not be able to exchange genes, that is, they will be reproductively isolated.

27.24   More haplotype diversity refers to the number of different haplotypes that are found in a population. If African populations of humans have the greatest haplotype diversity, then these populations appear to have had a longer time to accumulate different haplotypes—that is, they are older than other populations. Greater haplotype diversity in African populations of humans is therefore evidence that African populations were at the root of the modern human evolutionary tree.

✍ Notes!

✍ Notes!

✍ **Notes!**